AUTOMOTIVE EMISSION CONTROL AND COMPUTER SYSTEMS

AUTOMOTIVE EMISSION CONTROL AND COMPUTER SYSTEMS

Second Edition

DON KNOWLES

A Reston Book
Prentice Hall
Englewood Cliffs, New Jersey 07632

Library of Congress Cataloging-in-Publication Data

Knowles, Don.
    Automotive emission control and computer systems.

    "A Reston book."
    Includes index.
    1. Automobiles--Pollution control devices.
2. Automobiles--Motors--Control systems. 3. Automo-
biles--Electronic equipment. I. Title.
TL214.P6K56 1989   629.2'528   88-15237
ISBN 0-13-054156-7

Editorial/production supervision and
    interior design: WordCrafters Editorial Services, Inc.
Cover design: 20/20 Services, Inc.
Manufacturing buyer: Robert Anderson
Page layout: Meg VanArsdale

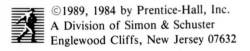
Printed in the United States of America

10 9 8 7 6 5 4 3 2 1

ISBN 0-13-054156-7

Prentice-Hall International (UK) Limited, *London*
Prentice-Hall of Australia Pty. Limited, *Sydney*
Prentice-Hall Canada Inc., *Toronto*
Prentice-Hall Hispanoamericana, S.A., *Mexico*
Prentice-Hall of India Private Limited, *New Delhi*
Prentice-Hall of Japan, Inc., *Tokyo*
Simon & Schuster Asia Pte. Ltd., *Singapore*
Editora Prentice-Hall do Brasil, Ltda., *Rio de Janeiro*

CONTENTS

6

ELECTRONIC IGNITION SYSTEMS   67

7

COMPUTER-CONTROLLED DISTRIBUTOR
ADVANCE SYSTEMS   111

8

COMPUTER-CONTROLLED CARBURETOR
SYSTEMS   135

PREFACE

Possibly the greatest challenge facing the automotive service industry in the 1980s is to keep service technicians informed and trained in the new automotive computer technology. With each model year, smarter chips provide a wider range of output control functions which produce improved performance, greater economy, and increased consumer appeal. On-board computers now control fuel economy, spark advance, exhaust emission devices, instrumentation, voice alert systems, anti-lock brakes, suspension systems, and transmission shifting. In the future, electronics will provide the car driver with such conveniences as road maps displayed on a cathode ray tube (CRT) and holographic instrument displays which appear several feet in front of the windshield.

A second challenge for the automotive industry is in the area of exhaust emission control. Some emission standards, such as diesel particulate emissions, are becoming more stringent, especially in California. The struggle against acid rain is continuing in the United States and Canada. Since automobiles are believed to be responsible for about 30 percent of the nitrous oxides (NOx) that contribute to acid rain, some bills have been presented to the U.S. Congress which call for stricter NOx standards in the automotive industry.

The second edition of this book was written to provide information on these two challenging areas of automotive computers and emission control. A history of California emission control programs is provided, as well as emission standards for the other 49 states and Canada. Most emission components and computer-controlled emission systems are clearly explained. Electronic and computer-controlled ignition systems are described in detail. The computer system explanations include the early systems which controlled spark advance, and the latest systems comprising several computers interconnected with serial data lines. Electronic instrumentation, voice alert systems, electronically controlled diesel injection systems, and import computer systems are included in the book. One chapter is devoted to scope pattern diagnosis of electronic ignition systems.

Electronic components are very expensive, and inaccurate diagnosis can be costly and time consuming. Therefore, special emphasis is placed on the service and diagnosis of each system. Many easy-to-follow diagnostic charts are provided to simplify diagnostic procedures. The self-diagnostic capabilities of many computer systems are explained thoroughly. In most cases, the diagnostic procedures require the use of basic test equipment such as the voltmeter and ohmmeter.

I would like to thank all the automotive manufacturers and other companies who gave us permission to use their diagrams in the book. I also express my appreciation to the publishing staff for their excellent cooperation in the production of the book.

1

AUTOMOTIVE EMISSION STANDARDS AND AIR POLLUTION

MOTOR VEHICLE EMISSIONS

Air Pollution

Programs for the control of motor vehicle emissions originated in California, where emission standards evolved as a result of Professor A. J. Haagen-Smit's reports in the early 1950s about automotive emissions. Haagen-Smit had discovered that two invisible automobile emissions, hydrocarbons and oxides of nitrogen, react together in the presence of sunlight to form oxidants such as ozone, a principal ingredient of Los Angeles "smog." Haagen-Smit also proclaimed that the petroleum and rubber industries, in addition to automobiles, were major sources of hydrocarbon and oxides of nitrogen emissions. Thirty years of further research have confirmed the basic soundness of Haagen-Smit's conclusions regarding air pollution.

The pollutants now known to be caused by each type of automotive emission are listed in Figure 1-1.

Of the automobile emissions, hydrocarbons (HC) and oxides of nitrogen (NOx) contribute to oxidants and total suspended particulate matter (TSP). Sulphur dioxide (SO_2) and particulate matter (PM) are known to contribute to smog formation because they are partly responsible for atmospheric particulate loading in the size range that scatters light.

Although carbon monoxide (CO) does not contribute to visible smog, it is recognized as a poisonous gas that can adversely affect human health. Some adverse short-term and long-term health effects caused by air pollution are listed in Table 1-1.

Regulated Motor Vehicle Emissions

Air pollution legislation usually regulates HC, CO, and NOx emissions from automobiles. These emissions have several sources in an automobile. One is gasoline, which is a hydrocarbon fuel; some unburned hydrocarbons (HC) are emitted in the exhaust because the combustion in the cylinders is never complete. Most emissions occur at the tailpipe of the vehicle. Evaporative sources such as fuel tanks and carburetors also contribute to HC emissions. Carbon monoxide (CO) is a by-product of the combustion process. Hydrogen and carbon in the gasoline combine with oxygen in the air to form a combustible air-fuel mixture in the engine cylinders. When the combustible mixture is burned in the cylinders, the resultant by-products are carbon monoxide (CO), carbon dioxide (CO_2), carbon (C), and water (H_2O). The atmosphere contains approximately 20 percent oxygen and 80 percent nitrogen, depending on the altitude and the amount of air pollution. When the temperature in the cylinders exceeds 2,500°F (1370°C) during the combustion process, the oxygen and nitrogen combine to form oxides of nitrogen (NOx).

1

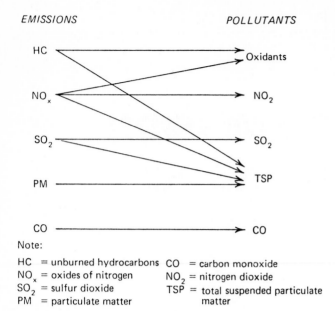

EMISSIONS   POLLUTANTS

Note:

HC = unburned hydrocarbons   CO = carbon monoxide
NO_x = oxides of nitrogen NO_2 = nitrogen dioxide
SO_2 = sulfur dioxide   TSP = total suspended particulate
PM = particulate matter   matter

FIGURE 1-1. Automotive Emissions and Air Pollution. (Reprinted with permission © 1981 Society of Automotive Engineers)

The state government of California became the first official body to regulate the emissions of new automobiles when in 1963 it passed legislation requiring positive crankcase ventilation (PCV) valves on new automobiles. By 1980 the air pollution control program in California progressed to the point where all new passenger cars had to be certified to standards representing 95 percent less HC, 89 percent less CO, and 72 percent less NOx in their exhaust compared with the uncontrolled cars of the early 1960s. In actual customer service, these vehicles are expected to achieve reductions of 82 percent HC, 88 percent CO, and 65 percent NOx. Most emission systems are designed by the car manufacturer to operate with a minimum amount of maintenance. Therefore, these levels of control will occur despite the fact that cars in customer service are sometimes improperly maintained.

The California Air Resources Board (CARB) is responsible for air quality in the state. The Environmental Protection Agency (EPA) is in charge of federal air pollution regulation in the United States. In Canada air quality is the responsibility of Environment Canada.

TABLE 1-1. Health Effects from Air Pollution

Pollutant	Short-Term Health Effects	Long-Term Health Effects
Oxidant (Ox) (ozone and others)	Difficulty in breathing, chest tightness, soreness, coughing	Impaired lung function, increased susceptibility to respiratory infection
Total suspended particulate matter (TSP)	Increased susceptibility to other pollutants	Many components of TSP are toxic and carcinogenic; contribute to silicosis, brown lung
Sulfate particles (SO_4)	Increased asthma attacks	Reduced lung function when oxidant is also present
Nitrogen dioxide (NO_2)	Similar to ozone, but at higher concentrations	Increased susceptibility to respiratory infection; adverse changes in cell structure of the lung wall

Reprinted with permission © 1981 Society of Automotive Engineers.

Passenger Car Emission Standards

Federal and California emission standards through the 1986 model year are provided in Table 1-2, and the 1987 federal, California, and Canadian emission standards are shown in Table 1-3.

Diesel Particulate Emission Standards

Diesel particulate emission standards for all vehicles from passenger cars to heavy-duty trucks are becoming more stringent. Diesel particulate emission standards for passenger cars through 1989 are listed in Table 1-4.

New federal diesel emission standards of 0.6 grams per horsepower-hour (g/bhp-h) of particulate emissions, and 6.0 g/bhp-h of NOx will apply to heavy-duty vehicles in 1988. Stricter standards of 0.25 g/bhp-h for particulate emissions, and 5.0 g/bhp-h for NOx emissions will be applied to heavy-duty diesel powered vehicles in 1991. These standards will apply to new heavy-duty vehicles, and the standards will not be retroactive. (Refer to Chapter 12 for discussion of diesel emission control systems.)

TABLE 1-2. Federal and California Emission Standards Through 1986

Model Year	Hydrocarbon (HC)		Carbon Monoxide (CO)		Oxides of Nitrogen (NO$_x$)	
	California	Federal	California	Federal	California	Federal
1978	0.41	1.5	9.0	15.0	1.5	2.0
1979	0.41	0.41	9.0	15.0	1.5	2.0
1980	0.39	0.41	9.0	7.0	1.0	2.0
1981	0.39	0.41	7.0	3.4	0.7	1.0
1982	0.39	0.41	7.0	3.4	0.4	1.0
1983	0.39	0.41	7.0	3.4	0.4	1.0
1984	0.39	0.41	7.0	3.4	0.4	1.0
1985	0.39	0.41	7.0	3.4	0.7	1.0
1986	0.39	0.41	7.0	3.4	0.7	1.0
1960 (No Control)		10.6		84		4.1

Courtesy of Chrysler Canada Ltd.

TABLE 1-3. Federal, California, and Canadian Emission Standards 1987

Gas Engines	HC (gm/mi) 1986 1987	CO (gm/mi) 1986 1987	NOx (gm/mi) 1986 1987
Federal	0.41	3.4	1.0
California	0.39*	7.0	0.7@
Canada	2.0	25.0	3.1

* = Non-methane hydrocarbons
@ = With 50,000 mile warranty and 75,000 mile recall liability

Courtesy of GM Product Service Training, General Motors Corporation

TABLE 1-4. Diesel Particulate Emission Standards Through 1989

1984 — 0.6	1985 — 0.4
1986 — 0.2	1989 — 0.08

Courtesy of GM Product Service Training, General Motors Corporation

Additional Emission Regulations

Federal emission standards since the 1975 model year, and California emission standards since the 1980 model year, apply both to gasoline- and diesel-powered cars.

The Sealed Housing Evaporative Determination (SHED) test is a method of determining evaporative emissions from motor vehicles. Effective for 1978 models, the evaporative emission standards are 6.0 grams per SHED test for all classifications of vehicles. Emission standards are 2.0 grams per SHED test for 1980 and later model vehicles.

California and federal motorcycle emission standards for 1984 and later model years are 1.0 g/km (HC) and 12 g/km (CO) for motorcycles with an engine displacement of 280 cubic centimeters (cc) or greater.

The U.S. Clean Air Act prescribes significant fines for any individual who tampers with, disconnects, or alters any emission device on a motor vehicle. A fine can be imposed for each device that is disconnected or altered.

In some states, such as California, vapor recovery systems on gasoline pumps are mandatory.

Emission Test Procedures

The "7 mode" test procedure was one of the early test methods used by the California Air Resources Board and the Environmental Protection Agency to determine the actual emissions from motor vehicles. Emission levels were monitored during a 7 mode, 140-second test cycle, as illustrated in Figure 1-2.

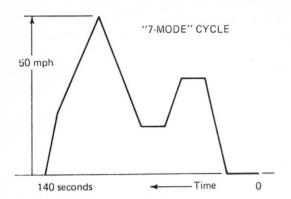

FIGURE 1-2. The "7 mode" Emission Test Procedure. (Reprinted with permission © 1981 Society of Automotive Engineers)

In order to develop a driving cycle that represented actual driving conditions more closely, California officials gathered test data on what is known as the "LA4" road route, which was considered typical of a home-to-work commuting trip in the Los Angeles area. This 12-mi (19.2 km) route took about 39 minutes to complete. A shortened version of the "LA4" test was subsequently developed by EPA. The Urban Dynamometer Driving Schedule (UDDS), which retained the characteristics of the "LA4" test, was thus adopted to replace the 7 mode test procedure. The UDDS follows a procedure known as constant volume sampling (CVS), which is the official U.S. and California certification test now used to measure HC, CO, and NOx emissions on new vehicles. The present CVS test procedure, developed in 1975, is referred to as CVS-75.

During this test, the vehicle is preconditioned by being driven on a simulated trip on a dynamometer. A 12-hour shutdown period at 78°F (26°C) allows the simulation of cold-start conditions at the beginning of the test. The test takes approximately 41 minutes, of which the first 23 minutes are devoted to a cold-start driving test, followed by a 10-minute "hot-soak" waiting period. A hot-start test is performed during the final 8 minutes of the test. This final test simulates a short trip situation in which a motorist might stop and start the car several times while the car is hot.

During these various stages of testing, the vehicle's exhaust is diluted with air to a constant volume, and a small amount of the diluted exhaust is collected in separate bags, each of which is analyzed for concentrations of HC, CO, and NOx. The mass of pollutants emitted by the vehicle can be calculated from the concentration of pollutants in the

sample bags and the total volume of exhaust air that was pumped through the CVS unit.

Durability Testing

Before manufacturers are allowed to sell a particular line of car, they must demonstrate that each "engine family" to be used will meet the emission standards. They do so by performing a durability demonstration with a prototype vehicle that involves testing the vehicle's emissions every 5,000 mi (8,000 km) for a driving distance of 50,000 mi (80,000 km). The mileage is logged by following a specific driving schedule, and maintenance on the vehicle is limited by regulation. The data from the durability vehicle are used to establish a deterioration factor.

Vehicles that will be equipped with the same "engine family" but that have different transmissions, weights, or other variables must demonstrate their compliance with the emission standards through the testing of "data vehicles." These vehicles are operated for 4,000 mi (6,400 km) and the deterioration factor from the durability vehicle is used to project the emission levels to 50,000 mi (80,000 km). In order to be certified for sale, all "data vehicles" must have emission levels below the federal or California emission standards. Federal emission standards are based on the 50,000-mi (80,000 km) levels. California emission standards may be based on durability standards of 50,000 mi (80,000 km) or 100,000 mi (166,000 km).

CALIFORNIA EMISSION CONTROL PROGRAM

New Car Certification

The California emission control program represents one of the most extensive efforts to date to reduce air pollution. As the new model year approaches, each car manufacturer must submit prototype emission data vehicles for test purposes. Emission testing of new vehicles is done by the EPA or CARB. Test procedures and durability test methods outlined earlier in the chapter are used to determine the emission levels of each engine family for the 50,000-mi (80,000-km) or 100,000-mi (166,000-km) periods. The car manufacturer receives certification of the engine family if the projected emissions for all emission data vehicles are equal to or lower than all of the emission standards. EPA administers a similar new car certification program nationwide.

Assembly Line Testing

Two percent of the vehicles produced for sale in California are selected at random for quality audit testing at the factory. This test involves the same CVS procedure that is used in new car certification. Every new car sold in California must also undergo assembly line testing. Each car is checked to ensure that it is tuned to factory specifications and that all its emission equipment is operating properly. Before leaving the assembly line, each car is given an idle exhaust emission test.

Confirmatory Testing

Approximately one-third of the confirmatory tests performed by the CARB are required because the certification tests appeared marginal, or the assembly line, quality audit tests indicated a problem of compliance. Approximately two-thirds of the confirmatory tests are ordered for vehicles that have the largest sales volumes in California. Confirmatory testing and new car certification tests are performed at CARB's laboratory in El Monte.

First, the vehicles are preconditioned on a dynamometer for 5 minutes at 50 MPH (80 KPH). Each vehicle is then placed in a cold-soak storage shed for 12–24 hours. Next, the vehicle is rolled onto the dynamometer without starting the engine and the official CVS test procedure is performed. After the test, the car is returned to the cold-soak storage shed. The test procedure is repeated for three days and the test data are averaged for the test series.

If two vehicles exceed one or more standards by 15 percent, or if one vehicle exceeds all three standards by 15 percent, they are considered failures. The manufacturer then has the option of supplying two additional vehicles for test purposes. If three out of five vehicles fail one or more standards, or if two out of five vehicles fail all three standards, Section 2109 of Title 13 may be invoked by the CARB officer.

Enforcement Action

When enforcement action is taken under Section 2109, the CARB executive officer requires the car manufacturer to submit a plan for bringing into compliance all vehicles that use the failed engine family. The plan may include correction of the vehicles at the factory and dealerships and a recall of the vehicles already sold. The enforcement action may also call for a halt in sales until all the affected vehicles are brought into compliance. In the event of a recall, manufacturers have ten days in which to request a public hearing to contest the recall order and specify their objections. Upon receipt of the hearing request, the executive officer may stay the recall until the CARB holds a hearing.

Under Section 43212, Part 5, Division 26 of the Health and Safety Code, any manufacturer or distributor failing to comply with standards or test procedures established under the subdivision shall be subject to a civil penalty of $50 for each vehicle sold in the state that does not comply with the regulations. Under Section 43211, any manufacturer who sells, or causes to be offered for sale, a new motor vehicle that fails to meet the applicable emission standards shall be subject to a civil penalty of $5,000 for each vehicle in violation.

Vehicle Inspection Program

Used cars or light trucks under 8,500 lb (3,825 kg) must pass an exhaust emission inspection if they are sold to residents of the California South Coast Air Basin. This basin includes parts of Los Angeles, Riverside, San Bernardino, and Santa Barbara counties, and all of Orange and Ventura counties. The mandatory inspection is also necessary if a car is being registered in California from another state. These vehicles cannot be registered until they have an exhaust emission inspection. The individual or dealer selling the vehicle is responsible for any repairs that are necessary to make the vehicle pass the emission standards. Emission testing is performed at seventeen test centers in the area. When a vehicle passes the emissions test, it is given a certificate of compliance that shows the test readings. If a vehicle fails the emission tests, it must be repaired and then returned for a retest.

Retrofit Programs

Some programs have been developed in California to bring 1955–1965 vehicles and 1966–1970 vehicles up to emission standards. For 1955–1965 cars, the program requires that an approved kit be installed in the vacuum advance hose to delay the advance under certain operating conditions. For 1966–1970 vehicles, the approved kits consist of an exhaust gas recirculation valve and a vacuum spark advance disconnect device. The retrofit programs are mandatory before vehicles in these age groups can be transferred and registered to new owners, or registered in California for the first time. The kits are installed by technicians who are licensed by the state, and a certificate of compliance is issued by the installer. The cost of the retrofit installations is limited by law.

Vehicle Inspection/Maintenance Programs

Many states have vehicle inspection/maintenance programs. These programs encourage vehicle owners to maintain their vehicles at approved emission standards. Information regarding state emission inspection/maintenance programs is provided in Table 1-5, and the abbreviations are explained in Table 1-6.

TABLE 1-5. State Emission Inspection/ Maintenance Programs

STATE	PROGRAM TYPE	TEST TYPE	VEHICLES INSP. TO GVW	LDV 1981 + STD. * HC	CO	TAMPER INSPECTION
Alaska (Anch)	D	2	12,000	NA	1.00	Y
Alaska (Fbks)	D	2	12,000	NA	1.00	Y
Arizona	CC	I	All	250	1.50	N
California	D	2	8,600	100	1.00	Y
Colorado	D	2	10,000	400	1.50	Y
Connecticut	CC	I	10,000	220	1.20	W
D.C.	CL	I	6,000	300	1.50	N
Delaware	CS	I	8,500	220	NA	N
Georgia	D	I	6,000	250	2.50	Y
Idaho	D	I	8,600	NA	1.20	Y
Indiana	CC	2	9,000	220	1.20	W
Kentucky (Lsvle)	CC	I	All	220	1.20	N
Maryland	CC	I	10,000	220	1.20	N
Massachusetts	D	I	8,500	220	1.20	Y
Michigan	D	I	8,500	220	1.20	N
Missouri	D	I	6,000	220	1.20	Y
Nevada	D	2	5,000	NA	3.00	W
New Jersey	CSD	I	All	220	1.20	Y
New York	D	I	8,500	220	1.20	N
North Carolina	D	I	All	NA	1.50	Y
Oklahoma	D	T	8,500	NA	NA	Y
Oregon	CS	2	All	225	1.00	Y
Pennsylvania	D	I	11,000	220	1.20	W
Rhode Island	D	I	8,000	300	3.00	N
Tennessee (Mem)	CL	I	8,500	NA	3.00	W
Tennessee (Nash)	CC	I	8,500	220	1.20	N
Texas (Dallas)	D	P	8,500	NA	NA	Y
Texas (Elpaso)	D	P	8,500	NA	NA	Y
Texas (Houston)	D	P	8,500	NA	NA	Y
Utah (Davis)	D	2	10,000	220	1.20	Y
Utah (S. Lake)	D	2	10,000	220	1.20	Y
Virginia	D	I	6,000	220	1.20	Y
Washington	CC	I	All	300	1.50	N
Wisconsin	CC	I	8,000	220	1.20	W

Courtesy of GM Product Service Training, General Motors Corporation

TABLE 1-6. Explanation of Abbreviations in Table   1-5

Program Type Key:		Test Type Key:	
D = Decentralized		2 = 2500 RPM + Idle	
CL = Central Local Run		I = Idle	
CC = Central Contractor		P = Parameter	
CS = Central State Run		T = Tamper Insp. Only	

W = Tamper inspection for waiver only
NA = Not Applicable
* = I/M stds. vary for pre-1981 MY LDV and other vehicle categories.

Courtesy of GM Product Service Training, General Motors Corporation

Questions

1. Name the two motor vehicle emissions that contribute to "smog."

2. Carbon monoxide (CO) is a by-product of the combustion process.　　　　　T　F

3. Oxides of nitrogen (NOx) emission levels increase at _____ cylinder temperature.

4. For the 1987 model year, federal emission standards are _____ g/mi HC.

5. In the 1989 model year, diesel particulate emission standards are _____ g/mi.

6. For the 1987 model year, federal emission standards are _____ g/mi CO.

2

EXHAUST GAS RECIRCULATION (EGR) SYSTEMS

EGR VALVES

Conventional EGR Valves

The conventional EGR valve contains a vacuum diaphragm linked to a tapered valve. One side of the tapered valve is connected to the exhaust crossover passage in the intake manifold, and the other side is connected to the fuel passages in the intake manifold. Ported vacuum from a vacuum outlet above the throttle valve is connected to the EGR valve vacuum chamber, as illustrated in Figure 2-1. A diaphragm spring holds the valve closed.

When the throttle is opened from the idle position, ported manifold vacuum applied to the EGR valve diaphragm opens the tapered valve. Exhaust gas flows into the intake manifold once the EGR valve is opened. Because of the lack of oxygen, the exhaust gas will not promote any combustion in the cylinders. Recirculating exhaust gas through the EGR valve into the intake manifold lowers combustion temperature and reduces oxides of nitrogen emission levels. Most EGR valves require 3 in Hg (10 kPa) of vacuum applied to the diaphragm to hold the valve open. At a wide throttle opening, the manifold vacuum will drop below 3 in Hg (10 kPa), and the diaphragm spring will close the EGR valve. Exhaust gas must not be allowed to recirculate into the intake manifold at idle speed. Rough idling will occur if the EGR valve sticks in the open position.

Back Pressure Transducer (BPT) EGR Valves

Some EGR valves contain a built-in BPT valve. Exhaust gas pressure is applied through the hollow EGR valve stem to the BPT diaphragm, as shown in Figure 2-2. A vacuum control valve is located above the BPT diaphragm.

When the throttle is opened from the idle position, ported vacuum will be available at the EGR valve. At low speeds, the vacuum applied to the EGR valve will be bled off through the control valve and vent opening, and the EGR valve will remain closed. Exhaust pressure will force the BPT diaphragm upward and close the control valve at 30–35 MPH (50–58 KPH). Once the control valve is closed (as pictured in the EGF. valve on the right side of Figure 2-2), the ported manifold vacuum will open the EGR valve.

EGR VALVE CONTROL

Coolant-controlled Ported Vacuum Switch (PVS)

Nitrous oxide emissions occur at high cylinder temperatures. While the engine is warming up, NOx emissions are low, and exhaust gas recirculation is not required. A ported vacuum switch operated by coolant temperature is connected in the EGR

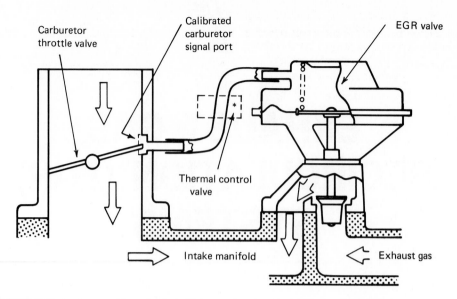

FIGURE 2-1. Conventional EGR Valve. (Courtesy of General Motors of Canada Ltd.)

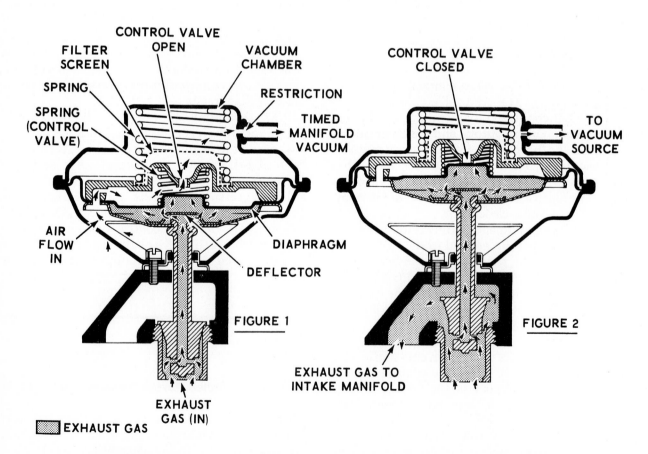

FIGURE 2-2. Back Pressure Transducer EGR Valve. (Courtesy of General Motors of Canada Ltd.)

vacuum hose, as pictured in Figure 2-3. The vacuum to the EGR valve is shut off by the PVS switch when engine coolant is below approximately 120°F (49°C). When the engine coolant reaches 120°F (49°C), the wax in the bottom of the PVS expands, moving the PVS plunger upward. Vertical movement of the PVS plunger opens the vacuum port to the EGR valve. Some EGR systems use a vacuum switch mounted in the air cleaner. The EGR vacuum switch may be referred to as a thermal vacuum switch (TVS) rather than a PVS.

EGR Vacuum Amplifier

Many engines have a vacuum amplifier to control the EGR valve. The intake manifold vacuum and carburetor venturi vacuum are both applied to the vacuum amplifier as shown in Figure 2-4. The intake manifold vacuum is created by the downward motion of the pistons on the intake strokes. The

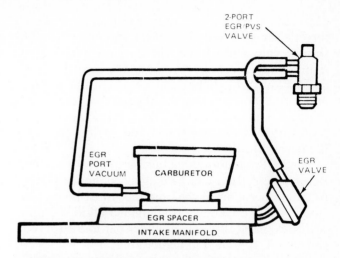

FIGURE 2-3. EGR Coolant-controlled Ported Vacuum Switch. (Courtesy of Ford Motor Co.)

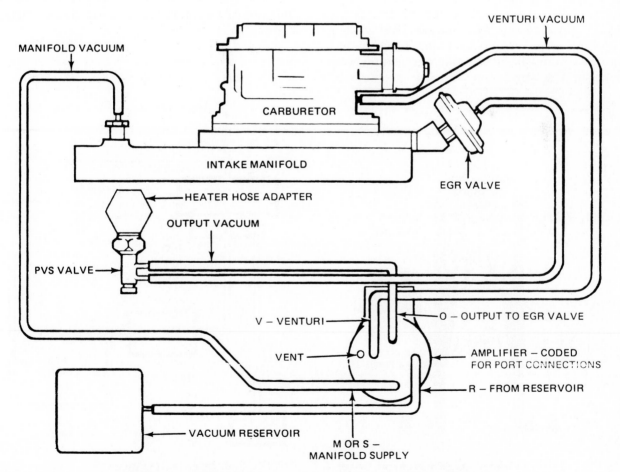

FIGURE 2-4. EGR Vacuum Amplifier. (Courtesy of Ford Motor Co.)

pistons pull air and fuel past the restriction of the throttle valves. Manifold vacuum is highest when the throttle valves are closed. The vacuum in the intake manifold decreases in relation to the throttle opening. Airflow through the carburetor venturi creates a pressure drop or vacuum in the narrow venturi, as illustrated in Figure 2-5. The venturi vacuum is proportional to engine speed and airflow through the carburetor.

The EGR vacuum amplifier is replaced as a unit. Before the system can be diagnosed, however, the internal operation of the amplifier must be understood. When the engine is idling, manifold vacuum will be available through the check ball and reservoir to valve A, as shown in Figure 2-6.

When the engine is idling, the manifold vacuum to the EGR valve will be shut off by valve A. The venturi vacuum at port B increases in relation to engine RPM. Diaphragms C and D are interconnected. At a vehicle speed of 30–35 MPH (50–58 KPH), the venturi vacuum will be sufficient to lift diaphragms C and D. As diaphragm D lifts up, port A opens and allows the manifold vacuum to open the EGR valve. The venturi vacuum is applied to the upper side of diaphragm E, and manifold vacuum becomes available at the lower diaphragm chamber. During idling or cruising speed, manifold vacuum is higher than venturi vacuum. The higher manifold vacuum holds diaphragm E downward and valve F remains closed.

At wide throttle openings, diaphragm E is lifted upward when venturi vacuum exceeds manifold vacuum. When valve F is open, the manifold vacuum in the EGR system is bled off through port G, and the EGR valve is allowed to close at wide throttle opening. Valve A is designed with a small amount of leakage. Vacuum applied to the EGR valve should not exceed 0.5 in Hg (1.6 kPa) at idle speed. Rough idling could result if excessive leakage at valve A held the EGR valve open during idle operation.

Back Pressure Transducer (BPT) Valves

Some EGR valves have an external BPT valve operated by exhaust pressure, as shown in Figure 2-7. The vacuum hose connected to the EGR valve is routed through the BPT valve. Exhaust pressure operates a diaphragm and vent port in the BPT valve. At vehicle speeds less than approximately 35 MPH (58 KPH), exhaust pressure is too low to move the diaphragm upward and the vent port is open. Vacuum applied to the EGR valve will be bled off by the open BPT valve port, as illustrated at the right of Figure 2-8.

Exhaust pressure will move the BPT valve diaphragm upward and close the vent port at vehicle speeds greater than 35 MPH (58 KPH), as indicated

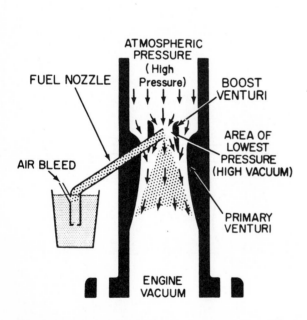

FIGURE 2-5. Venturi Vacuum. (Courtesy of General Motors of Canada Ltd.)

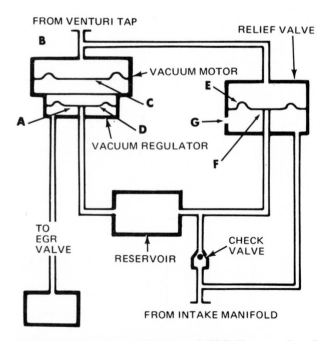

FIGURE 2-6. Internal Design of EGR Vacuum Amplifier. (Courtesy of Ford Motor Co.)

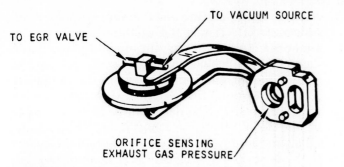

FIGURE 2-7. BPT Valve. (Courtesy of General Motors of Canada Ltd.)

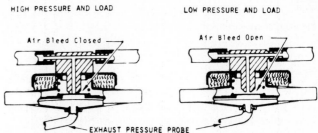

FIGURE 2-8. BPT Valve Operation. (Courtesy of General Motors of Canada Ltd.)

at the left of Figure 2-8. When the BPT valve vent port is closed, vacuum is applied to the EGR valve in the normal manner. The BPT valve improves drivability by keeping the EGR valve inoperative at low speeds.

Wide-Open Throttle (WOT) Valve

The EGR valve vacuum hose is routed through the WOT valve. The WOT valve and the BPT valve may be connected together in series. Figure 2-9 illustrates

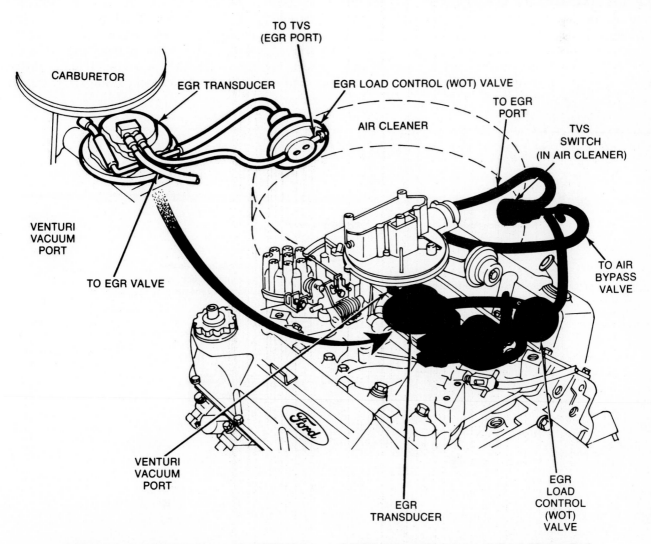

FIGURE 2-9. EGR System with WOT Valve and BPT Valve. (Courtesy of Ford Motor Co.)

an EGR system with a WOT valve, BPT valve, and air cleaner temperature switch.

The WOT valve is operated by venturi vacuum. At a normal cruising speed, venturi vacuum is insufficient to lift the WOT valve diaphragm. When the WOT valve vent port remains closed, ported manifold vacuum is allowed to pass through the WOT valve to the EGR valve, as shown in Figure 2-10.

At a wide throttle opening, the increased venturi vacuum lifts the WOT valve diaphragm and opens the vent port, as shown in Figure 2-11. The open WOT valve port bleeds off the vacuum applied to the EGR valve. When the WOT valve causes the EGR valve to close at a wide throttle opening, engine power is improved.

ERG TIME-DELAY SYSTEMS

Design

The EGR time-delay system uses a conventional EGR valve with a vacuum amplifier and a coolant-

controlled temperature valve. A vacuum solenoid is connected in series in the vacuum hose from the intake manifold to the vacuum amplifier. One end of the vacuum solenoid winding is connected to the ignition switch, and the other end of the winding is grounded through the EGR delay timer. Solid state circuitry is used in the delay timer. Three terminals on the delay timer are connected to the vacuum solenoid winding, the ignition switch, and the starter relay, as indicated in Figure 2-12. The starter relay connection supplies power to the delay timer while cranking the engine.

Operation

The EGR delay timer allows current to flow through the vacuum solenoid winding and the delay timer for 35 seconds after the engine is started. The energized vacuum solenoid winding moves the solenoid plunger to the left, as indicated in Figure 2-12. A tapered needle valve is located on each end of the solenoid plunger. When the plunger is held to the left, vacuum to the EGR amplifier is shut off and vacuum to the idle enrichment system is turned on.

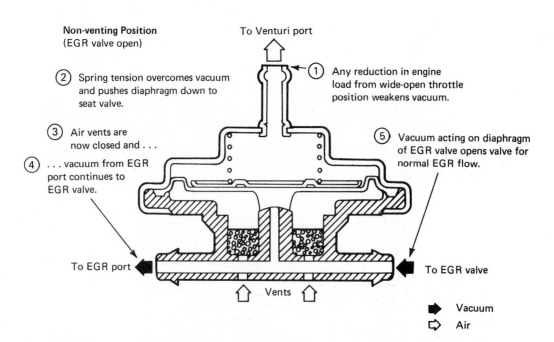

FIGURE 2-10. WOT Valve with Vent Port Closed. (Courtesy of Ford Motor Co.)

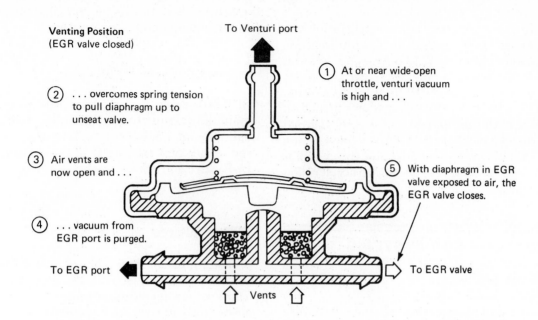

Venting Position
(EGR valve closed)

To Venturi port

① At or near wide-open throttle, venturi vacuum is high and . . .

② . . . overcomes spring tension to pull diaphragm up to unseat valve.

③ Air vents are now open and . . .

⑤ With diaphragm in EGR valve exposed to air, the EGR valve closes.

④ . . . vacuum from EGR port is purged.

To EGR port

To EGR valve

Vents

FIGURE 2-11. WOT Valve with Vent Port Open. (Courtesy of Ford Motor Co.)

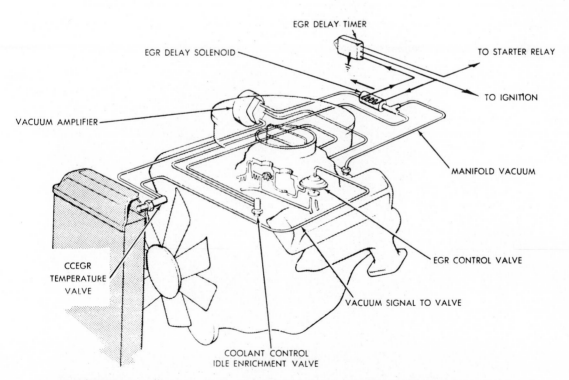

EGR DELAY TIMER

EGR DELAY SOLENOID

TO STARTER RELAY

TO IGNITION

VACUUM AMPLIFIER

MANIFOLD VACUUM

CCEGR TEMPERATURE VALVE

EGR CONTROL VALVE

VACUUM SIGNAL TO VALVE

COOLANT CONTROL IDLE ENRICHMENT VALVE

FIGURE 2-12. EGR Time-Delay System up to 35 Seconds After the Engine Is Started. (Courtesy of Chrysler Canada Ltd.)

(The idle enrichment system is discussed in Chapter 4.)

Thirty-five seconds after the engine is started, the delay timer turns off the current flow through the vacuum solenoid winding. With the vacuum solenoid winding deenergized, the plunger return spring is allowed to move the plunger to the right. As a result, manifold vacuum to the EGR amplifier is turned on and vacuum to the idle enrichment system is turned off, as illustrated in Figure 2-13. If the EGR system is kept inoperative for 35 seconds after the engine is started, drivability improves on sudden acceleration after a restart.

DIAGNOSIS OF EGR SYSTEMS

Conventional EGR Valve

An EGR valve that is stuck open will cause extremely rough idling. The conventional EGR valve may be diagnosed by applying full manifold vacuum to the valve at idle speed. The engine should slow down 150 RPM or stall completely when the EGR valve is opened by full manifold vacuum at idle speed. If there is very little change in idle speed when full manifold vacuum is applied to the EGR valve, the valve is stuck open.

Internal BPT—EGR Valves

BPT EGR valves may be diagnosed by applying full manifold vacuum to the EGR valve with the engine operating at 1,500 RPM. A slight change in engine speed indicates normal EGR valve operation. The EGR valve is stuck open when there is no change in engine speed.

EGR Amplifiers

A vacuum gauge should be connected into the hose from the amplifier to the EGR valve by a "T" fitting. The vacuum gauge hose must be long enough to allow the gauge to be mounted in the passenger compartment. A vacuum greater than 0.5 in Hg (1.6 kPa) at idle speed indicates a defective amplifier. The next two checks must be done during a road test, or on a dynamometer. At normal cruising speeds greater than 35 MPH (58 KPH), the gauge reading

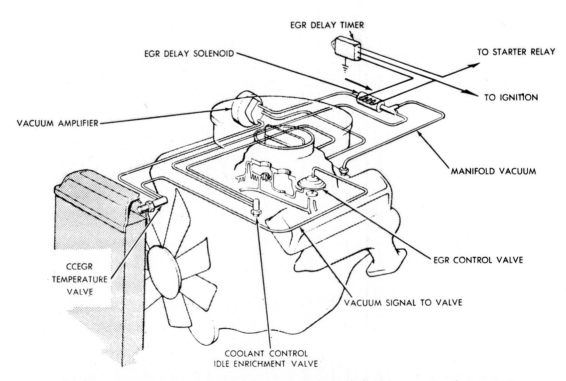

FIGURE 2-13. EGR Time-Delay System More than 35 Seconds After the Engine is Started. (Courtesy of Chrysler Canada Ltd.)

should be 5–15 in Hg (16.5–49.5 kPa). The gauge reading should drop to zero when the throttle is momentarily pushed to the wide-open position. Amplifier replacement is necessary if gauge readings are abnormal. Check all vacuum hoses for proper connections and leaks before replacing the amplifier.

EGR Time-Delay Systems

Connect a vacuum gauge to the hose between the delay solenoid and the EGR amplifier with a "T" fitting, as shown by gauge 1 in Figure 2-14. The vacuum gauge should read zero for 35 seconds after the engine is started. Full manifold vacuum should be indicated on the gauge once the engine has been running for more than 35 seconds.

If the vacuum gauge readings are incorrect, the following steps should be taken.

1. Check all vacuum hoses for proper routing and leaks.
2. With the ignition switch on, use a 12-V test light to check for power at the two lower delay timer terminals.
3. The wire from the ignition switch to the center terminal is open if there is no power at the center terminal. If there is no power at the bottom delay timer terminal, the delay solenoid winding or connecting wires are open. Connect an ohmmeter to the delay solenoid terminals. An infinite reading indicates an open winding.
4. Check for power at the delay timer top terminal with the ignition switch in the start position. If no power is available, the starter relay or connecting wire is open.
5. Check the delay solenoid. Vacuum ports A and B in Figure 2-14 should be connected when the solenoid winding is deenergized. When the solenoid winding is energized, vacuum ports A and C should be connected. The delay solenoid winding may be energized by grounding the solenoid terminal connected to the delay timer with the ignition switch on.
6. Check the delay timer ground. If all tests are satisfactory and the vacuum gauge reading is incorrect, replace the delay timer.

Some models have a charge temperature switch (CTS) and a CTS timer to control the EGR vacuum solenoid, as pictured in Figure 2-15.

The CTS senses the temperature of the air-fuel mixture charge in the intake manifold. When the

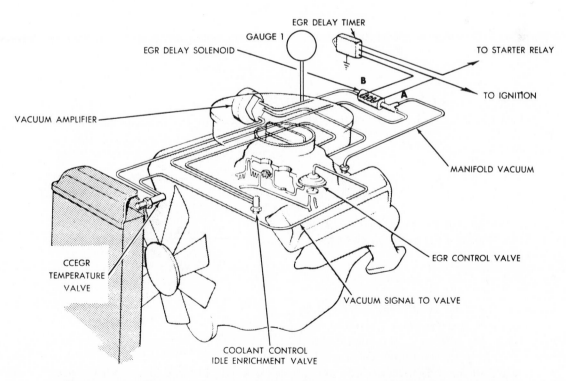

FIGURE 2-14. EGR Delay System Diagnosis. (Courtesy of Chrysler Canada Ltd.)

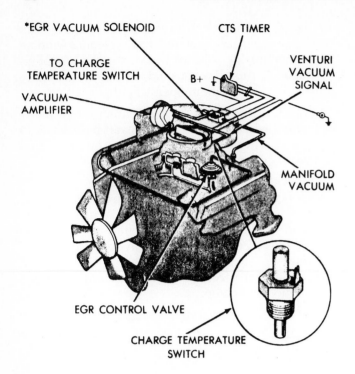

the timer will be able to energize the vacuum solenoid and shut off the vacuum to the amplifier for 35 seconds each time the engine is started. After the engine has been running for 35 seconds, the timer will deenergize the vacuum solenoid and vacuum will be applied to the amplifier.

EGR Maintenance Reminder System

The maintenance reminder system contains a switch-counting device in the speedometer cable, as illustrated in Figure 2-16. The switch-counting mechanism completes the dash reminder bulb circuit from the bulb to ground at 15,000-mile (24,900-km) intervals. The dash indicator bulb reminds the operator to have the EGR system checked. The reminder bulb stays illuminated until a reset screw is rotated in the switch counter, as shown in Figure 2-16.

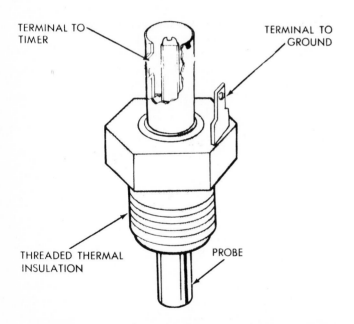

FIGURE 2-15. Charge Temperature Switch (CTS) and Timer. (Courtesy of Chrysler Canada Ltd.)

intake charge temperature is less than 60°F (16°C) the CTS will be closed and the timer will not be able to operate. Under this condition, the solenoid will shut off the vacuum to the amplifier and the EGR valve. When the intake charge temperature is greater than 60°F (16°C), the CTS will be open and

Questions

1. The purpose of the EGR valve is to reduce _____ _____ emission levels.

2. When a conventional EGR valve is used without a vacuum amplifier, the EGR valve vacuum source is always located _____ _____ _____.

3. A BPT EGR valve may be opened at idle speed by applying full manifold vacuum to the EGR valve diaphragm.   T   F

4. An EGR valve stuck in the open position would result in _____ _____ _____.

5. The BPT valve is operated by _____ _____ _____.

6. The WOT valve is controlled by manifold vacuum.   T   F

7. In the EGR delay system, the delay timer and delay solenoid turn on manifold vacuum to the amplifier _____ after the engine is started.

8. A defective EGR amplifier could cause rough idling of the engine.   T   F

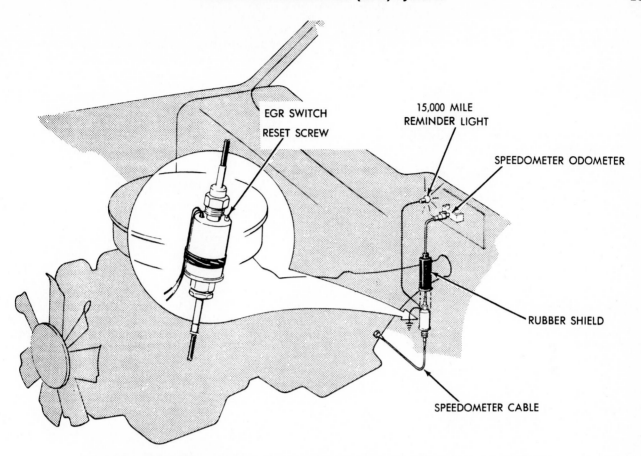

FIGURE 2-16. EGR Maintenance Reminder System. (Courtesy of Chrysler Canada Ltd.)

3

AIR INJECTION EMISSION SYSTEMS

AIR INJECTION REACTOR (AIR) SYSTEMS

Air Pump

The main component in the AIR system is a belt-driven air pump (see Figure 3-1). Air is moved through the pump by two rotating vanes, as outlined in Figure 3-2.

Air enters the pump through the centrifugal filter behind the pulley, and exits via the outlet at the rear of the pump. Dirt particles are cleaned from the air by the centrifugal filter, as shown in Figure 3-3.

AIR System

Airflow from the pump moves through the diverter valve, hoses, one-way check valves, and piping into each exhaust port, as shown in Figure 3-4. Some cylinder heads have internal air passages rather than external air pipes. Vehicles without catalytic converters use air injection to burn hydrocarbon (HC) emissions at the exhaust ports. Oxygen is necessary for the catalytic converter to oxidize carbon monoxide (CO) and HC into carbon dioxide (CO_2) and water vapor. In cars equipped with catalytic converters, the AIR pump supplies oxygen to the converter. (Various types of catalytic converters are discussed in detail in Chapter 8.) In some converter-equipped vehicles, air is injected downstream into the exhaust manifold flange or converter rather than into the exhaust ports. Some air injection systems use a switching arrangement to change air injection from the exhaust ports to a downstream location under specific operating conditions. Several types of air injection systems are covered later in this chapter.

FIGURE 3-1. AIR Pump. (Courtesy of General Motors of Canada Ltd.)

21

The vane is travelling from a small area into a larger area—consequently a vacuum is formed that draws fresh air into the pump.

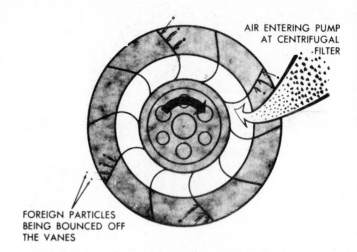

FIGURE 3-3. Centrifugal Filter Operation. (Courtesy of General Motors of Canada Ltd.)

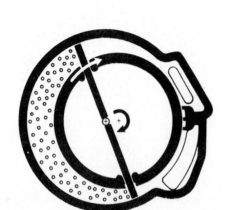

As the vane continues to rotate, the other vane has rotated past the inlet opening. Now the air that has just been drawn in is entrapped between the vanes. This entrapped air is then transferred into a smaller area and thus compressed.

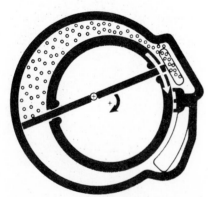

As the vane continues to rotate it passes the outlet cavity in the pump housing bore and exhausts the compressed air into the remainder of the system.

FIGURE 3-2. Airflow through AIR Pump. (Courtesy of General Motors of Canada Ltd.)

Air flow through the air injection system is pictured in Figure 3-5. The one-way check valves prevent exhaust from flowing into the air system should the pump become inoperative. Notice the vacuum connection from the intake manifold to the diverter valve.

The diverter valve assembly contains a pressure relief valve and a diverter valve operated by a vacuum diaphragm. The pressure relief valve limits pump pressure to 2–6 PSI (14–42 kPa). Under idle, cruise, or wide-open throttle conditions, the manifold vacuum is insufficient to move the diverter valve diaphragm. With the diaphragm and diverter valve in the downward position, air is allowed to flow from the pump to the exhaust ports, as illustrated in Figure 3-6. On sudden deceleration, manifold vacuum will be approximately 21–22 in Hg (69–72 kPa) and will lift the diverter valve and diaphragm that directs air from the pump through the diverter valve muffler to the atmosphere.

Sudden deceleration causes the air-fuel mixture to become richer and thus HC and CO emissions increase. If the AIR pump continued pumping air into the exhaust ports on deceleration, excessive burning of HC would occur in the exhaust manifolds. Manifold backfiring would result from the excessive burning. The diverter valve prevents manifold backfiring on deceleration by momentarily directing AIR pump flow to the atmosphere.

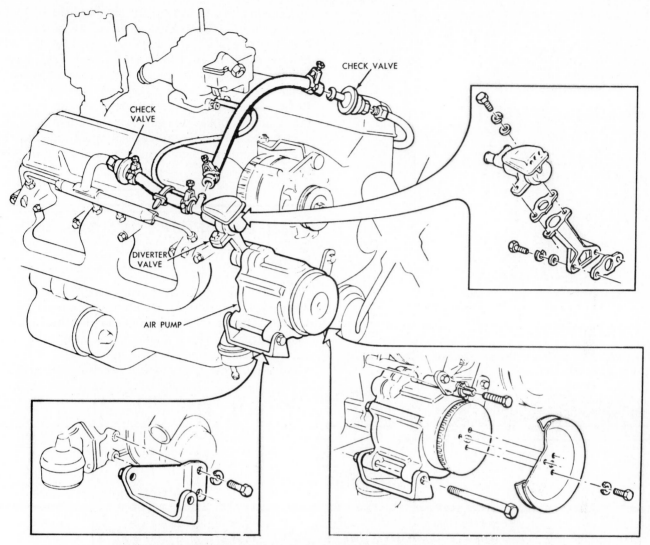

FIGURE 3-4. AIR System. (Courtesy of General Motors of Canada Ltd.)

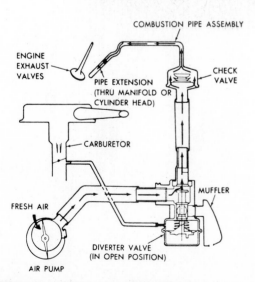

FIGURE 3-5. Air Flow through AIR System. (Courtesy of General Motors of Canada Ltd.)

PULSE AIR INJECTION REACTOR (PAIR) SYSTEMS

Design and Operation

The PAIR system draws clean air from the air cleaner and injects the air into each exhaust port. A one-way check valve and pipe are connected from a small air tank to each exhaust port, as pictured in Figure 3-7. Clean air is delivered from the air cleaner to the air storage tank by a connecting hose.

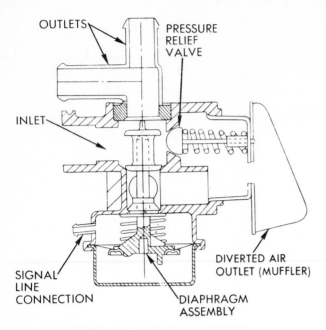

EXTERNAL MUFFLER TYPE

FIGURE 3-6. Diverter Valve Design. (Courtesy of General Motors of Canada Ltd.)

As each exhaust valve opens, high (positive) pressure occurs at the exhaust port. At the same time, low, negative pressure occurs between each positive high-pressure peak at the exhaust ports. The negative pressure opens the one-way check valve and pulls air into the exhaust port. Air injected by the PAIR system ignites and burns HC at the exhaust ports and supplies oxygen for the catalytic converter. The PAIR system is advantageous because it does not require engine power to turn a belt-driven pump.

THERMACTOR AIR PUMP SYSTEMS

Design

The air pump and air injection lines in thermactor air pump systems are similar to the ones used in the AIR system. The main components in the thermactor system are the bypass valve, idle vacuum valve, and vacuum delay valve. Ported manifold vacuum from above the throttles is supplied through the exhaust

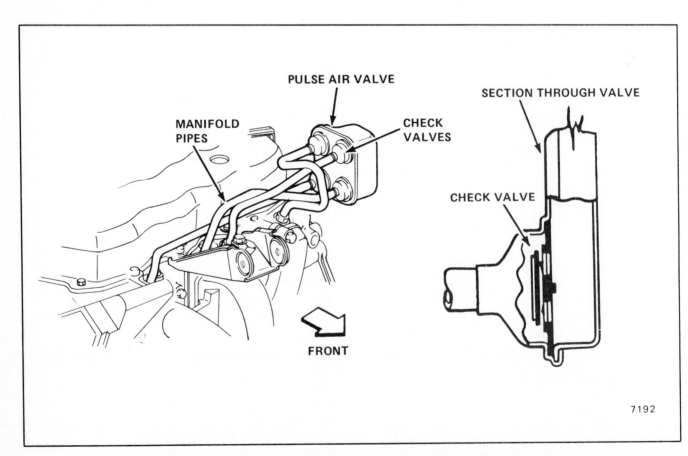

FIGURE 3-7. PAIR System. (Courtesy of General Motors of Canada Ltd.)

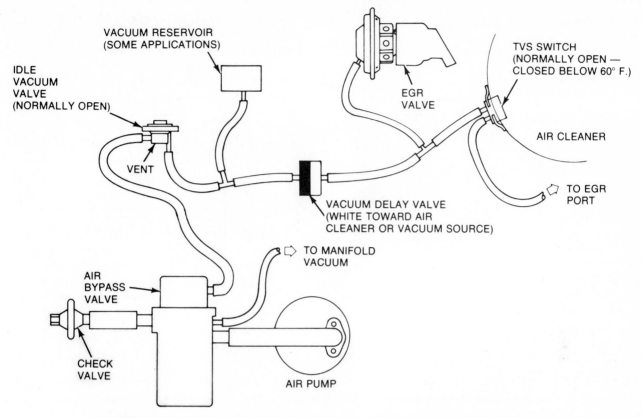

FIGURE 3-8. Thermactor Air Pump System. (Courtesy of Ford Motor Co.)

gas recirculation (EGR) thermal vacuum switch (TVS) and vacuum delay valve to the idle vacuum valve. The idle vent valve outlet is connected to the upper port on the air bypass valve. The intake manifold vacuum is connected directly to the air bypass valve lower port, as illustrated in Figure 3-8.

Idle Vacuum Valve

Ported vacuum from the EGR TVS is applied to the diaphragm in the idle vacuum valve. When such vacuumn is supplied, the movement of the diaphragm and valve seals the vent port connected to the upper air bypass valve port, as shown in Figure 3-9.

At idle speed, ported vacuum to the idle vacuum valve is zero, and consequently the diaphragm and valve move upward. The outlet from the upper air bypass valve is vented to the atmosphere when the idle vacuum valve diaphragm is in the upward position, as indicated in Figure 3-10.

Cold-Engine Operation

During cold-engine conditions, vacuum to the idle vacuum valve is shut off by the EGR TVS. The upper air bypass diaphragm chamber is vented to the atmosphere through the idle vacuum valve. Manifold vacuum under the air bypass valve diaphragm holds the diaphragm and valve assembly downward, and thus air from the thermactor pump is able to flow through the air bypass valve to the atmosphere, as pictured in Figure 3-11.

Normal Warm-Engine Operation

When the throttle is opened, ported vacuum is supplied to the idle vacuum valve through the EGR TVS. The idle vacuum valve remains in the nonventing position. The bleed port of the air bypass valve diaphragm allows manifold vacuum to equalize on both sides of the diaphragm. Diaphragm spring pressure moves the air bypass valve diaphragm and valve assembly upward. Thermactor pump air then flows through the air bypass valve to the exhaust ports, as shown in Figure 3-12.

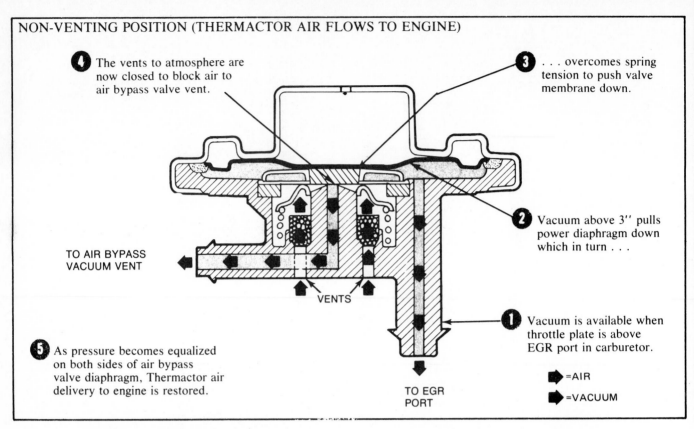

NON-VENTING POSITION (THERMACTOR AIR FLOWS TO ENGINE)

4 The vents to atmosphere are now closed to block air to air bypass valve vent.

3 ... overcomes spring tension to push valve membrane down.

2 Vacuum above 3" pulls power diaphragm down which in turn ...

TO AIR BYPASS VACUUM VENT

VENTS

5 As pressure becomes equalized on both sides of air bypass valve diaphragm, Thermactor air delivery to engine is restored.

1 Vacuum is available when throttle plate is above EGR port in carburetor.

TO EGR PORT

➡ =AIR
➡ =VACUUM

FIGURE 3-9. Nonventing Position of Idle Vacuum Valve. (Courtesy of Ford Motor Co.)

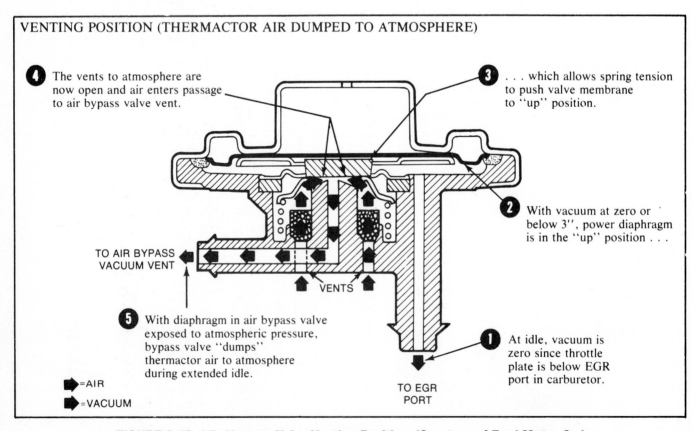

VENTING POSITION (THERMACTOR AIR DUMPED TO ATMOSPHERE)

4 The vents to atmosphere are now open and air enters passage to air bypass valve vent.

3 ... which allows spring tension to push valve membrane to "up" position.

2 With vacuum at zero or below 3", power diaphragm is in the "up" position ...

TO AIR BYPASS VACUUM VENT

VENTS

5 With diaphragm in air bypass valve exposed to atmospheric pressure, bypass valve "dumps" thermactor air to atmosphere during extended idle.

➡ =AIR
➡ =VACUUM

1 At idle, vacuum is zero since throttle plate is below EGR port in carburetor.

TO EGR PORT

FIGURE 3-10. Idle Vacuum Valve Venting Position. (Courtesy of Ford Motor Co.)

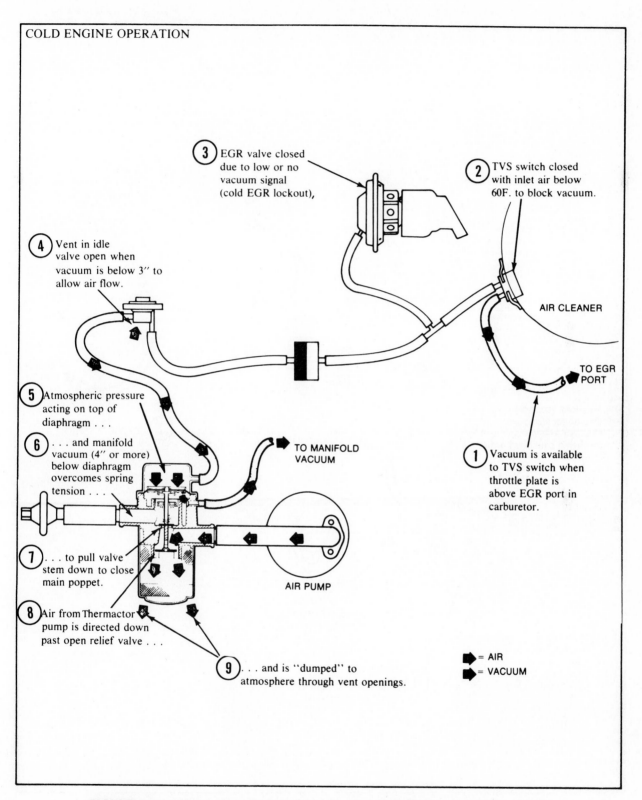

COLD ENGINE OPERATION

3 EGR valve closed due to low or no vacuum signal (cold EGR lockout),

2 TVS switch closed with inlet air below 60F. to block vacuum.

4 Vent in idle valve open when vacuum is below 3″ to allow air flow.

AIR CLEANER

5 Atmospheric pressure acting on top of diaphragm . . .

TO EGR PORT

6 . . . and manifold vacuum (4″ or more) below diaphragm overcomes spring tension . . .

TO MANIFOLD VACUUM

1 Vacuum is available to TVS switch when throttle plate is above EGR port in carburetor.

7 . . . to pull valve stem down to close main poppet.

AIR PUMP

8 Air from Thermactor pump is directed down past open relief valve . . .

9 . . . and is "dumped" to atmosphere through vent openings.

= AIR
= VACUUM

FIGURE 3-11. Thermactor System Operating in a Cold Engine. (Courtesy of Ford Motor Co.)

WARM ENGINE OR NORMAL OPERATION

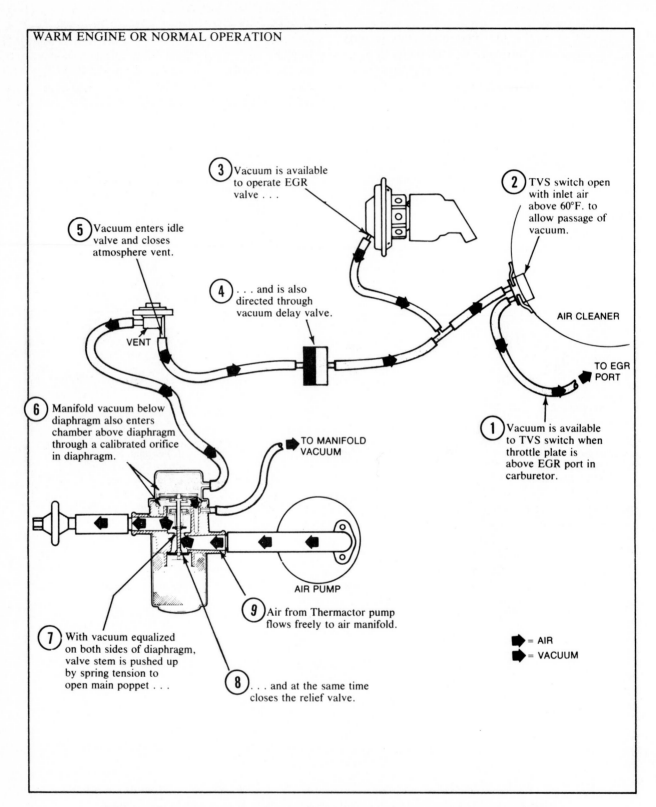

3 Vacuum is available to operate EGR valve . . .

2 TVS switch open with inlet air above 60°F. to allow passage of vacuum.

5 Vacuum enters idle valve and closes atmosphere vent.

4 . . . and is also directed through vacuum delay valve.

VENT

AIR CLEANER

TO EGR PORT

6 Manifold vacuum below diaphragm also enters chamber above diaphragm through a calibrated orifice in diaphragm.

1 Vacuum is available to TVS switch when throttle plate is above EGR port in carburetor.

TO MANIFOLD VACUUM

AIR PUMP

9 Air from Thermactor pump flows freely to air manifold.

7 With vacuum equalized on both sides of diaphragm, valve stem is pushed up by spring tension to open main poppet . . .

8 . . . and at the same time closes the relief valve.

= AIR
= VACUUM

FIGURE 3-12. Thermactor System Operating in a Warm Engine. (Courtesy Ford Motor Co.)

Extended Idle Operation

Ported vacuum to the idle vacuum valve is zero when the throttle is returned to idle. The idle vacuum valve slowly returns to the venting position as ported vacuum is bled off through the vacuum delay valve.

Atmospheric pressure is available to the upper air bypass valve diaphragm chamber through the idle vacuum valve. Manifold vacuum below the air bypass valve diaphragm holds the diaphragm and valve assembly downward. The air bypass valve directs thermactor pump air to the atmosphere, as outlined in Figure 3-13.

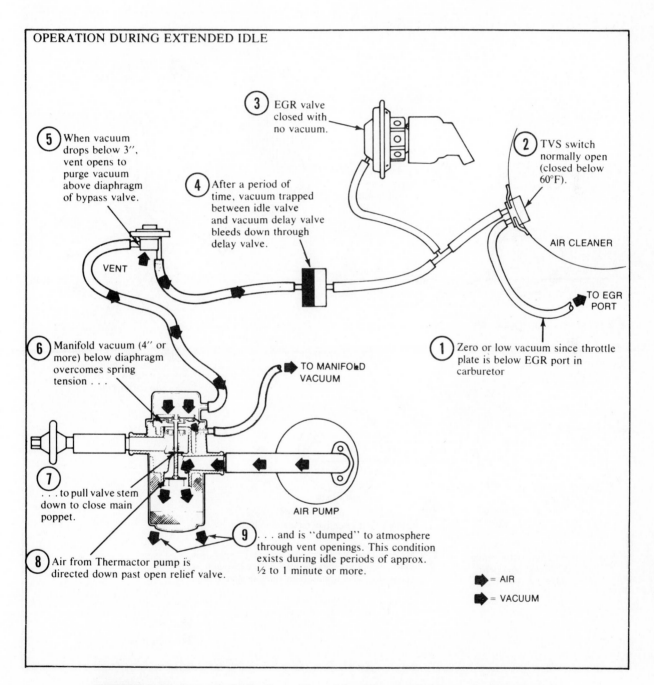

OPERATION DURING EXTENDED IDLE

3) EGR valve closed with no vacuum.

5) When vacuum drops below 3″, vent opens to purge vacuum above diaphragm of bypass valve.

2) TVS switch normally open (closed below 60°F).

4) After a period of time, vacuum trapped between idle valve and vacuum delay valve bleeds down through delay valve.

AIR CLEANER

VENT

TO EGR PORT

6) Manifold vacuum (4″ or more) below diaphragm overcomes spring tension . . .

TO MANIFOLD VACUUM

1) Zero or low vacuum since throttle plate is below EGR port in carburetor

7)
. . . to pull valve stem down to close main poppet.

AIR PUMP

9) . . . and is "dumped" to atmosphere through vent openings. This condition exists during idle periods of approx. ½ to 1 minute or more.

8) Air from Thermactor pump is directed down past open relief valve.

= AIR

= VACUUM

FIGURE 3-13. Extended Idle Operation of the Thermactor System. (Courtesy of Ford Motor Co.)

Operation during Deceleration

When the throttle is returned to the idle position, ported vacuum applied to the idle vacuum and vacuum delay valves drops to zero. Vacuum applied to the idle vacuum valve is trapped temporarily by the vacuum delay valve. The vent of the idle vacuum valve remains closed. Manifold vacuum applied to the lower air bypass valve port reaches 22 in Hg (72 kPa) on sudden deceleration. Extremely high vacuum under the air bypass valve diaphragm momentarily pulls the diaphragm and valve assembly downward. Thermactor pump air is directed to the atmosphere past the lower valve in the air bypass valve assembly, as pictured in Figure 3-14.

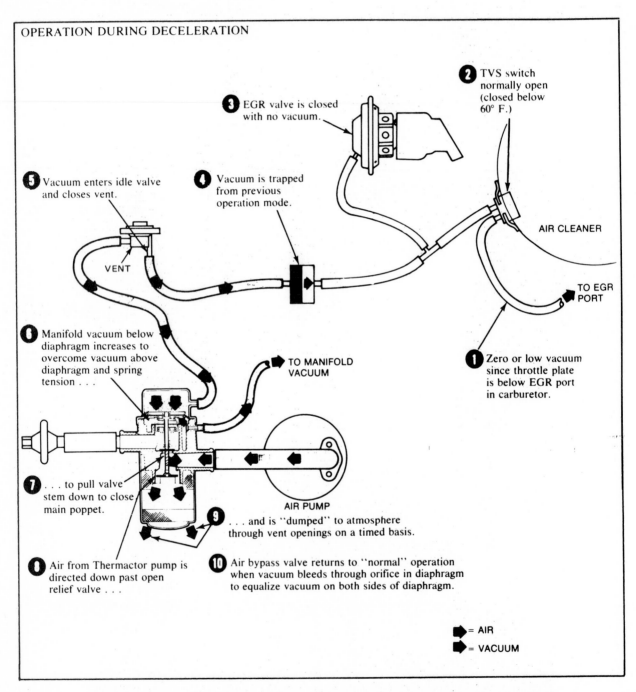

FIGURE 3-14. Thermactor System during Deceleration. (Courtesy of Ford Motor Co.)

Pressure Relief

During normal warm-engine operation, pump pressure greater than 6 PSI (42 kPa) applied to the lower air bypass valve forces the air bypass diaphragm and valve assembly downward slightly. Thermactor pump pressure is limited by directing partial airflow to the atmosphere, as shown in Figure 3-15.

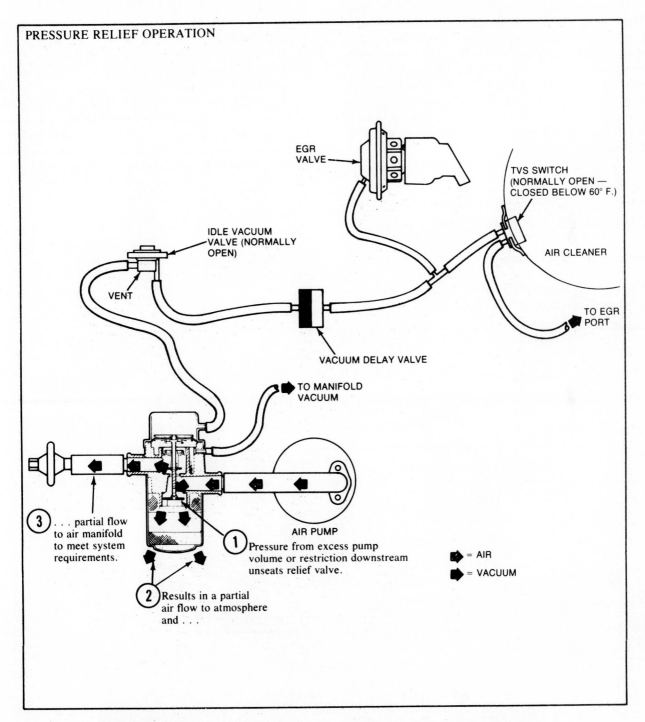

PRESSURE RELIEF OPERATION

EGR VALVE

TVS SWITCH (NORMALLY OPEN — CLOSED BELOW 60° F.)

IDLE VACUUM VALVE (NORMALLY OPEN)

AIR CLEANER

VENT

TO EGR PORT

VACUUM DELAY VALVE

TO MANIFOLD VACUUM

③ . . . partial flow to air manifold to meet system requirements.

AIR PUMP

① Pressure from excess pump volume or restriction downstream unseats relief valve.

= AIR

= VACUUM

② Results in a partial air flow to atmosphere and . . .

FIGURE 3-15. Pressure Relief Operation of the Thermactor System. (Courtesy of Ford Motor Co.)

Vacuum Differential Valve

Some thermactor air pump systems have a vacuum differential valve (VDV) connected in the vacuum hose to the air bypass valve. At normal cruising speed the VDV remains closed, as shown in Figure 3-16.

When the VDV is closed at cruising speed, manifold vacuum will be applied to the air bypass valve, and thermactor air will be directed to the exhaust ports. If the engine is decelerated, the sudden increase in manifold vacuum will open the VDV valve, as illustrated in Figure 3-17. Under this condition, the manifold vacuum applied to the air bypass valve will be vented through the VDV, and the air bypass valve will bypass the thermactor air to the atmosphere.

CHRYSLER AIR INJECTION SYSTEM

Design and Operation

The Chrysler air injection system contains a diverter valve and an air-switching valve (see Figure 3-18).

Airflow from the pump is directed to the exhaust ports or the exhaust manifold flange by the air-switching valve. A coolant control vacuum switch is connected in the vacuum circuit to the air-switching valve. Airflow is directed into the air system or bypassed to the atmosphere by the diverter valve.

Diverter Valve

Under normal operating conditions, the diverter valve directs airflow from the pump into the air system, as illustrated in Figure 3-19. When the engine is suddenly decelerated, high manifold vacuum pulls the diaphragm and valve assembly upward. Airflow is then directed from the pump to the atmosphere by the upward valve position. The pressure relief valve limits pump pressure to a maximum of 6 PSI (42 kPa) by directing excess airflow to the atmosphere. As mentioned earlier, airflow into the exhaust ports during deceleration would cause severe exhaust manifold backfiring owing to excessive burning of hydrocarbon emissions in the exhaust manifold.

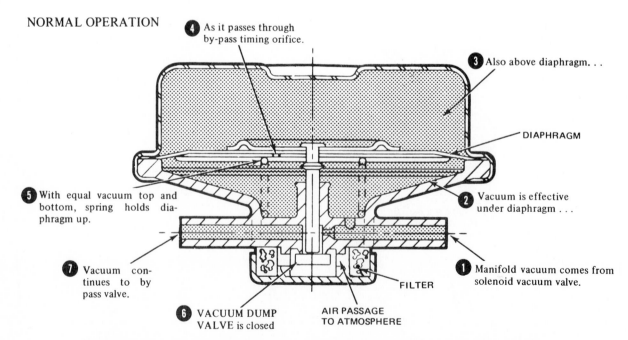

FIGURE 3-16. Vacuum Differential Valve, Closed Position. (Courtesy of Ford Motor Co.)

DECELERATION OR VACUUM FAILURE

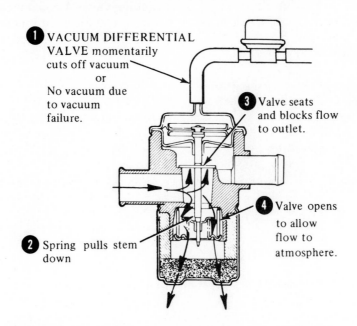

1 VACUUM DIFFERENTIAL VALVE momentarily cuts off vacuum
or
No vacuum due to vacuum failure.

3 Valve seats and blocks flow to outlet.

4 Valve opens to allow flow to atmosphere.

2 Spring pulls stem down

FIGURE 3-17. Vacuum Differential Valve (VDV), Open Position. (Courtesy of Ford Motor Co.)

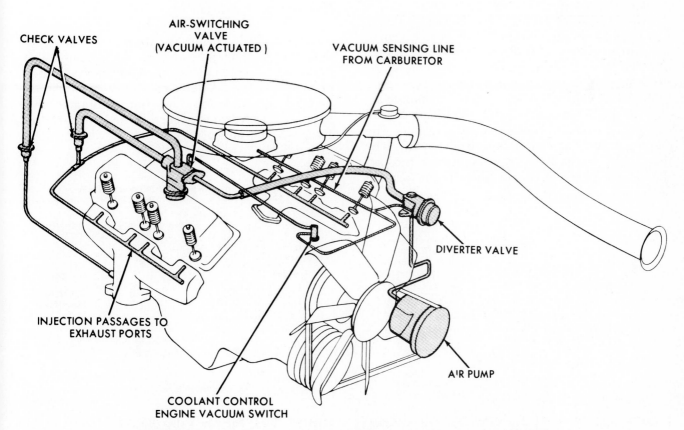

CHECK VALVES

AIR-SWITCHING VALVE (VACUUM ACTUATED)

VACUUM SENSING LINE FROM CARBURETOR

DIVERTER VALVE

INJECTION PASSAGES TO EXHAUST PORTS

AIR PUMP

COOLANT CONTROL ENGINE VACUUM SWITCH

FIGURE 3-18. Air Injection System. (Courtesy of Chrysler Canada Ltd.)

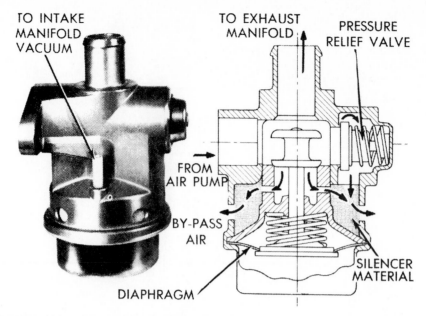

FIGURE 3-19. Diverter Valve Operation. (Courtesy of Chrysler Canada Ltd.)

Air-Switching Valve

When the engine coolant is cold, manifold vacuum is applied through the coolant control vacuum switch to the air-switching valve. The air-switching valve and diaphragm move upward when manifold vacuum is applied to the diaphragm. Airflow is then directed past the lower side of the valve to the exhaust ports, as shown in Figure 3-20.

When the engine coolant is warm, manifold vacuum to the air-switching valve diaphragm is shut off by the coolant control vacuum switch. The diaphragm and valve assembly is held downward by the diaphragm spring. Airflow is directed through the air-switching valve to the downstream location at the exhaust manifold flange.

JET VALVE SYSTEMS

Design and Operation

Some Mitsubishi engines use the jet valve air injection system. The jet valve is a small air valve mounted in the cylinder head. A jet valve is located in each cylinder. Jet valve closing is accomplished by a light return spring. Each time the intake valve opens, the jet valve is opened by an adjustable set screw on the intake rocker arm, as illustrated in Figure 3-21.

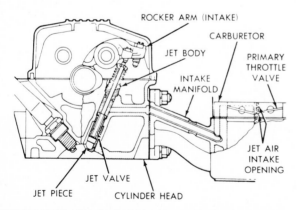

FIGURE 3-20. Air-Switching Valve Operation. (Courtesy of Chrysler Canada Ltd.)

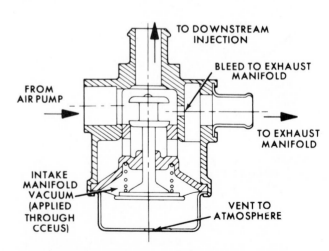

FIGURE 3-21. Jet Air Valve System. (Courtesy of Chrysler Canada Ltd.)

The jet valve air passage extends through the cylinder head and intake manifold to ports at the edge of the throttle. When the piston moves down on the intake stroke, vacuum in the cylinder will be higher than vacuum at the carburetor ports. Opening the jet valve on the intake stroke allows high cylinder vacuum to pull air through the jet valve system into the cylinder. Air injected into the cylinder through the jet valve increases cylinder turbulence and combustion efficiency, and thus reduces the HC and CO emissions.

Jet Air Control Valve and Thermo Valve

When the jet air control valve is opened, additional air flows through the valve into the jet valve system. The jet air control diaphragm and valve are opened by ported manifold vacuum above the throttle, as shown in Figure 3-22.

The thermo valve is operated by coolant temperature. When the engine coolant is cold or hot, the jet air control valve will be inoperative because the open thermo valve vents the vacuum in the system, as outlined in Figure 3-22. When the engine coolant is warm, the thermo valve will close and thus allow ported manifold vacuum to be applied to the jet air control valve. While the choke is on, additional air can pass into the jet valve system during engine warm-up via the jet air control valve and help to reduce HC and CO emissions.

Adjusting Jet Air Valve Clearance

Proceed as follows when adjusting the jet air valve clearance:

1. Back off the intake valve adjusting screw at least two turns, with the piston at top dead center (TDC) compression and the engine at normal temperature.
2. Loosen the lock nut of the jet valve adjusting screw.
3. Rotate the jet valve adjusting screw counterclockwise.
4. Place a 0.006-in (0.15-mm) feeler gauge between the adjusting screw and the jet valve stem.
5. Turn the adjusting screw clockwise until the screw touches the feeler gauge.
6. Tighten the lock nut on the jet air valve adjusting screw.

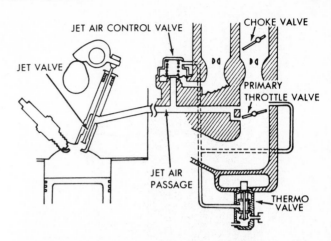

FIGURE 3-22. Jet Air Control Valve and Thermo Valve. (Courtesy of Chrysler Canada Ltd.)

7. Adjust the intake valve clearance to 0.006 in (0.15 mm), as shown in Figure 3-23.

The spring tension on the jet air valve is fairly weak. When completing step 5, be sure the jet air valve is not forced open.

ASPIRATOR SYSTEMS

Design and Operation

The aspirator system uses a one-way check valve and hose connected from the air cleaner to the exhaust system, as pictured in Figure 3-24. Between cylinder firings clean air is moved through the aspirator hose and valve into the exhaust system by

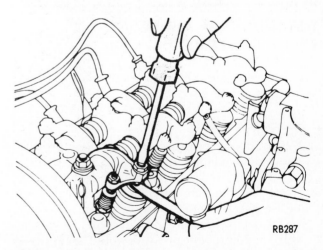

FIGURE 3-23. Jet Air Valve and Intake Valve Adjustment. (Courtesy of Chrysler Canada Ltd.)

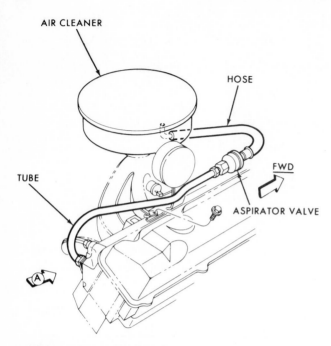

FIGURE 3-24. Aspirator System. (Courtesy of Chrysler Canada Ltd.)

negative exhaust manifold pressures. The aspirator system operates on the same principle as the PAIR system described earlier in this chapter. Exhaust is prevented from entering the air cleaner during exhaust manifold positive pressures by the one-way aspirator valve. Aspirator air may be used to burn hydrocarbons in the exhaust manifold, or to supply oxygen to the catalytic converter.

DIAGNOSIS OF AIR INJECTION PUMP SYSTEMS

The procedure for diagnosing air injection pump systems is as follows:

1. Check belt tension.
2. Check pump pressure by installing pressure gauge and adapter downstream from diverter valve or air bypass valve. Gauge adapter must fit tightly in air system hose. Pump pressure should be 2–6 PSI (14–42 kPa) on most air injection pump systems. Low pressure indicates a defective relief valve or pump. High pressure would be caused by a defective relief valve. Most air pumps are replaced as a unit.

3. On sudden deceleration, airflow should be directed to the atmosphere by the diverter valve or bypass valve. If air flows into the exhaust ports on decleration, exhaust manifold backfiring will occur.
4. Thermactor systems should divert air from the bypass valve to the atmosphere during extended idle conditions.
5. Check systems with an air-switching valve for air flow to the exhaust ports when the engine is cold. When the engine is warm, airflow should be directed to the downstream location.

Questions

1. Most air pumps use a _____ filter.
2. Burned air system hoses would indicate defective _____.
3. Air injection into the exhaust ports lowers _____ _____ emissions.
4. In vehicles equipped with a catalytic converter, the air pump may be used to supply _____ to the converter.
5. An air pump pressure of 11 PSI (77 kPa) would be normal.   T   F
6. Manifold backfiring could be caused by a defective _____ _____.
7. The PAIR system requires a belt-driven pump.   T   F
8. In a thermactor system, airflow is directed from the pump to the exhaust ports when the engine is cold.   T   F
9. Ported EGR vacuum is applied to the idle vacuum valve in the thermactor system.   T   F
10. The Chrysler air injection system directs airflow to the _____ _____ when the engine is warm.

4

FUEL SYSTEM EMISSION CONTROL EQUIPMENT

EARLY FUEL EVAPORATION (EFE) SYSTEMS

Design and Operation

A vacuum-operated power actuator is linked to the heat riser valve in the EFE system, as illustrated in Figure 4-1.

Manifold vacuum is routed through a thermal vacuum switch (TVS) to the power actuator. The EFE TVS may be combined with the EGR TVS, as pictured in Figure 4-2.

When engine coolant is cold, the TVS allows vacuum to be applied to the power actuator and closes the heat riser valve. Additional exhaust gas is forced through the intake manifold crossover passage by the closed heat riser valve. Intake manifold fuel vaporization and drivability are improved by the EFE system during cold-engine operation. When the coolant is warm, the TVS shuts off the vacuum to the power actuator and thus allows the heat riser valve to open.

Grid-type EFE System

Some EFE systems have an electrical grid mounted in a spacer block between the carburetor and the intake manifold, as indicated in Figure 4-3. A two-stage two-barrel carburetor with a primary and secondary throttle is used in these systems. The EFE electrical grid is located under the primary throttle.

The EFE grid receives electrical power from the ignition switch through a coolant temperature switch. When the coolant is cold, electrical power is supplied through the closed temperature switch contacts to the EFE grid, as indicated in Figure 4-4.

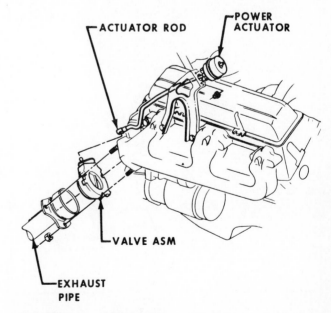

FIGURE 4-1. EFE System. (Courtesy of General Motors of Canada Ltd.)

Automotive Emission Control and Computer Systems

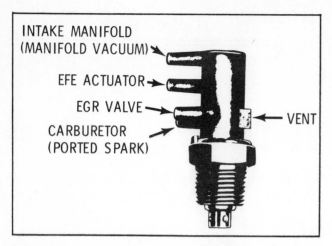

FIGURE 4-2. Combined EFE and EGR TVS. (Courtesy of General Motors of Canada Ltd.)

Heat from the EFE grid warms the fuel vapor, improves cold drivability, and reduces emission levels. Warm-engine coolant opens the EFE temperature switch contacts and stops current flow through the grid. A ground wire on the engine completes the electrical circuit from the grid to ground.

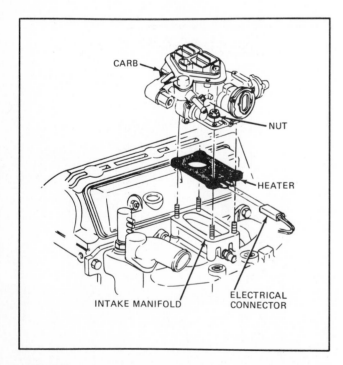

FIGURE 4-3. Grid-type EFE System. (Courtesy of General Motors of Canada Ltd.)

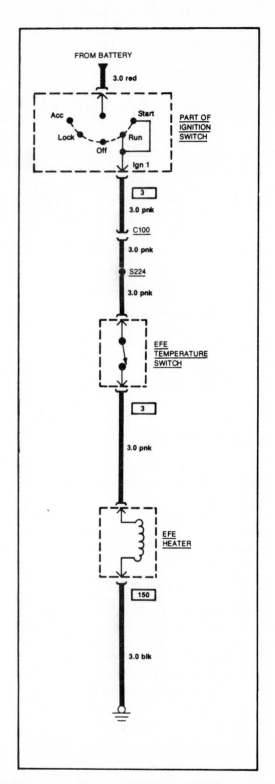

FIGURE 4-4. Grid-type EFE Electrical Circuit. (Courtesy of General Motors of Canada Ltd.)

HEATED AIR INLET SYSTEMS

Design and Operation

The heated air inlet system contains a thermostatic air bleed and a vacuum-operated air control valve mounted in the air cleaner. An exhaust manifold heat stove is connected to the air control valve by a flexible hose, as shown in Figure 4-5.

When the air cleaner temperature is below 125°F (52°C), the thermostatic air bleed supplies manifold vacuum to the air control valve diaphragm. The air control valve is lifted upward by the manifold vacuum; this upward movement shuts off the flow of cold underhood air and allows warm air from the manifold stove into the air cleaner. Heated intake air improves cold-engine drivability. At temperatures above 125°F (52°C), the thermostatic air bleed vents the vacuum applied to the air control valve diaphragm and thus allows the air control valve to move downward. Cooler underhood air is supplied to the air cleaner by the downward position of the air control valve.

CRANKCASE "BLOW-BY" GASES

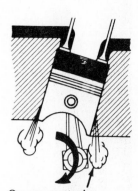

On compression stroke, unburned air and HC "blow by" the piston into the crankcase.

On power stroke, combustion (exhaust) gases "blow-by" the piston into the crankcase.

FIGURE 4-6. Origin of Crankcase Emissions. (Courtesy of Ford Motor Co.)

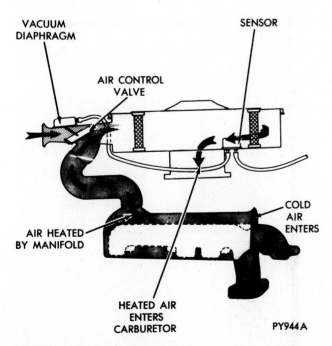

FIGURE 4-5. Heated Air Inlet System. (Courtesy of Chrysler Canada Ltd.)

POSITIVE CRANKCASE VENTILATION (PCV) SYSTEMS

Design

The PCV system prevents crankcase emissions from escaping to the atmosphere. Unburned HC emissions "blow by" the piston rings on the compression stroke. When the piston is on the power stroke, small quantities of exhaust gases escape past the rings and piston into the crankcase, as outlined in Figure 4-6.

The PCV valve is mounted in one of the rocker arm covers and is connected through a hose to the intake manifold. A fresh air hose is connected from the air cleaner to the other rocker arm cover, as pictured in Figure 4-7. Clean air is drawn from the air cleaner through the clean air hose to the rocker arm cover, and into the crankcase. Emissions in the crankcase mix with the clean air. The manifold vacuum moves crankcase emissions and air through the PCV valve into the intake manifold, as illustrated in Figure 4-7.

NORMAL ENGINE OPERATION

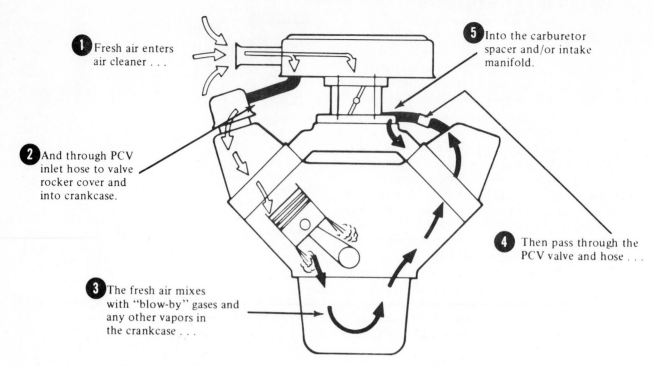

1 Fresh air enters air cleaner . . .

5 Into the carburetor spacer and/or intake manifold.

2 And through PCV inlet hose to valve rocker cover and into crankcase.

4 Then pass through the PCV valve and hose . . .

3 The fresh air mixes with "blow-by" gases and any other vapors in the crankcase . . .

FIGURE 4-7. PCV System. (Courtesy of Ford Motor Co.)

The PCV valve contains a tapered flow control valve. Manifold vacuum and crankcase pressure act as closing forces on the valve. Spring pressure pushes the valve toward the open position, as illustrated in Figure 4-8.

PCV System under Normal Loads

During idle or normal cruising speeds, the PCV valve is held in a reduced flow position by high manifold vacuum, as pictured in Figure 4-9. At idle or part throttle, crankcase emissions are reduced because of lower cylinder pressure.

PCV System under Heavy Engine Loads

Higher cylinder pressure that develops when the engine is operating under heavy loads or high-speed conditions increases crankcase emissions. Reduced manifold vacuum at wide-throttle opening allows the spring to increase PCV valve opening, as outlined in Figure 4-10. If the engine is in normal condition, the PCV flow rating will adequately handle the

CONTROL FACTORS

1 This end of the PCV valve is subject to crankcase pressure . . . tending to close the valve.

2 This end is subject to intake manifold vacuum . . . also tending to close the valve.

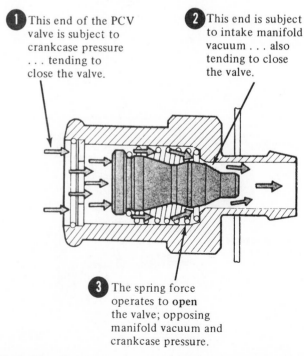

3 The spring force operates to **open** the valve; opposing manifold vacuum and crankcase pressure.

FIGURE 4-8. Internal Design of PCV Valve. (Courtesy of Ford Motor Co.)

NORMAL OPERATION

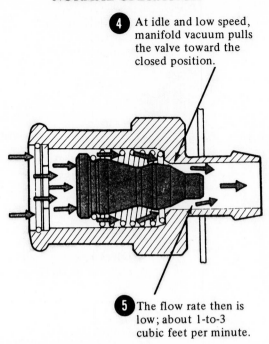

4 At idle and low speed, manifold vacuum pulls the valve toward the closed position.

5 The flow rate then is low; about 1-to-3 cubic feet per minute.

FIGURE 4-9. Normal Operation of PCV Valve. (Courtesy of Ford Motor Co.)

HIGH-SPEED OR LOAD OPERATION

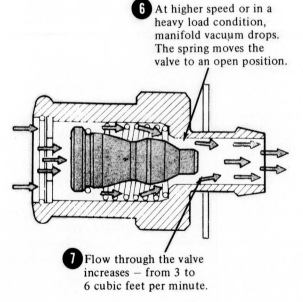

6 At higher speed or in a heavy load condition, manifold vacuum drops. The spring moves the valve to an open position.

7 Flow through the valve increases — from 3 to 6 cubic feet per minute.

NOTE:

Any flow more than 6 cubic feet per minute reverses to the carburetor inlet as shown on the page before.

FIGURE 4-10. PCV Valve Operation at Heavy Engine Loads. (Courtesy of Ford Motor Co.)

crankcase emissions at heavy load. Crankcase emissions may exceed the PCV valve flow rating if piston ring "blow by" is excessive. When the flow rating of the PCV valve is exceeded, crankcase pressure will force crankcase emissions through the clean air hose into the air cleaner.

An intake manifold backfire will seat the tapered valve against the PCV valve housing and prevent backfiring into the crankcase, as indicated in Figure 4-11. Most PCV systems have a screen on the end of the clean air hose in the air cleaner as a backfire safety device.

ADJUSTABLE PART-THROTTLE (APT) EMISSION SYSTEMS

Design and Operation

Fuel flows past the APT metering rod and fixed metering jet into the primary main wells, as illustrated in Figure 4-12. The APT adjusting screw moves the metering rod up or down in the fixed

BACKFIRE DURING CRANKING

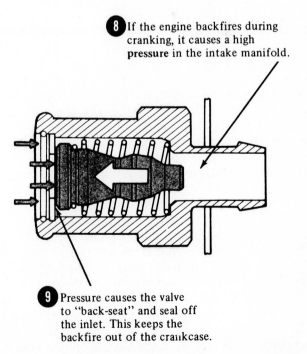

8 If the engine backfires during cranking, it causes a high **pressure** in the intake manifold.

9 Pressure causes the valve to "back-seat" and seal off the inlet. This keeps the backfire out of the crankcase.

FIGURE 4-11. PCV Valve during Backfire Conditions. (Courtesy of Ford Motor Co.)

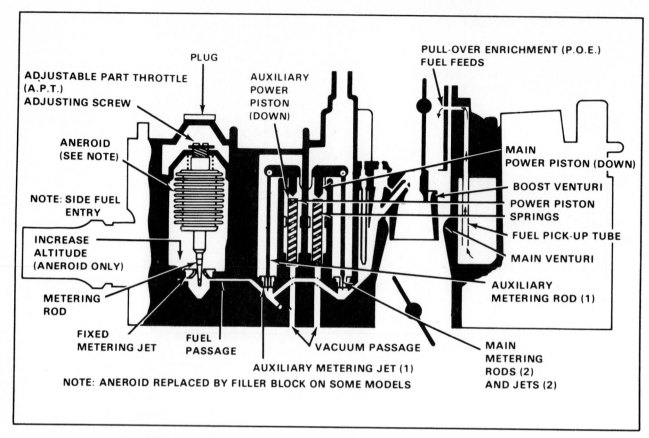

FIGURE 4-12. APT System. (Courtesy of General Motors of Canada Ltd.)

metering jet. The carburetor is flow-tested at the factory by rotating the APT screw until the correct air-fuel ratio and emission levels are obtained.

An aneroid, or a solid screw, may be used in the APT system. The aneroid expands and contracts in response to atmospheric pressure changes at different altitudes. Adjustment of the APT screw is not recommended in service. If the APT screw must be removed, count the number of turns required to bottom the C clip on the screw retainer before removing the screw. Reinstall the APT screw and back it off the original number of turns from the position where the C clip is bottomed.

The auxiliary power piston is used to obtain more accurate air-fuel ratios over the entire speed range. The auxiliary power piston metering rod and jet allow additional fuel to flow into the primary main wells. Manifold vacuum is applied to the main and the auxiliary power pistons. The auxiliary power piston spring is stronger than the spring on the main power piston. When the primary throttles are approximately half open, the auxiliary power piston spring overcomes the manifold vacuum and moves the piston upward. Mixture enrichment occurs

when the auxiliary power piston and metering rod move upward. A further reduction in manifold vacuum at approximately 75 percent primary throttle opening will allow the main power piston spring to move the piston and metering rods upward. Additional mixture enrichment is provided by the upward position of the main power piston metering rods in the main jets. The power piston springs must not be interchanged.

The pull-over enrichment (POE) system allows extra fuel to flow into the carburetor air horn at wide throttle opening. The high venturi vacuum at wide throttle opening moves fuel through the POE system which provides the necessary mixture enrichment to obtain increased engine power. Leaner mixtures can be used at moderate cruising speeds when engine power requirements are reduced.

Another type of APT adjustment is pictured in Figure 4-13. When the power piston is in the downward position, the power piston stop pin contacts the factory-metering adjustment screw. Correct part-throttle air-fuel ratios are obtained during the manufacturing process by rotating the factory-metering adjustment screw. This rotation adjusts

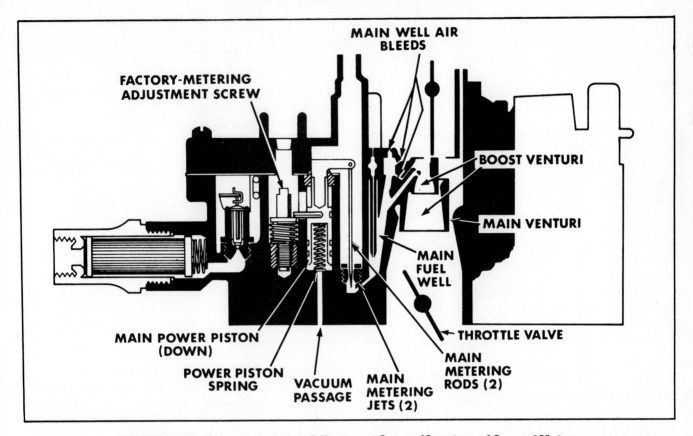

FIGURE 4-13. Factory-metering Adjustment Screw. (Courtesy of General Motors of Canada Ltd.)

the primary metering rod location in the main metering jets. The factory-metering adjustment screw is not adjustable in service.

IDLE ENRICHMENT EMISSION SYSTEMS

Idle Enrichment Valve

When the engine is operating at normal temperature, air flows through the idle enrichment valve into the idle circuit, as pictured in Figure 4-14. Additional airflow into the idle circuit provides a leaner air-fuel ratio and lower emission levels. During cold operation the idle enrichment valve will be closed by manifold vacuum to prevent drivability problems.

Idle Enrichment System

The EGR system (see Chapter 2) and the idle enrichment system are interconnected. The vacuum sole-

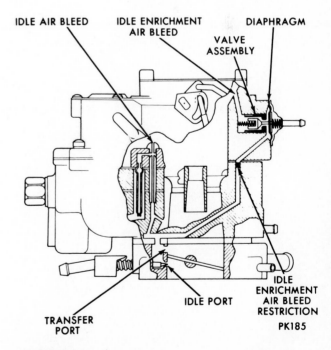

FIGURE 4-14. Idle Enrichment Valve. (Courtesy of Chrysler Canada Ltd.)

noid and the EGR delay timer (see Figure 4-15) are used in both systems. Current flow is maintained through the delay timer and vacuum solenoid winding for 35 seconds after the engine is started. The energized vacuum solenoid winding moves the solenoid plunger in the direction of the arrow, as pictured in Figure 4-15. Solenoid plunger movement opens the vacuum passage through the solenoid to the coolant control valve and the idle enrichment valve. The coolant control valve in the idle enrichment system will be open if coolant temperature is below 150°F (66°C).

Thirty-five seconds after the engine is started, the delay timer turns off the current flow through the vacuum solenoid winding. The solenoid plunger is moved in the opposite direction by a plunger return spring. This movement turns off the vacuum to the idle enrichment system and opens the vacuum passage through the solenoid to the EGR amplifier. The idle enrichment valve will be closed for 35 seconds each time a cold engine is started.

DECEL EMISSION SYSTEMS

Decel Valve

The decel valve clean air inlet is connected to the air cleaner. When the decel valve is open, clean air flows through the decel valve into the intake manifold. Under normal driving conditions, spring pressure holds the tapered valve in the closed position. On deceleration, high manifold vacuum applied to the decel valve diaphragm moves the diaphragm and valve assembly downward, as pictured in Figure 4-16. Downward valve movement allows air to flow through the open decel valve into the intake manifold. Additional airflow into the intake manifold provides leaner mixtures and lower emissions on deceleration.

Electrically Operated Decel Systems

A solid state speed switch operates a decel solenoid in the electrically controlled system. Engine RPM signals are sent to the speed switch by two primary ignition circuit connections, as illustrated in Figure 4-17. The speed switch energizes the solenoid winding at 1,500 RPM. When the engine is decelerated, the extended solenoid plunger holds the throttle open until RPM drops below 1,500. Deceleration emission levels are improved by holding the throttle open as the engine slows down. High hydrocarbon emissions on deceleration could overheat and damage the catalytic converter. The electrically operated decel system protects the catalytic converter.

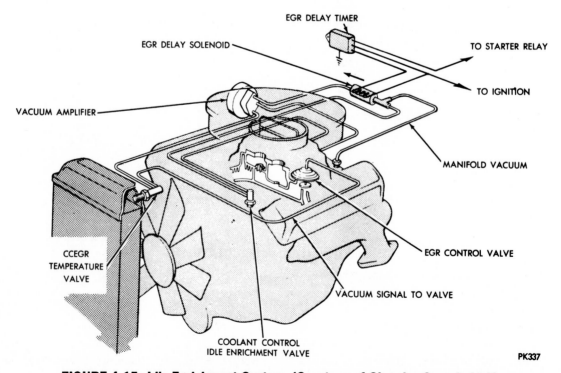

FIGURE 4-15. Idle Enrichment System. (Courtesy of Chrysler Canada Ltd.)

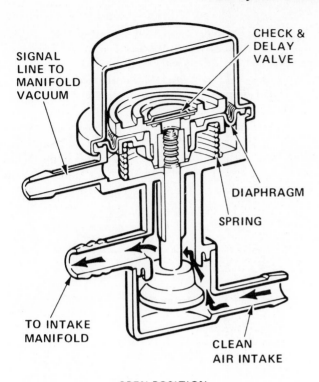

SIGNAL LINE TO MANIFOLD VACUUM

CHECK & DELAY VALVE

DIAPHRAGM

SPRING

TO INTAKE MANIFOLD

CLEAN AIR INTAKE

OPEN POSITION

FIGURE 4-16. Decel Valve. (Courtesy of General Motors of Canada Ltd.)

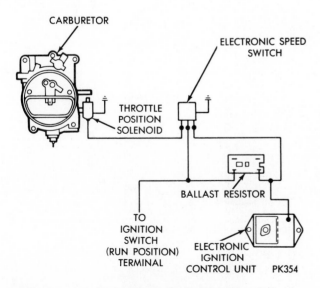

CARBURETOR

ELECTRONIC SPEED SWITCH

THROTTLE POSITION SOLENOID

BALLAST RESISTOR

TO IGNITION SWITCH (RUN POSITION) TERMINAL

ELECTRONIC IGNITION CONTROL UNIT

PK354

FIGURE 4-17. Electrically Operated Decel System. (Courtesy of Chrysler Canada Ltd.)

ELECTRIC ASSIST CHOKE SYSTEMS

Operation

Some electric assist chokes use a heating element located in the choke spring cover. Current flows from the alternator stator terminal through the silver contacts and heating element to the gound strap, as illustrated in Figure 4-18. The contacts will complete the electric circuit when choke cover temperature exceeds 60°F (16°C). Manifold vacuum moves hot air through the heat pipe into the choke housing in the normal manner. The electric assist choke opens the choke faster and reduces HC and CO emissions at choke cover temperatures above 60°F (16°C).

Electric Assist Choke with External Control Switch

The external control switch is connected between the ignition switch and the choke heating element, as illustrated in Figure 4-19. A single or a dual control switch may be used. Dual control switches have a resistor mounted beside the switch.

The single control switch operates as follows:

1. Below 60°F (16°C), the control switch circuit is open.

2. Between 60°F (16°C) and 130°F (54°C), the control switch supplies full power to the choke heating coil.

3. Above 130°F (54°C), the control switch circuit is open.

The dual control switch operates slightly differently:

1. Below 60°F (16°C), the control switch supplies partial voltage to the choke heating coil.

2. When the control switch temperature is between 60°F (16°C) and 130°F (54°C), full voltage is supplied to the choke heating coil.

3. Above 130°F (54°C), the control switch is open. (Figure 4-20 pictures a single and dual control switch.)

The later model electric assist choke system is shown in Figure 4-21. A set of oil pressure switch

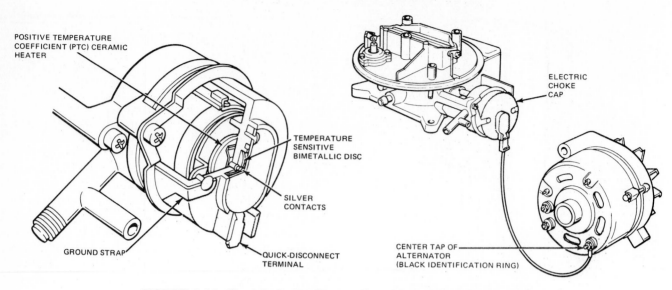

FIGURE 4-18. Electric Assist Choke. (Courtesy of Ford Motor Co.)

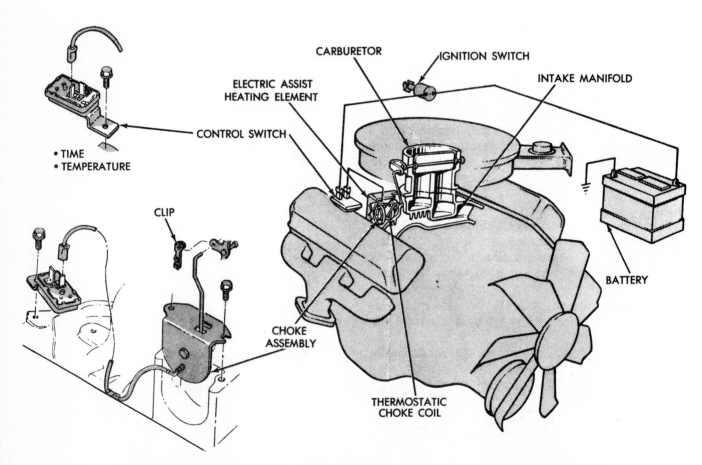

FIGURE 4-19. Electric Assist Choke with External Control Switch. (Courtesy of Chrysler Canada Ltd.)

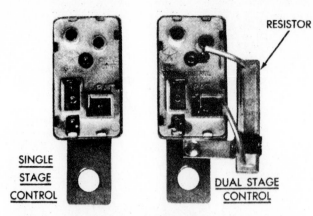

FIGURE 4-20. Electric Assist Choke Control Switches. (Courtesy of Chrysler Canada Ltd.)

Different single- and dual-control switches are used in the later model system. Single-control switch A in Figure 4-21 is closed above 80°F (27°C) and open below 55°F (13°C). The dual-control unit, item B, supplies partial voltage to the choke heating coil below 55°F (13°C), and full power above 80°F (27°C).

EVAPORATIVE EMISSION CONTROL SYSTEMS

Design and Operation

Fuel vapors from the fuel tank and carburetor are vented into a charcoal canister in the evaporative emission control system. The canister absorbs the fuel vapors while the engine is not running. A vacuum hose moves fuel vapors out of the canister into the intake manifold when the engine is running. The rollover valve in the vapor line of the fuel tank will prevent liquid fuel from entering the canister if the vehicle is overturned in an accident. Some systems use two canisters, as illustrated in Figure 4-22. A special filter-separator is used in some systems with a return line to the fuel tank. Fuel vapors are separated and returned to the fuel tank by the filter-

contacts is connected in series between the ignition switch and the control switch. Oil pressure will not be available unless the engine is running and thus can close the contacts and supply power to the control switch. For systems without an oil pressure switch circuit, the tension of the choke spring could be relaxed on a cold engine if the ignition switch was left on without starting the engine.

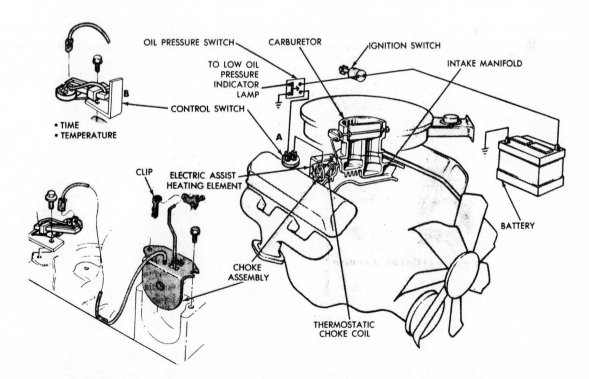

FIGURE 4-21. Later Model Electric Assist Choke System. (Courtesy of Chrysler Canada Ltd.)

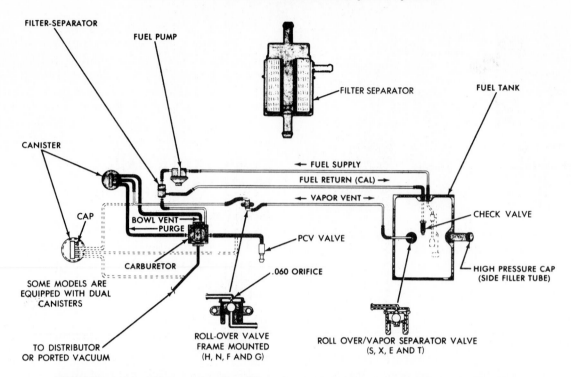

FIGURE 4-22. Evaporative Emission Control System. (Courtesy of Chrysler Canada Ltd.)

separator and return line. The evaporative emission control system prevents the escape of HC emissions from the fuel tank and carburetor to the atmosphere.

The canister purge port in the carburetor is usually located above the throttle. In this type of system canister purging will only occur when the throttle is opened. A few systems have a small, constant-bleed canister purge port below the throttles to provide some canister purging at idle speed. A vacuum-operated canister purge valve, as shown in Figures 4-23 and 4-24, is used in some systems.

The purge control valve operates as follows:

1. During engine shutdown, vapors from the fuel tank and carburetor are routed through the purge control valve into the canister.

2. When the engine is operating at normal cruising speed, the vacuum opens the valve and enables fuel vapors to be purged from the canister into the PCV hose or carburetor spacer. Spark port, or EGR port vacuum is usually applied to the purge control valve.

3. At idle speed, the vacuum at the spark port or EGR port is insufficient to open the purge control

valve. Vapors are routed through the valve into the canister. In models in which the manifold vacuum is applied to the purge control valve, a small amount of vapor will be purged from the canister at idle speed.

Carburetor Bowl Vents

Mechanically operated bowl vents (see Figure 4-25) are used to vent the carburetor float bowl to the canister when the engine is idling or shut off. Opening the throttle causes the bowl vent to close. Fuel vapors will be vented from the float bowl to the air horn under the air cleaner when the bowl vent to the canister is closed.

Some carburetors have an electrically operated bowl vent valve. When the engine is shut off, the vent valve winding is deenergized and fuel vapors from the float bowl are allowed to enter the canister, as pictured in Figure 4-26. With the engine running, the vent valve winding is energized and the valve positioned to allow fuel bowl vapors to escape to the vent tube under the air cleaner.

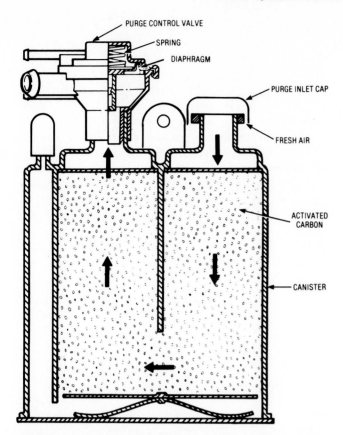

FIGURE 4-23. Vapor Canister with Purge Valve. (Courtesy of Ford Motor Co.)

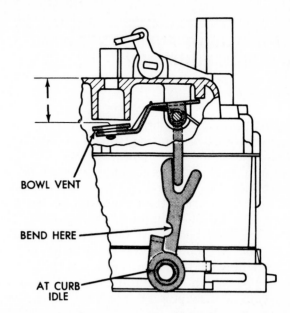

FIGURE 4-25. Mechanically Operated Carburetor Bowl Vent Valve. (Courtesy of Chrysler Canada Ltd.)

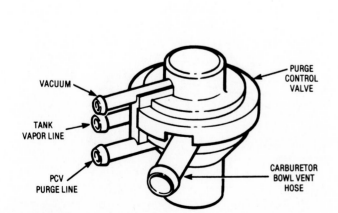

FIGURE 4-24. Purge Control Valve. (Courtesy of Ford Motor Co.)

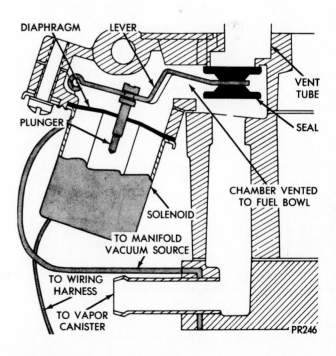

FIGURE 4-26. Electrically Operated Bowl Vent Valve. (Courtesy of Chrysler Canada Ltd.)

DIAGNOSIS OF FUEL SYSTEM EMISSION EQUIPMENT

EFE System

The EFE actuator diaphragm and heat riser valve may be checked by applying 10 in Hg (33 kPa) to the diaphragm, as shown in Figure 4-27. When the heat riser is fully closed, there should be 0.065 in (1.65 mm) between the actuator arm and stop. Free movement of the heat riser valve from fully open to fully closed is essential. Vacuum through the EFE TVS should be zero when the engine is at normal temperature. The TVS should allow vacuum to the actuator diaphragm when the engine coolant is cold. When the engine coolant is cold if the heat riser valve is open continuously, power loss or acceleration stumbles can be expected.

Grid-type EFE

A 12-V test light should be used to check power at the grid terminal connected to the temperature switch with the ignition on. Power should be available with a cold engine. Warm engine coolant should cause the temperature switch to open the circuit to the grid. The grid may be checked for continuity by connecting an ohmmeter across the grid terminals. A satisfactory grid will register low resistance, and an open grid will provide an infinite ohmmeter reading. Loss of power and acceleration stumbles can be expected from an inoperative grid system. (The grid system is shown in Figure 4-4.)

FUNCTIONAL TEST/MAINTENANCE

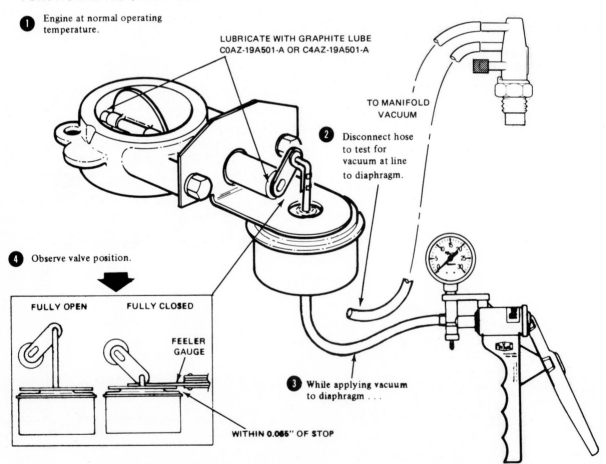

❶ Engine at normal operating temperature.

LUBRICATE WITH GRAPHITE LUBE
C0AZ-19A501-A OR C4AZ-19A501-A

TO MANIFOLD VACUUM

❷ Disconnect hose to test for vacuum at line to diaphragm.

❹ Observe valve position.

FULLY OPEN   FULLY CLOSED

FEELER GAUGE

❸ While applying vacuum to diaphragm . . .

WITHIN 0.065" OF STOP

FIGURE 4-27. Diagnosis of EFE System. (Courtesy of Ford Motor Co.)

Heated Air Inlet Systems

The heated air inlet system is diagnosed as follows:

1. With the engine idling, check for full manifold vacuum at the thermostatic air bleed.
2. When the thermostatic bleed is below 125°F (52°C), vacuum should be available at the vacuum diaphragm. The thermostatic bleed may be cooled for testing purposes.
3. Apply 10 in Hg (33 kPa) to the vacuum diaphragm and make sure that the air control valve is closed. Free movement of the air control valve is essential.
4. Check the condition of the heat stove flex tube. (Refer to Figure 4-5 for heated air inlet system design.)

An air control valve that is continuously open when the engine is cold can cause poor performance and acceleration stumbles. Detonation may occur when the air control valve remains in the hot air position with the engine at normal temperature.

PCV System

The engine should slow down 50 RPM when the PCV valve is plugged with the engine idling. If very little change occurs in RPM when the PCV valve is covered, the PCV valve or hose is restricted. An excessive change in RPM when the PCV valve is plugged indicates a sticking PCV valve. The antibackfire screen on the clean air hose must be cleaned or replaced when the air cleaner is serviced.

Idle Enrichment System

A slight decrease in engine idle speed should occur when 10 in Hg (33 kPa) is applied to the idle enrichment valve. Lack of RPM change indicates an inoperative idle enrichment valve. (Refer to Figure 4-14 for idle enrichment valve design.) Vacuum should be available at the idle enrichment valve for 35 seconds each time a cold engine is started. The coolant control idle enrichment valve should be open below 150°F (66°C). Connect a vacuum gauge between the vacuum solenoid and the coolant control idle enrichment valve with a "T" fitting. Full manifold vacuum should be indicated on the gauge for 35 seconds each time the engine is started. (Refer to Figure 4-15 for idle enrichment system vacuum and electric circuit.) An incorrect vacuum gauge

reading indicates a defective vacuum solenoid or EGR delay timer. (Diagnosis of the vacuum solenoid and delay timer was explained in Chapter 2.)

Vacuum-Operated Decel Valves

The vacuum-operated decel valve may be diagnosed as follows:

1. Check for full manifold vacuum at the decel valve signal hose. (Refer to Figure 4-16.)
2. Disconnect the clean air hose at the decel valve or air cleaner.
3. An audible rush of air should be taken into the decel valve when the engine is rapidly accelerated and decelerated. Replace the decel valve if air cannot be heard entering the valve on deceleration.

Electrically Operated Decel Valves

Follow the steps outlined below when diagnosing the electric decel system:

1. With the ignition switch on, check for power at both connections from the primary ignition circuit to the electronic speed switch. Lack of power at either terminal indicates an open wire. (Refer to Figure 4-17 for circuit connections.)
2. Connect an ohmmeter from the throttle position solenoid terminal at the speed switch to ground. A low ohmmeter reading is satisfactory, while an infinite reading indicates an open wire or solenoid winding.
3. If no defects have been located in steps 1 and 2, replace the electronic speed switch.

Integral Electric Assist Choke

The integral electric assist choke should be diagnosed with an ohmmeter and a voltmeter, as outlined below:

1. With the engine running, connect a voltmeter from the choke terminal to ground. Voltage should be approximately 7 V. If the voltage is below 6, check alternator output and the stator wire from the alternator to the choke terminal. (See Figure 4-18 for an electric assist integral choke.)

2. Connect an ohmmeter from the choke terminal to the ground strap. Continuity should only be available above 60°F (16°C).

Electric Assist Choke with External Control Switch

Use the following sequence to trouble-shoot the external electric choke (see Figure 4-19):

1. With the ignition switch on, check for power at the control switch ignition terminal using a 12-V test light. If no power is available, the wire from the ignition switch to the control switch is open.
2. With a cold engine operating at idle speed, connect a 12-V test light from the choke heating element terminal at the control switch to ground.
3. A single- or dual-control switch should supply power when the control switch temperature is 60°F to 130°F (16°C to 54°C). Partial voltage should be available through the dual-control unit below 60°F (16°C). Later model single-control switches should only supply power above 80°F (27°C). Dual-control switches on later model systems should supply partial voltage below 55°F (13°C) and full voltage above 80°F (27°C). (See Figure 4-21 for the later model system.)
4. Connect an ohmmeter from the heating coil lead to ground. A low ohmmeter reading indicates a satisfactory heating coil. Replace the heating coil if an infinite reading is obtained.

Evaporative Control Systems

The purge control valve shown in Figure 4-24 may be diagnosed as follows:

1. Disconnect the vacuum signal hose from the purge valve.
2. Remove the purge valve from the canister. A slight hissing noise should be heard from the small orifice of the purge valve. If the hissing noise is strong, however, the purge valve should be replaced.
3. Reconnect the purge valve vacuum signal line and accelerate the engine to 2,000 RPM. A strong hissing noise should now be heard from the purge valve.

FIGURE 4-28. Canister Air Filter Cleaning. (Courtesy of Chrysler Canada Ltd.)

4. Return the engine to idle speed. If a ported vacuum signal is used at the purge valve, only a slight purge valve hissing should be detected.

Periodic cleaning of the canister air filter is necessary on some canisters, as outlined in Figure 4-28.

Questions

1. An acceleration stumble during cold-engine operation could be caused by loss of vacuum to the EFE actuator diaphragm.   T   F
2. Current flows through the EFE grid at normal operating engine temperature.   T   F
3. If the air control valve in a heated air inlet system is in the cold air position continuously, the engine may _____ on cold _____.
4. When vacuum is applied to the air control valve diaphragm, the valve will be in the _____ air position.
5. PCV valve flow decreases when the engine is accelerated from idle to 2,000 RPM.   T   F
6. A defective PCV valve is indicated if the engine slows down 150 RPM when the valve is covered.   T   F
7. The APT screw is adjustable in service. T   F
8. The aneroid in the APT system provides _____ _____.

9. Additional air flows through the idle enrichment valve into the idle circuit with the engine at normal temperatures.   T   F

10. The coolant control idle enrichment valve is closed when engine coolant is below 150°F (66°C).   T   F

11. A dual external control switch in an electric assist choke system supplies partial power to the choke heating coil at temperatures below 55°F (13°C).   T   F

12. On later models of external control switch-type electric assist chokes, power is supplied to the control switch when the ignition switch is on and the engine is not running.   T   F

5

DISTRIBUTOR SPARK RETARD EMISSION SYSTEMS

SPARK DELAY VALVE SYSTEMS

Coolant Spark Control (CSC) System

Two ported vacuum switches (PVS), a spark delay valve, and a check valve are used in the coolant spark control system. The CSC system is interconnected with the EGR system, as illustrated in Figure 5-1. Ported EGR vacuum is connected to the distributor vacuum advance through one PVS switch and the check valve.

Cold-Engine Operation

Ported manifold vacuum is available at the EGR port as soon as the throttle is opened from the idle position. Spark port vacuum is not available until a higher RPM is reached because the spark port is located a considerable distance above the throttles. When engine coolant is below 85°F (30°C), the EGR PVS allows EGR vacuum to be applied through the check valve to the distributor advance as soon as the throttle is opened. This provides faster vacuum advance than vacuum supplied from the spark port, as shown in Figure 5-2. Engine drivability is improved during cold operation by providing vacuum advance sooner.

Warm-Engine Operation

When the coolant reaches 85°F (30°C), the EGR PVS shuts off vacuum to the distributor advance and turns on vacuum to the EGR valve. Vacuum for the distributor advance is supplied from the spark port, 225° PVS, and the spark delay valve. A restricted port in the spark delay valve delays vacuum to the distributor advance for approximately 20 seconds after vacuum is available at the spark port, as pictured in Figures 5-3 and 5-4. Under warm-engine conditions the check valve prevents spark port vacuum from leaking into the EGR system.

Some spark delay valves have a "dump" valve that vents the vacuum advance through the filter to the atmosphere when there is no vacuum signal to the valve, as shown in Figure 5-4. The action of the "dump" valve prevents fuel vapors from condensing in the vacuum advance. When the engine is decelerated, the spark delay valve releases the vacuum instantly to prevent the vacuum advance from locking in the advanced position, as pictured in Figure 5-5.

If the engine begins to overheat, the 225° PVS will supply manifold vacuum to the distributor advance, as outlined in Figure 5-6. Delayed vacuum advance in the low-speed range reduces emission levels. The 85° PVS shuts off vacuum to the dual area diaphragm of the transmission modulator when the engine is cold. Loss of vacuum to the dual area diaphragm changes the transmission shift points slightly and improves cold engine drivability.

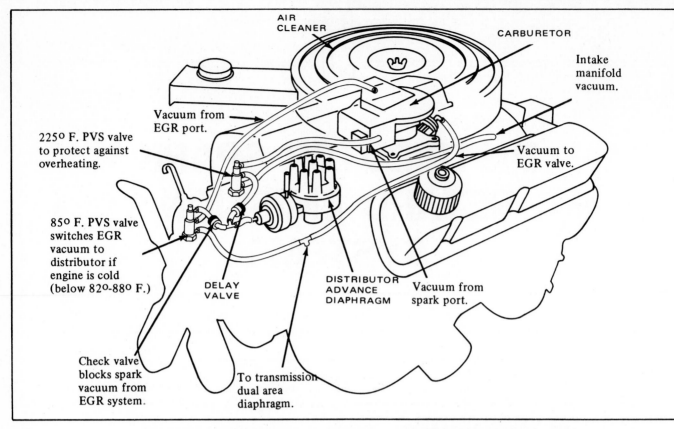

FIGURE 5-1. Coolant Spark Control System. (Courtesy of Ford Motor Co.)

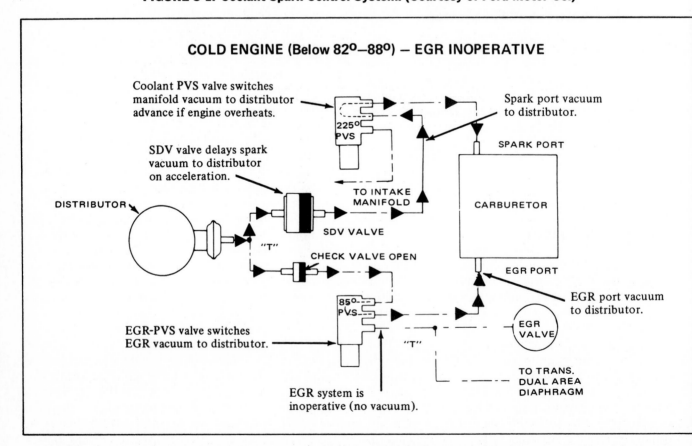

FIGURE 5-2. Coolant Spark Control System during Cold-Engine Operation. (Courtesy of Ford Motor Co.)

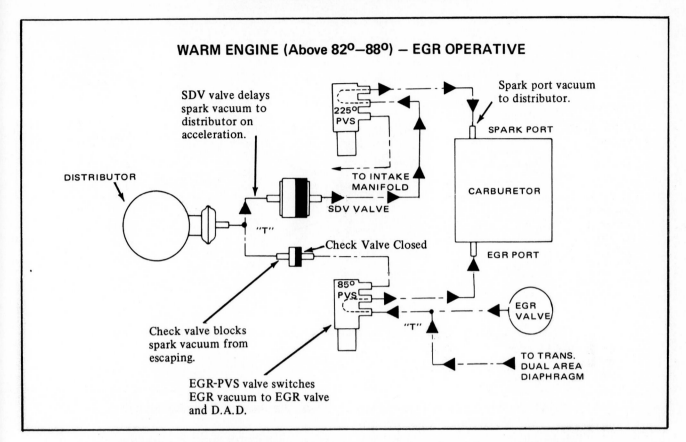

FIGURE 5-3. Coolant Spark Control System during Warm-Engine Operation. (Courtesy of Ford Motor Co.)

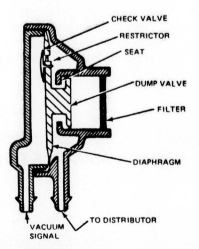

FIGURE 5-4. Internal Design of Spark Delay Valve. (Courtesy of Ford Motor Co.)

Cold-start Spark Advance System (CSSA), Cold Operation

Two PVS switches, a spark delay valve, and a check valve are connected in the CSSA system. Manifold vacuum is available through the top ports of the 125° PVS to the vacuum advance when coolant temperature is below 125°F (52°C). On hard acceleration, manifold vacuum decreases instantly. Vacuum advance is maintained for a few seconds on hard acceleration because the check valve momentarily traps vacuum in the distributor advance, as shown in Figure 5-7. Cold engine drivability is improved during cold operation by applying full manifold vacuum to the vacuum advance.

Cold-start Spark Advance System (CSSA), Warm Operation

When the coolant reaches 125°F (52°C), the 125° PVS turns off the manifold vacuum applied to the distributor advance. Spark port vacuum is now

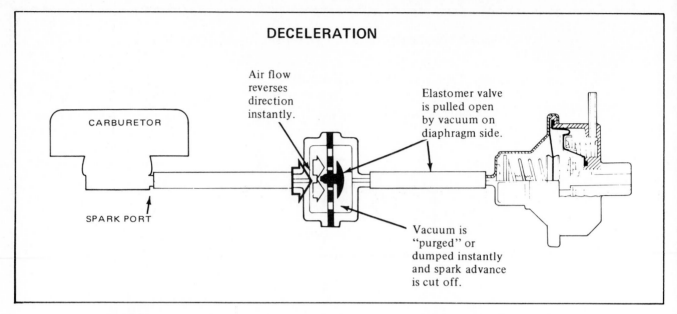

DECELERATION

CARBURETOR

SPARK PORT

Air flow reverses direction instantly.

Elastomer valve is pulled open by vacuum on diaphragm side.

Vacuum is "purged" or dumped instantly and spark advance is cut off.

FIGURE 5-5. Spark Delay Valve Operation on Deceleration. (Courtesy of Ford Motor Co.)

available through the cooling PVS, spark delay valve, and lower 125° PVS ports to the vacuum advance, as illustrated in Figure 5-8. The spark delay valve will delay the vacuum applied to the distributor advance for approximately 20 seconds each time the throttle is opened. Extremely hot coolant temperatures will cause the cooling PVS to supply manifold vacuum through the spark delay valve to the vacuum advance.

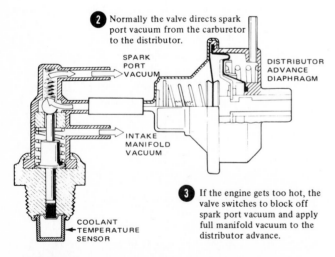

❷ Normally the valve directs spark port vacuum from the carburetor to the distributor.

SPARK PORT VACUUM

DISTRIBUTOR ADVANCE DIAPHRAGM

INTAKE MANIFOLD VACUUM

COOLANT TEMPERATURE SENSOR

❸ If the engine gets too hot, the valve switches to block off spark port vacuum and apply full manifold vacuum to the distributor advance.

FIGURE 5-6. PVS Operation. (Courtesy of Ford Motor Co.)

Dual Diaphragm Distributors

Many distributor spark retard emission systems use a dual diaphragm vacuum advance. Ported vacuum is supplied to the outer diaphragm through the spark delay valve or other emission devices. Full manifold vacuum is supplied to the inner diaphragm as illustrated in Figure 5-9. High manifold vacuum at idle speed or on deceleration will overcome the inner diaphragm spring and move the stop toward the distributor housing. The outer diaphragm and vacuum advance arm will move toward the distributor housing until the slot in the advance arm contacts the retard stop. Inward movement of the vacuum advance arm rotates the distributor plate and points, or pickup coil, to retard the timing 6°–8°. Both vacuum hoses must be disconnected and plugged whenever ignition timing is checked.

ORIFICE SPARK ADVANCE CONTROL (OSAC) SYSTEMS

Design and Operation

The OSAC valve is connected in series between the carburetor vacuum port and the distributor advance. A restricted port in the OSAC valve delays

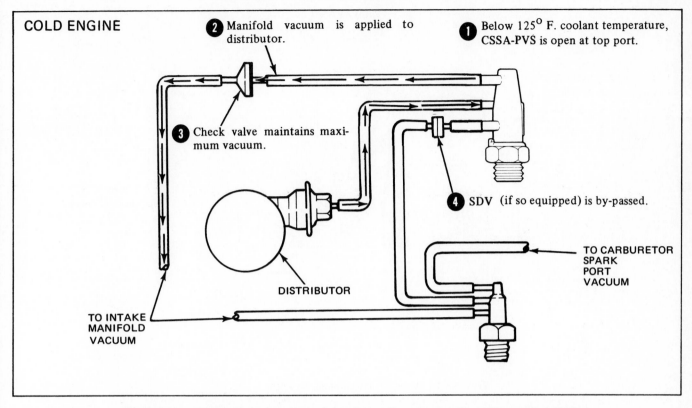

FIGURE 5-7. CSSA System during Cold Operation. (Courtesy of Ford Motor Co.)

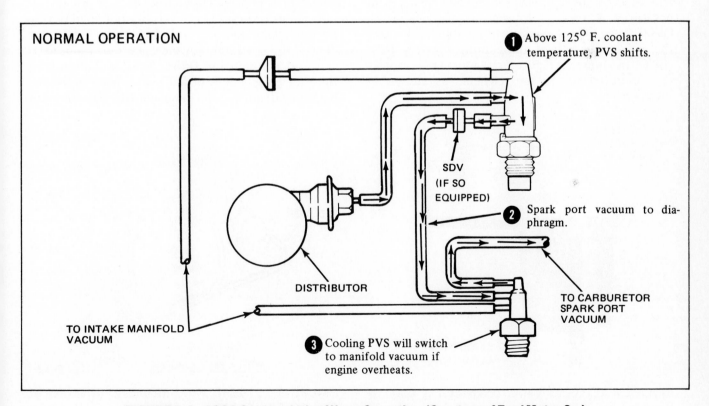

FIGURE 5-8. CSSA System during Warm Operation. (Courtesy of Ford Motor Co.)

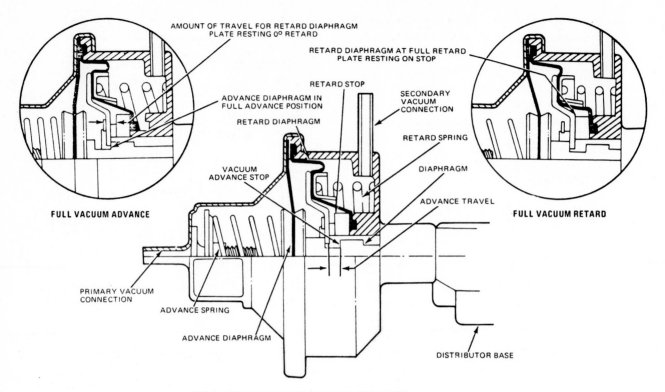

VACUUM ADVANCE AND RETARD DIAPHRAGMS AT REST

FIGURE 5-9. Dual Diaphragm Vacuum Advance. (Courtesy of Ford Motor Co.)

vacuum advance for approximately 20 seconds, when the throttle is opened at normal engine temperatures. A temperature sensing element in the OSAC valve allows vacuum to be applied through the valve instantly at temperatures below 60°F (16°C), as shown in Figure 5-10. Most OSAC valves are mounted in the air cleaner, as shown in Figure 5-11.

The OSAC valve provides instant vacuum advance and improved drivability during cold-engine operation. Improved emission levels are obtained during normal engine temperatures when the OSAC valve delays the vacuum advance. Ported vacuum is supplied through the OSAC valve and the upper thermal ignition control (TIC) valve ports to the distributor advance at normal engine temperatures. A

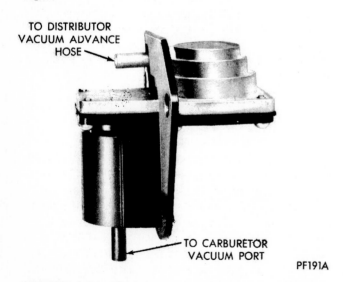

FIGURE 5-10. OSAC Valve. (Courtesy of Chrysler Canada Ltd.)

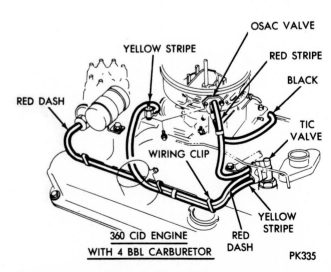

FIGURE 5-11. OSAC Valve Mounting. (Courtesy of Chrysler Canada Ltd.)

manifold vacuum hose is connected to the lower TIC valve port, as shown in Figure 5-11. The TIC valve supplies full manifold vacuum to the distributor advance if the engine starts overheating.

TRANSMISSION-CONTROLLED SPARK (TCS) SYSTEMS

Low-gear Operation

The main components in the TCS system are the vacuum advance solenoid, transmission switch, and temperature switch. Contacts in the transmission switch are normally open. The transmission switch contacts will close when pressure from the transmission hydraulic circuit is applied to the switch. Manifold vacuum applied to the vacuum advance is routed through the vacuum advance solenoid. In low or second gear, the transmission switch contacts are open and the vacuum advance solenoid is allowed to remain deenergized. The vacuum advance solenoid plunger closes the vacuum advance passage, as shown in Figure 5-12. When vacuum to the distributor advance is shut off, the vacuum advance is vented through the air filter on the solenoid.

High-gear Operation

When the transmission shifts into high gear, the transmission switch contacts are closed by hydraulic

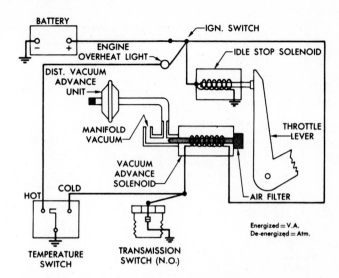

FIGURE 5-13. TCS System during High-gear Operation. (Courtesy of General Motors of Canada Ltd.)

pressure. The vacuum advance solenoid electrical circuit is completed to ground through the transmission switch. Manifold vacuum is applied past the end of the solenoid plunger to the vacuum advance, as pictured in Figure 5-13. The TCS system improves emission levels in the low-speed range by keeping the vacuum advance inoperative until the transmission shifts into high gear.

Cold Operation

When the engine coolant is below 85°F (30°C), the closed temperature switch contacts will ground the solenoid winding. Vacuum advance will be provided continuously when the coolant is below 85°F (30°C), as illustrated in Figure 5-14. The hot contacts in the temperature switch operate the engine overheat light.

Hot Operation

Some systems have a separate temperature switch to operate the engine overheat light. The TCS temperature switch contains two contacts. One set of temperature switch contacts will close if the coolant is below 85°F (30°C), and the other contacts will close if the coolant is above 225°F (107°C), as outlined in Figure 5-15. When either set of temperature switch contacts are closed, the solenoid winding is grounded and thus provides vacuum advance continuously.

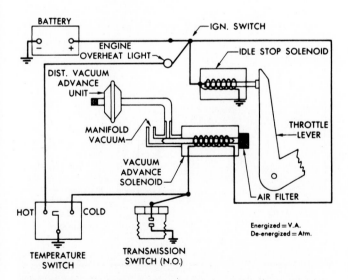

FIGURE 5-12. TCS System during Low-gear Operation. (Courtesy of General Motors of Canada Ltd.)

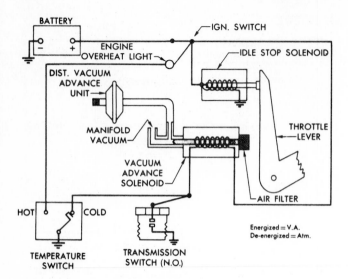

FIGURE 5-14. TCS System, Cold Operation. (Courtesy of General Motors of Canada Ltd.)

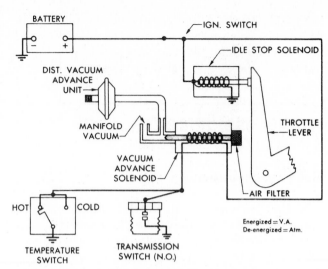

FIGURE 5-15. TCS System, Hot Operation. (Courtesy of General Motors of Canada Ltd.)

MODULATED DISTRIBUTOR ADVANCE SYSTEMS

Spark Retard Delay Valve Systems

A thermal vacuum valve (TVV) is used with the spark retard delay valve to control vacuum advance, as illustrated in Figure 5-16.

When the engine coolant is cold, the TVV remains closed. The only vacuum path to the distributor is through the spark retard delay valve. On hard acceleration, the manifold vacuum decreases instantly, and the spark retard delay valve traps vacuum in the distributor advance for 4 seconds. Cold drivability is improved by gradual retarding the vacuum advance. Warm engine coolant opens the TVV and supplies manifold vacuum directly to the vacuum advance without going through the spark retard delay valve.

Spark Advance Vacuum Modulator (SAVM) Valve Systems

Some vacuum advance systems use a SAVM valve with the spark retard delay valve and the thermal vacuum valve, as shown in Figure 5-17. The TVV is connected parallel to the spark retard delay valve.

Ported vacuum is supplied directly to the SAVM valve. Manifold vacuum is available through the TVV or spark retard delay valve to the SAVM valve. The SAVM valve operates as follows:

1. Full manifold vacuum is available through the modulator valve to the distributor advance when manifold vacuum is below 10 in Hg (33 kPa) at wide-throttle opening.

2. When manifold vacuum is greater than 10 in Hg (33 kPa) and ported vacuum is less than 10 in Hg

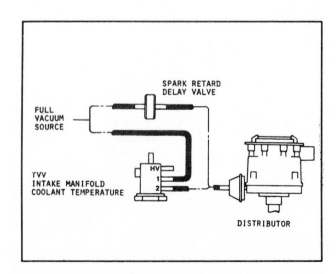

FIGURE 5-16. Spark Retard Delay Valve. (Courtesy of General Motors of Canada Ltd.)

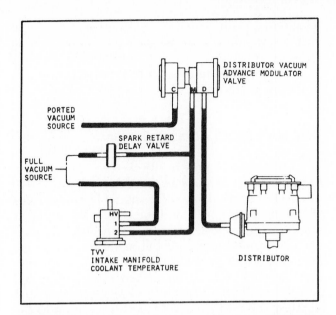

FIGURE 5-17. Spark Advance Vacuum Modulator Valve System. (Courtesy of General Motors of Canada Ltd.)

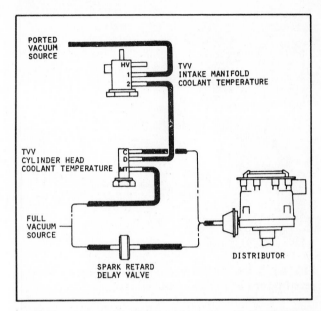

FIGURE 5-18. Dual Thermal Vacuum Valve System. (Courtesy of General Motors of Canada, Ltd.)

(33 kPa), the modulator valve supplies a constant 10 in Hg (33 kPa) to the distributor advance. This condition would occur at idle speed.

3. Ported vacuum is supplied to the vacuum advance from the modulator valve when both ported and manifold vacuum signals exceed 10 in Hg (33 kPa) at normal cruising speed.

Vacuum Advance Systems with Dual Thermal Vacuum Valves

Dual TVVs are connected with the spark retard delay valve in some applications, as pictured in Figure 5-18. One TVV senses intake manifold coolant temperature, and the other TVV is operated by cylinder head coolant temperature.

The dual TVV system operates as follows:

1. When the coolant is cold, the only vacuum path to the distributor advance is through the spark retard delay valve, because of the closed TVVs. A 4-second delay in retard on hard acceleration is supplied by the spark retard delay valve.

2. During warm coolant temperatures, the intake manifold TVV opens, and all three ports on the cylinder head TVV are interconnected. Manifold vacuum and ported vacuum are available at the vacuum advance. At idle speed, vacuum advance

is lost because the ported vacuum outlets above the throttles vent manifold vacuum. Ported vacuum will be supplied to the distributor advance once the throttle is opened.

3. When the engine coolant reaches approximately 225°F (107°C), the cylinder head TVV connects full manifold vacuum to the vacuum advance. Ported vacuum is shut off by the cylinder head TVV when the engine coolant is hot.

Diagnosis of Spark Advance Vacuum Modulator Valve (SAVM)

The SAVM valve should be diagnosed using the following procedure.

1. Connect a vacuum gauge to the distributor port. Apply 10 in Hg (33 kPa) to the manifold vacuum port. The vacuum gauge reading should increase to 10 in Hg (33 kPa) and remain constant. Vacuum specifications will vary depending on the application. (See Figure 5-19 for vacuum port on SAVM valve.)

2. Leave the vacuum gauge connected to the distributor port. Connect the vacuum pump to the carburetor port and plug the manifold vacuum port. Gradually increase the carburetor port vacuum. The vacuum gauge should read zero

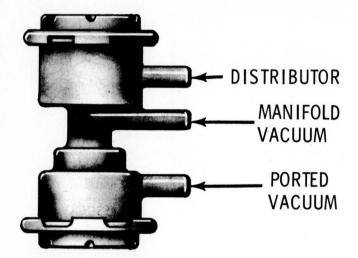

FIGURE 5-19. SAVM Valve Ports. (Courtesy of General Motors of Canada Ltd.)

until pump output reaches 10 in Hg (33 kPa). As pump output increases above 10 in Hg (33 kPa), the gauge reading should equal pump output. Replace the SAVM valve if it fails any of the checks in step 1 or 2.

DIAGNOSIS OF DISTRIBUTOR SPARK RETARD EMISSION SYSTEMS

Spark Delay Valve

Spark delay valves must be installed correctly. The carburetor and distributor outlets are easily identified (see Figure 5-20). Spark delay valves may be

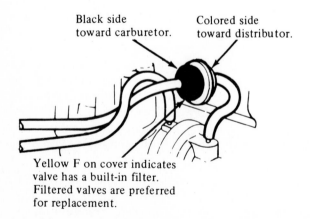

Black side toward carburetor.   Colored side toward distributor.

Yellow F on cover indicates valve has a built-in filter. Filtered valves are preferred for replacement.

FIGURE 5-20. Installation of Spark Delay Valve. (Courtesy of Ford Motor Co.)

tested by applying 10 in Hg (33 kPa) of vacuum to the carburetor side of the valve and recording the time required to obtain 8 in Hg (26 kPa) on a vacuum gauge at the distributor side of the valve, as outlined in Figure 5-21.

Cold-start Spark Advance System

The CSSA system may be checked by connecting a vacuum gauge at the vacuum advance with a "T" fitting. The spark delay valve should be removed as shown in Figure 5-22, and the engine should be at normal temperature. When the engine is accelerated and decelerated, a rise and fall on the vacuum gauge indicates normal system operation. Check all vacuum hoses carefully. In any system using a restricted port type valve in the vacuum advance hose, the slightest vacuum leak will vent the vacuum to the distributor advance.

Transmission-controlled Spark System

When diagnosing a TCS system, use a "T" fitting to connect a vacuum gauge at the vacuum advance. With the engine at normal temperature, lift the vehicle on a hoist and shift the transmission into high gear. Vacuum at the distributor advance should be present in high gear. Some TCS systems will also have vacuum advance in reverse because transmission hydraulic pressure closes the TCS switch in high or reverse. If vacuum is not available at the vacuum advance in high gear, follow the diagnostic steps listed below:

1. Be sure all vacuum hoses and electrical connections are tightened securely.
2. Check for vacuum at the vacuum advance solenoid. For TCS systems using ported vacuum, the throttle will have to be partly opened before vacuum is available at the solenoid.
3. Check power at both vacuum solenoid terminals with the ignition switch on. Lack of power at the ignition switch terminal on the solenoid indicates an open fuse, or an open wire from the solenoid to the ignition switch. If power is available at the solenoid ignition switch terminal and no power is present at the solenoid transmission terminal, the solenoid winding is open.
4. Disconnect the transmission wire from the vacuum solenoid. With the vehicle operating in high gear, connect a 12 V test light from the battery positive terminal to the disconnected transmis-

VACUUM DELAY TEST

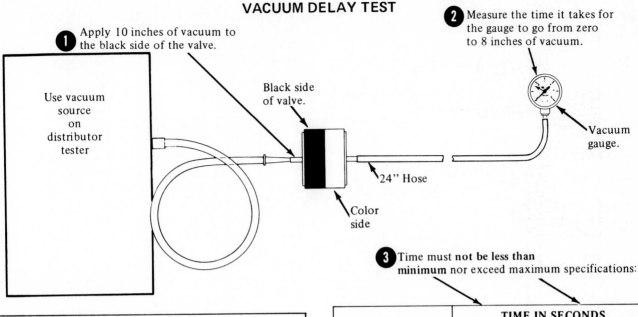

① Apply 10 inches of vacuum to the black side of the valve.

② Measure the time it takes for the gauge to go from zero to 8 inches of vacuum.

Use vacuum source on distributor tester

Black side of valve.

Vacuum gauge.

24" Hose

Color side

③ Time must not be less than minimum nor exceed maximum specifications:

RESULTS: (Refer to table)

● **Less than minimum time specified** – replace valve.
● **Longer than maximum time specified** – replace valve.
● **As specified** – delay valve is okay. Check spark vacuum to source.

TYPE VALVE	TIME IN SECONDS	
	MINIMUM	MAXIMUM
Black & Gray	1	4
Black & Brown	2	5
Black & White	4	12
Black & Yellow	5,8	14
Black & Blue	7	16
Black & Green	9	20
Black & Orange	13	24
Black & Red	15	28

FIGURE 5-21. Testing Spark Delay Valve. (Courtesy of Ford Motor Co.)

sion wire. If the test light is on, the circuit through the transmission switch is satisfactory. If the test light is off, the transmission switch, or the wire from the switch to the solenoid, is open.

5. If the tests in steps 1–4 are satisfactory and there is no vacuum at the distributor advance in high gear, replace the vacuum advance solenoid.

Questions

1. The spark delay valve momentarily traps vacuum in the distributor vacuum advance when the engine is decelerated.   T   F

2. In a CSC system with the coolant below 85°F (30°C), EGR port vacuum is applied to the vacuum advance   T   F

3. The PVS in the vacuum advance system supplies manifold vacuum to the distributor advance when the engine is _____.

4. Vacuum is supplied through the spark delay valve to the vacuum advance when the coolant is 75°F (24°C) in a CSSA system.   T   F

5. When the basic timing of a dual diaphragm distributor is being checked, the vacuum hoses should be connected.   T   F

6. The OSAC valve provides instant vacuum advance when the throttle is opened and the air cleaner temperature is 150°F (66°C).   T   F

7. In a TCS system, vacuum advance is provided in high gear with the engine at normal operating temperature.   T   F

8. A spark delay valve will operate normally with the vacuum hoses reversed.   T   F

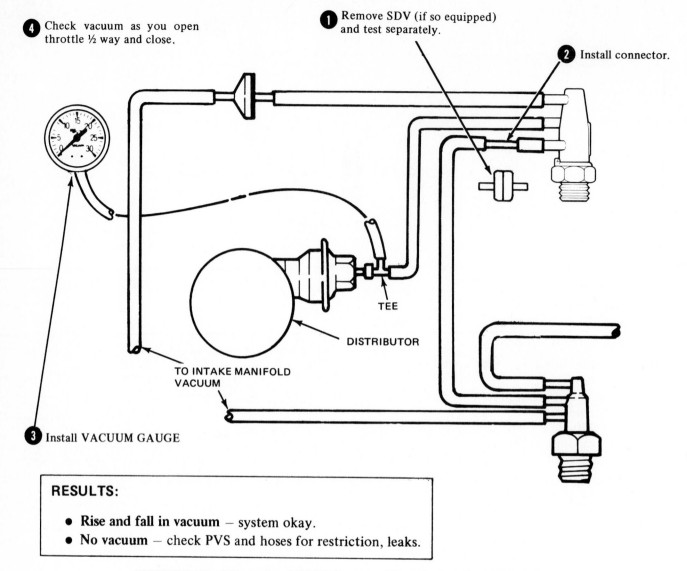

4 Check vacuum as you open throttle ½ way and close.

1 Remove SDV (if so equipped) and test separately.

2 Install connector.

TEE

DISTRIBUTOR

TO INTAKE MANIFOLD VACUUM

3 Install VACUUM GAUGE

RESULTS:

- **Rise and fall in vacuum** – system okay.
- **No vacuum** – check PVS and hoses for restriction, leaks.

FIGURE 5-22. Diagnosis of CSSA System. (Courtesy of Ford Motor Co.)

---

9. The spark retard delay valve provides a delay of vacuum advance when the engine is accelerated from idle to cruising speed.   T   F

10. When an engine equipped with a spark advance vacuum modulator valve is operating at normal temperatures and idle speed, vacuum applied to the distributor advance would be _____ _____ in Hg.

6

ELECTRONIC IGNITION SYSTEMS

ELECTRONIC IGNITION SYSTEM COMPONENTS

Spark Plug

Spark plugs contain a center electrode surrounded by an insulator. This center electrode and insulator assembly is mounted in a steel shell. A ground electrode is attached to the steel shell, and this electrode is positioned directly below the center electrode. Since the center electrode must withstand high temperatures, it is manufactured from a nickel alloy.

The spark plug electrodes provide the gap within the combustion chamber for high voltage to arc across and ignite the air-fuel mixture. When combustion takes place in the cylinder, the heat is concentrated on the center spark plug electrode. This heat is dissipated through the insulator and the shell of the spark plug into the cylinder head. Coolant is circulated through a passage in the cylinder head which surrounds the spark plug seat. This coolant conducts the heat away from the spark plug.

The heat range of a spark plug is determined by the depth of the insulator before it contacts the lower end of the spark plug shell. Spark plugs with normal, cold, and hot heat ranges are pictured in Figure 6-1.

If an engine is driven continuously at idle and low speeds, the spark plugs may become carbon fouled because their operating temperature is too low. Under this type of driving condition, hotter range spark plugs may be required to prevent carbon fouling. When an engine is operated continuously under heavy load or at high speed, extremely high temperatures may burn the spark plug electrodes. If these types of operating conditions are encountered, spark plugs with a colder heat range may be required. The heat range of the spark plugs recommended by the manufacturer for a specific engine will be adequate for average driving conditions.

Spark plug gaps should always be set to the manufacturer's specifications. When spark plugs are installed, they should be tightened to the specified torque.

Spark Plug Wires

Many spark plug wires have a core which contains glass threads coated with carbon. This type of conductor in the core reduces electromagnetic interference (EMI). Vehicles equipped with computer systems must be equipped with this type of spark plug wire. (See Chapters 8 and 9 for an explanation of computer systems.) Older type spark plug wires had cores with stranded copper conductors. This type of spark plug wire could produce EMI, which

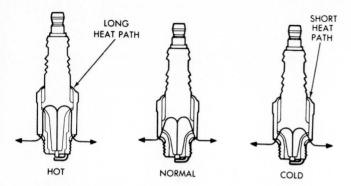

FIGURE 6-1. Spark Plug Heat Range. (Courtesy of Chrysler Canada Ltd.)

can interfere with computer input or output signals. The core of the spark plug wire is surrounded by insulation and a hypalon or silicone jacket, as indicated in Figure 6-2.

When spark plug wires are being removed, they should not be stretched. The spark plug boot should be rotated back and forth on the spark plug before the wire is removed from the plug. Spark plug wires may be tested with an ohmmeter on the X1000 scale. The wires are defective if the resistance exceeds manufacturer's specifications.

Distributors

The distributor is usually positioned in the engine block. In some overhead-cam engines, the distributor is located in the cylinder head. A bushing, or bushings, in the distributor housing support the distributor shaft. The distributor drive gear is attached to the shaft with a roll pin. In many engines this drive gear is driven by a gear on the camshaft. The centrifugal advance mechanism is mounted on the upper end of the distributor shaft,

and a reluctor is held on the upper end of the shaft with a roll pin. An aligning lug inside the rotor fits into a slot on top of the distributor shaft.

The pickup coil assembly is mounted on a plate in the distributor housing. A linkage from the vacuum advance is connected to the pickup plate. The pickup coil plate is designed to rotate with vacuum advance diaphragm movement. A permanent magnet is positioned in the pickup coil. An aligning notch in the distributor cap fits into a notch in the distributor housing, so the distributor cap can only be installed in one position. Spring clips are used to retain the distributor cap on top of the housing. A complete distributor is illustrated in Figure 6-3.

FIGURE 6-3. Distributor Assembly. (Courtesy of Chrysler Canada Ltd.)

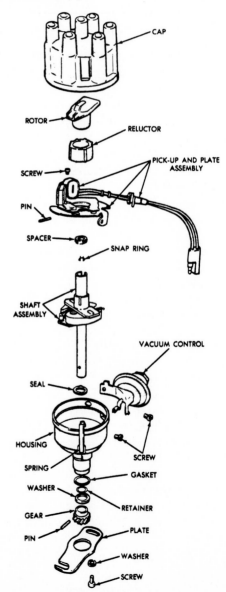

FIGURE 6-2. Spark Plug Wire. (Courtesy of Chrysler Canada Ltd.)

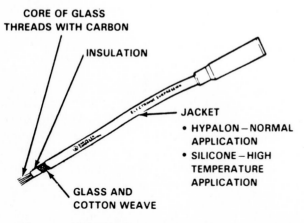

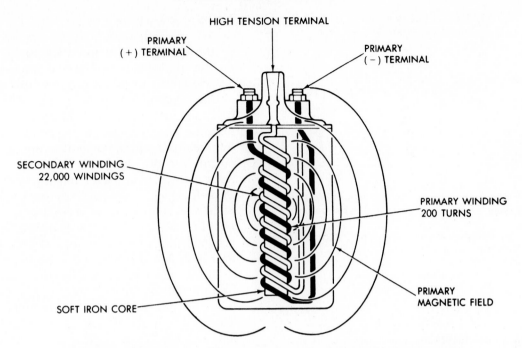

FIGURE 6-4. Ignition Coil. (Courtesy of Chrysler Canada Ltd.)

Ignition Coil

A laminated iron core is located at the center of the coil windings. The secondary coil winding is wound around this iron core. This winding contains thousands of turns of very fine wire. A coating of insulating material on the wire prevents the turns of wire from touching each other. One end of the secondary winding is connected to the secondary high-tension terminal in the coil tower, and the other end of the winding is often connected to one of the primary terminals. A primary terminal winding is wound on top of the secondary winding. This primary winding contains much heavier wire than the secondary winding, and has approximately 200 turns of wire. The ends of the primary winding are connected to the two primary terminals on top of the coil. These terminals are identified with positive (+) and negative (−) symbols.

The coil tower and winding assembly is sealed in a round metal container. The assembled coil is filled with oil to help cool the windings and to prevent air space inside the coil, which would allow the formation of moisture. A typical ignition coil is shown in Figure 6-4.

Ignition Module

The ignition module is an electronic device that is completely sealed and therefore must be replaced as a unit. Many ignition modules are bolted to the

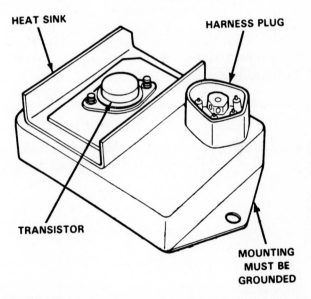

FIGURE 6-5. Ignition Module. (Courtesy of Chrysler Canada Ltd.)

fender shield or firewall. Some ignition modules must be grounded where they are mounted, or the ignition system will be inoperative. Various wires from the electronic ignition system are connected to the module wiring harness plug. An ignition module is pictured in Figure 6-5.

1.2 OHM RESISTANCE IN PRIMARY CIRCUIT
"RUN" POSITION

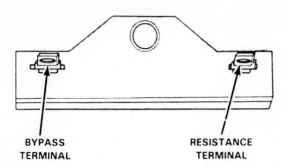

BYPASS
TERMINAL

RESISTANCE
TERMINAL

FIGURE 6-6. Primary Circuit Resistor. (Courtesy of Chrysler Canada Ltd.)

Primary Circuit Resistor

A primary circuit resistor is used in some ignition systems. This resistor is connected between the ignition switch and the positive primary coil terminal to provide the correct voltage and current flow in the primary circuit. A primary circuit resistor is pictured in Figure 6-6.

Some ignition systems have a resistance wire in place of the block-type primary circuit resistor.

Complete Ignition System

In the ignition system shown in Figure 6-7, the pickup coil leads are connected to the number 4 and 5 terminals on the ignition module.

The negative (−) primary coil terminal is connected to the number 2 terminal on the ignition module. Another wire is connected from ignition 1 (I1) terminal on the ignition switch to the primary circuit resistor, and this same wire is also connected to the number 1 terminal on the ignition module. The other end of the primary circuit resistor is connected to the positive (+) primary coil terminal. A wire from the ignition 2 (I2) terminal is connected to the primary circuit resistor terminal that is attached to the coil + primary connection.

The I2 terminal of the ignition switch supplies battery voltage directly to the ignition coil + primary terminal while the engine is being cranked. This action maintains a higher voltage at the coil and provides an increase in primary current while starting the engine. If the I2 ignition switch terminal did not perform this function, voltage and current would be supplied through the primary circuit resistor to the coil + primary terminal. Under this condition the voltage at the coil + primary would be reduced and primary current flow would decrease. If the primary current flow is decreased, the magnetic field in the coil will be weak and the maximum secondary coil voltage will be low.

FIGURE 6-7. Complete Ignition System. (Courtesy of Chrysler Canada Ltd.)

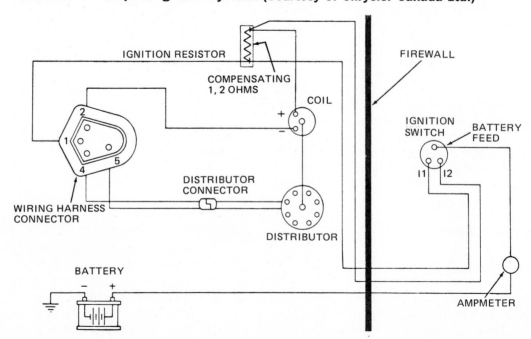

ELECTRONIC IGNITION SYSTEM OPERATION

Ignition Module and Pickup Coil Operation

One high point is located on the distributor reluctor for each engine cylinder. When the reluctor high points are out of alignment with the pickup coil, a weak magnetic field exists around the pickup coil, as indicated in Figure 6-8. Under this condition, the module will close the primary ignition circuit and current will flow through the ignition switch, resistor, primary coil winding, and the module to ground. This current flow through the primary circuit will create a magnetic field around both coil windings.

As the distributor shaft rotates, a reluctor high point moves into alignment with the pickup coil. Since the metal reluctor tip is a better conductor for magnetic lines of force than air, the magnetic field around the pickup coil is strengthened, as pictured in Figure 6-9.

The instant that the reluctor tip begins to move out of alignment with the pickup coil, the magnetic field around the pickup coil suddenly collapses. When this action takes place, a voltage is induced in the pickup coil, which causes the module to open the primary circuit. When the primary circuit is opened, the magnetic field collapses across the coil

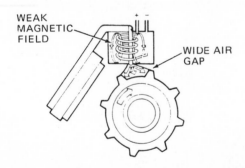

FIGURE 6-8. Reluctor High Point Out of Alignment with the Pickup Coil. (Courtesy of Chrysler Canada Ltd.)

FIGURE 6-9. Reluctor Tip in Alignment with the Pickup Coil. (Courtesy of Chrysler Canada Ltd.)

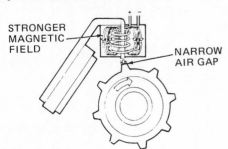

windings. This action induces a very high voltage in the secondary coil winding because of the large number of secondary turns. This high induced secondary voltage forces current flow out the secondary coil wire to the center distributor cap terminal. From this point the secondary current flows through the rotor, distributor cap terminal, spark plug wire, and spark plug electrodes to ground. This spark across the plug electrodes ignites the air-fuel mixture in the cylinder, which forces the piston down in the power stroke. The purpose of the ignition system is to create a spark at each plug electrode gap at the right instant.

Some ignition manufacturers refer to the ignition module as an electronic control unit (ECU). The operation of the ignition system is outlined on the diagram in Figure 6-10.

The purpose of the ignition module is to open and close the primary ignition circuit. When the reluctor tip moves a very short distance out of alignment with the pickup coil, the induced voltage in the pickup coil decreases. As this occurs, the ignition module closes the primary circuit and primary current flow resumes. The ignition module must keep the primary circuit turned on long enough to allow the magnetic field to build up in the ignition coil. This "on time" for the primary circuit is referred to as dwell time. In most electronic ignition systems, the dwell time is determined electronically by the ignition module.

In the older point-type ignition systems, the ignition points and the cam lobe opened and closed the primary circuit in place of the ignition module. A point-type distributor and ignition coil are shown in Figure 6-11.

Secondary Coil Voltage

When the engine is operating at low speeds, the normal required secondary ignition coil voltage is approximately 10,000 V. However, many ignition coils are capable of producing a maximum secondary coil voltage of 25,000 V or more. The difference between required secondary voltage and maximum secondary voltage is referred to as reserve voltage. When the engine is operated with a wide-open throttle at high speeds or under heavy load conditions, cylinder pressure increases. This pressure increase between the spark electrodes requires an increase in secondary voltage to keep firing the spark plugs.

Spark plug electrodes wear gradually from continual arcing and extreme heat. When the electrode gap becomes wider, the required secondary voltage must increase to fire the spark plug. This increase

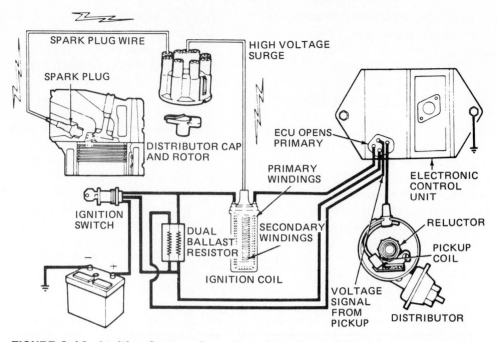

FIGURE 6-10. Ignition System Operation. (Courtesy of Chrysler Canada Ltd.)

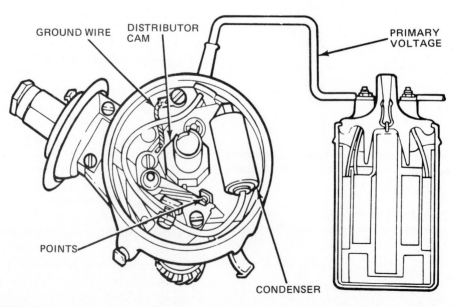

FIGURE 6-11. Point-Type Distributor and Ignition Coil. (Courtesy of Chrysler Canada Ltd.)

in required secondary coil voltage will also occur if additional resistance develops in the spark plug wires or rotor gap. Therefore, secondary reserve voltage is necessary to compensate for higher cylinder pressures under wide-open throttle conditions, and for any increase in secondary resistance.

The maximum available secondary voltage must always exceed the normal required secondary voltage. Excessive resistance at spark plug electrodes or in spark plug wires may reduce the secondary voltage reserve and cause misfiring under hard acceleration. The maximum secondary coil voltage may be reduced by a defective coil, low primary current, or a cracked distributor cap and rotor. Any of these defects could cause secondary misfiring under hard acceleration.

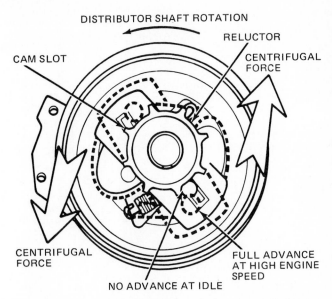

FIGURE 6-12. Centrifugal Advance Mechanism. (Courtesy of Chrysler Canada Ltd.)

DISTRIBUTOR ADVANCES

Centrifugal Advance Mechanism

The centrifugal advance mechanism is attached to the distributor shaft. This advance mechanism contains pivoted advance weights which move outward when the engine and distributor shaft speed increase. The advance weight movement is controlled by two springs. When the advance weights move outward, they turn the reluctor in the same direction that the distributor shaft is rotating. This action causes the reluctor tips to be aligned sooner with the pickup coil, which results in earlier spark at the spark electrodes. When the piston speed increases, the spark must occur sooner at the spark plug electrodes to sustain maximum pressure on the piston from combustion. If this spark advance did not occur in relation to piston speed, the piston would move down in the power stroke before the air-fuel mixture had time to start burning, which would result in reduced engine power.

The centrifugal advance mechanism advances the spark at the spark plug electrodes in relation to engine speed as indicated in Figure 6-12.

Vacuum Advance

The vacuum advance contains a diaphragm in a sealed chamber. On some engines, this chamber is connected to a ported vacuum source above the throttle; in other engines, the vacuum source for the vacuum advance is directly from the intake manifold. A linkage from the vacuum advance diaphragm is connected to the pickup coil plate. When the engine is operating at part throttle, a relatively high vacuum is applied to the vacuum advance diaphragm. This vacuum overcomes the spring tension on the diaphragm and moves the diaphragm toward the vacuum outlet. When this occurs, the diaphragm link rotates the pickup coil plate opposite to the distributor shaft rotation, which causes the pickup coil to be aligned sooner with the reluctor tips. This pickup coil movement causes the spark to occur sooner at each spark plug electrode.

At part-throttle, light-load operating conditions, additional spark advance provides improved fuel economy and performance. If the throttle is moved to the wide-open position, cylinder pressure and heat increase, which results in faster burning of the air-fuel mixture. Under this condition, the spark advance must be retarded to prevent detonation. At wide-open throttle, manifold vacuum decreases and the vacuum advance return spring moves the diaphragm toward the distributor housing, which moves the pickup coil to the retarded position.

The vacuum advance controls spark advance in relation to engine load. On most current production cars, the distributor advances have been discontinued and the spark advance is computer controlled. (Computer-controlled spark advance systems are explained later in this chapter.) The operation of a vacuum advance mechanism is illustrated in Figure 6-13.

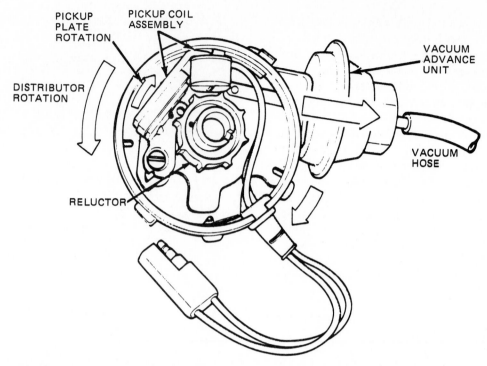

FIGURE 6-13. Vacuum Advance Operation. (Courtesy of Chrysler Canada Ltd.)

GENERAL MOTORS HIGH-ENERGY IGNITION (HEI)

Design

Point-type ignition systems are subject to changes in point setting and timing, which result in high exhaust emissions. Electronic ignition systems provide more stable operation because the dwell is determined by a solid state module. Exhaust emissions are reduced by the increased stability of electronic ignition systems.

The high-energy ignition system is self-contained in the distributor. Most HEI coils are mounted on top of the distributor cap. A spring contact connects the secondary coil terminal to the center distributor cap terminal. High-voltage leakage is prevented by the seal around the center cap terminal, as shown in Figure 6-14. Some HEI coils are mounted externally from the distributor cap.

The ground terminal dissipates induced voltages from the coil frame to ground on the distributor housing. Early model HEI coils have the secondary winding connected from the center cap terminal to the primary winding. The secondary windings on later model coils are connected from the center cap terminal to the coil frame. Four screws attach the coil to the distributor cap. The primary coil leads,

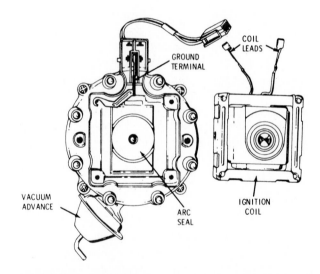

FIGURE 6-14. HEI Ignition Coil. (Courtesy of General Motors of Canada Ltd.)

identified as "bat" and "tach" terminals, are mounted in the distributor cap. A double connector is used on each primary terminal. One connection on the "bat" terminal is connected to the ignition switch. The "bat" terminal is also connected to the HEI module. "Tach" inner terminals are connected to the module, and the outer "tach" connection extends to the dash-mounted tachometer or diagnostic connector. (The diagnostic connector is discussed under "High-Energy Ignition Diagnosis.") A cover is located on top of the coil, as illustrated in Figure 6-15. Four latches attach the distributor cap to the housing.

Two latches hold the spark plug wire retaining ring to the coil cover. The spark plug wire retaining ring cannot be improperly installed because locating notches are present on the coil cover, as pictured in Figure 6-16.

Two screws attach the electronic module to the distributor housing. The pickup coil is mounted on top of the distributor bushing. Two pickup coil leads are connected to the module. Two pickup coil terminals—one wide and one narrow—prevent incorrect pickup lead connections, as shown in Figure 6-17. Heat-dissipating silicone grease must be applied between the module and the distributor housing. The "bat" and "tach" primary lead wires are connected to the module. Radio interference is eliminated by the capacitor connected from the "bat" wire to ground.

A waved retaining ring holds the pickup coil on the distributor bushing. Three screws attach the pole-piece assembly to the pickup coil. Mounted between the pole piece and the pickup coil is a flat permanent magnet. The number of pole-piece teeth is matched to the engine cylinders. Figure 6-18 pictures a pole piece from an eight-cylinder engine.

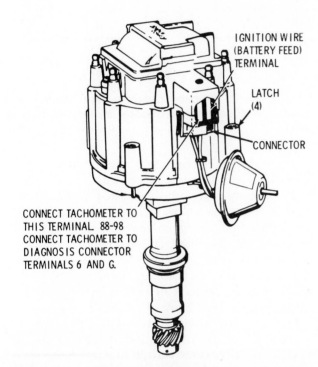

FIGURE 6-15. HEI Primary Coil Connection. (Courtesy of General Motors of Canada Ltd.)

FIGURE 6-16. Installation of HEI Spark Plug Wire. (Courtesy of General Motors of Canada Ltd.)

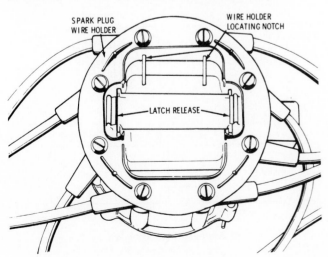

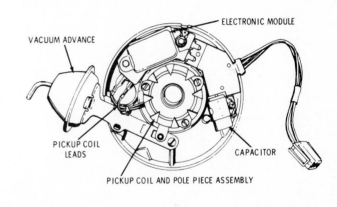

FIGURE 6-17. HEI Module and Pickup Coil. (Courtesy of General Motors of Canada Ltd.)

FIGURE 6-18. HEI Pickup Coil Assembly. (Courtesy of General Motors of Canada Ltd.)

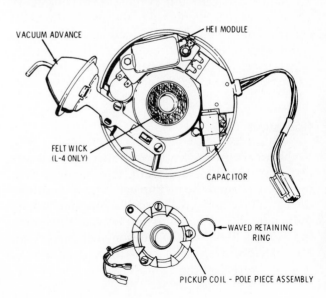

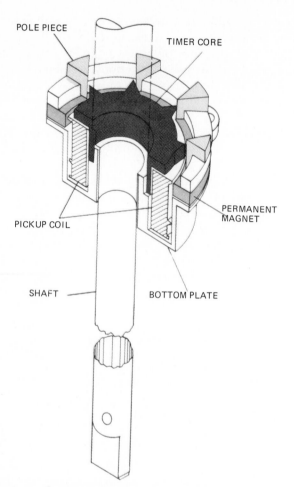

FIGURE 6-19. HEI Timer Core. (Courtesy of General Motors of Canada Ltd.)

A timer core rotates inside the pickup coil assembly. Eight high points on the timer core match the eight pole-piece teeth, as pictured in Figure 6-19. The timer core is connected to the distributor shaft. Outward movement of the centrifugal advance weights rotates the timer core ahead of the distributor shaft to advance the timing in relation to engine speed.

Operation

As the timer core teeth approach alignment with the teeth on the pole-piece teeth, the magnetic field builds up around the pickup coil. The resulting induced voltage in the pickup coil signals the module to turn on the primary current flow, as illustrated in Figure 6-20. When the ignition switch is on and the distributor shaft is not turning, there is no primary current flow.

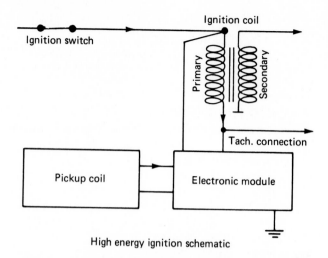

High energy ignition schematic

FIGURE 6-20. Primary Current Flow during HEI Operation. (Courtesy of General Motors of Canada Ltd.)

When the timer core teeth begin moving out of alignment with the pole-piece teeth, the magnetic field suddenly collapses across the pickup coil. The module opens the primary circuit when the pickup coil induced voltage signal is received, as indicated in Figure 6-21. Magnetic collapse occurs across the ignition coil windings when the module opens the primary circuit. High voltage required to fire the spark plug is induced in the secondary winding when the primary circuit is opened by the module.

Primary dwell time is the length of time that the primary circuit remains closed by the module. The module extends primary dwell time as engine speed increases, providing higher primary magnetic strength and maximum secondary voltage. Other

FIGURE 6-21. HEI Operation, Primary Circuit Open. (Courtesy of General Motors of Canada Ltd.)

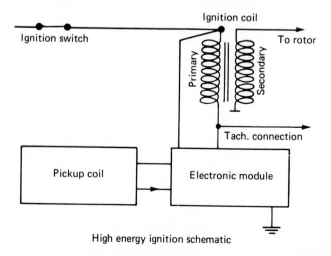

High energy ignition schematic

electronic ignition systems operate on the same basic principle, but many systems provide a constant dwell time regardless of engine RPM.

DIAGNOSIS OF HEI SYSTEMS

Ohmmeter, Voltmeter Tests

With the ignition switch on, connect a voltmeter from the coil "bat" terminal to ground. The voltmeter reading should exceed 12 V. A zero voltage reading indicates an open circuit between the ignition switch and the coil "bat" terminal. When the voltmeter registers less than 12 V, the battery is discharged or a high-resistance problem exists between the ignition switch and the coil "bat" terminal. The pickup coil may be checked for open circuits and shorts by connecting an ohmmeter across the disconnected pickup coil leads, as indicated by ohmmeter 2 in Figure 6-22. An infinite reading indicates an open pickup coil. If the ohmmeter reading is below the specification range of 650-850 Ω, the pickup coil is shorted. Ohmmeter 1 in Figure 6-23 is testing the pickup coil for a grounded condition. A low ohmmeter reading indicates a grounded pickup coil. The pickup coil is satisfactory if ohmmeter 1 displays an infinite reading.

Pull lightly on the pickup leads when testing for open circuits. A fluctuating ohmmeter reading indicates an intermittent open circuit. Severe engine surging usually occurs if an intermittent open circuit exists in the pickup coil.

FIGURE 6-22. Testing HEI Pickup Coil. (Courtesy of General Motors of Canada Ltd.)

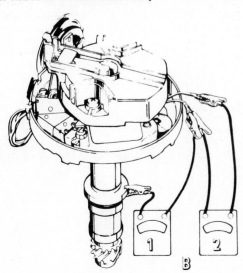

TESTING IGNITION COIL

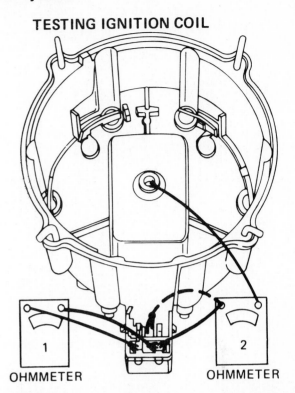

FIGURE 6-23. Testing HEI Ignition Coil Windings. (Courtesy of General Motors of Canada Ltd.)

The primary ignition coil winding should be tested by connecting an ohmmeter across the primary terminals, as indicated by ohmmeter 1 in Figure 6-23. A reading below the specified value of 0.5 Ω indicates a shorted winding. The primary winding is open if an infinite reading is obtained. Ohmmeter 2 is connected from the center cap terminal to the ground terminal to test the secondary winding. The secondary winding should have 12,000–20,000 Ω. Early model HEI secondary windings are connected from the center cap terminal to one of the primary terminals. When early model coils are being tested, ohmmeter leads must be connected from the center cap terminal to one of the primary terminals.

The distributor cap and rotor should be inspected for cracks, any sign of leakage, and corroded terminals. Inspect the underside of the rotor for burn marks that indicate leakage problems. If all the voltmeter and ohmmeter tests are satisfactory but there is no spark at the spark plugs while the engine is being cranked, the module or the ignition coil is defective. Ohmmeter tests will not indicate insulation leakage defects in the ignition coil. Scope testing of modules and coils are detailed later in this chapter.

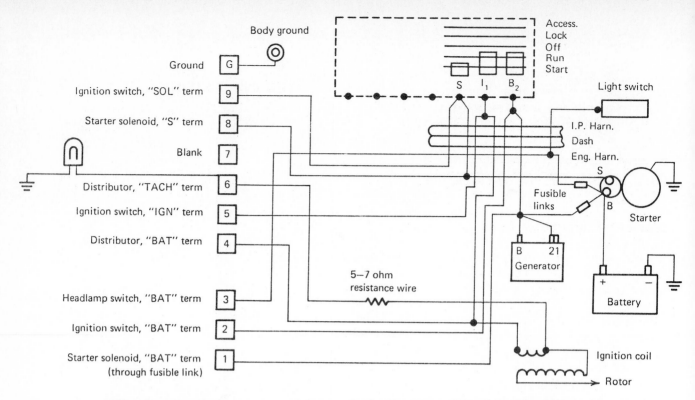

FIGURE 6-24. HEI Test Lamp Diagnosis. (Courtesy of General Motors of Canada Ltd.)

Test Lamp Method

When the engine fails to start, a 12 V test lamp may be connected from the "tach" terminal to ground. In the case of vehicles with a diagnostic connector, the test lamp may be connected from terminal 6 in the diagnostic connector to ground, as pictured in Figure 6-24. Failure of the test lamp to light with the ignition switch on indicates an open circuit in the coil primary or in the circuit from the ignition switch to the coil "bat" terminal. The test lamp should flash on and off while the engine is being cranked. The module or pickup coil is defective if the test lamp is on continuously while the engine is being cranked.

If the test lamp diagnosis indicates a satisfactory pickup coil and module, the no-start condition is caused by a defect in the secondary circuit. The HEI ST 125 tester may be used to test the secondary circuit. A spark plug boot is used to connect the ST 125 tester to the center distributor cap terminal, as shown in Figure 6-25. The ST 125 tester is a special-

FIGURE 6-25. ST 125 HEI Tester. (Courtesy of General Motors of Canada Ltd.)

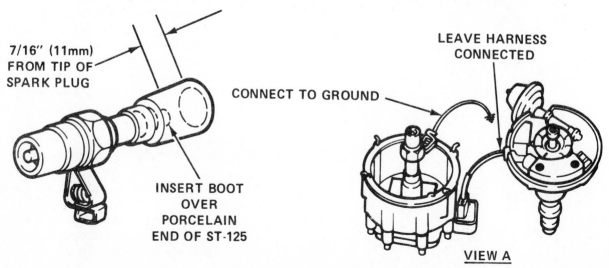

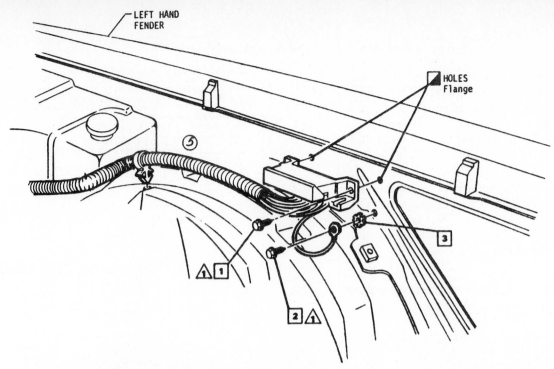

FIGURE 6-26. Location of Diagnostic Connector. (Courtesy of General Motors of Canada Ltd.)

ly designed spark plug with the ground electrode and part of the center electrode removed. Voltage requirement of the tester is 25,000 V. The spark plug wires must be removed from the cap and the primary coil leads must be connected as illustrated in Figure 6-25.

If the module and pickup coil were satisfactory in the test lamp diagnosis, failure of the ST 125 tester to fire during engine cranking indicates a defective ignition coil. With the distributor cap and wires installed, the ST 125 tester may be connected from one of the spark plug wires to ground. If the tester does not fire while the engine is being cranked but did fire when connected to the center cap terminal, a leakage exists in the rotor, cap, or plug wire.

Various tests may be performed by connecting a 12-V test lamp from the diagnostic connector terminals to ground. The circuit or component that may be tested at each diagnostic connector terminal is identified in Figure 6-24. Figure 6-26 shows the underhood location of the diagnostic connector.

ELECTRONIC MODULE RETARD (EMR) HEI SYSTEMS

Design and Operation

Conventional HEI systems use a four-terminal module. EMR HEI systems have a five-terminal module. The extra module terminal is connected to an EMR vacuum switch containing a set of normally open contacts operated by a vacuum diaphragm. Manifold vacuum is applied to the vacuum switch through a 6-port TVS switch when coolant temperature is below 120°F (49°C), as outlined in Figure 6-27.

FIGURE 6-27. EMR HEI System. (Courtesy of General Motors of Canada Ltd.)

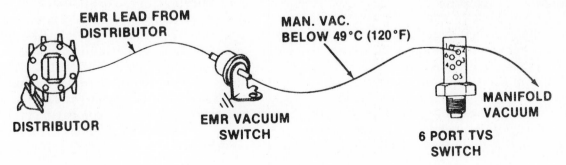

79

The vacuum switch contacts will close when manifold vacuum applied to the switch exceeds 4 in Hg (13.2 kPa). Closing the vacuum switch contacts signals the HEI module to retard the ignition timing 10°. When the engine coolant reaches 120°F (49°C), the vacuum to the switch is shut off by the TVS switch. The HEI system operates normally when the vacuum switch contacts are open. Retarded ignition timing during cold-engine operation helps the engine and catalytic converter to warm up faster and reduces emissions. The EMR system is used in some Oldsmobile models.

DIAGNOSIS OF EMR SYSTEMS

Diagnosis of the EMR system should be performed as follows:

1. Connect a vacuum gauge at the vacuum switch with a "T" fitting. The vacuum reading should be zero when engine coolant is above 120°F (49°C). If vacuum is available at the switch with a warm engine, the TVS switch is defective.

2. Disconnect the lead wire from the vacuum switch. Connect an ohmmeter from the vacuum switch terminal to ground. An infinite ohmmeter reading indicates satisfactory switch contacts. A low ohmmeter reading should be obtained when vacuum in excess of 4 in Hg (13.2 kPa) is applied to the switch with a vacuum pump. The vacuum switch is leaking if it does not hold the applied vacuum.

3. Ground the wire from the module to the vacuum switch while observing the timing marks with a timing light. A 10° retard should occur when the vacuum switch wire is grounded.

ELECTRONIC SPARK CONTROL (ESC) HEI SYSTEMS

Design and Operation

Turbocharged engines and light-duty trucks with certain engine transmission combinations use the ESC system to prevent engine detonation. The pickup coil leads are connected to the solid state ESC controller and to the ignition module as illustrated in Figure 6-28. A black wire is connected from the ignition module to the ESC controller. Electrical ground for the controller is provided by

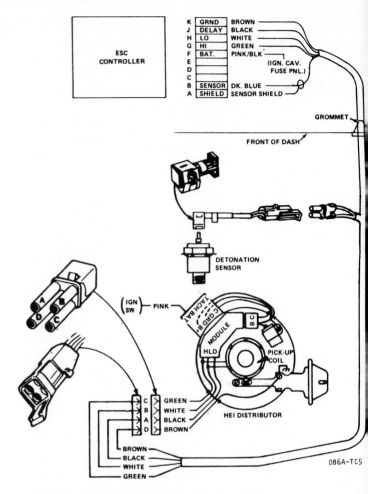

FIGURE 6-28. ESC HEI System. (Courtesy of General Motors of Canada Ltd.)

the brown wire attached to the distributor housing. A detonation sensor is located in the intake manifold on turbocharged engines. In light-duty trucks the detonation sensor is located in the coolant drain plug on the right side of the engine.

When detonation occurs in the cylinders, a signal is sent from the detonation sensor to the controller. The controller changes the pickup coil signal and retards the spark advance. Maximum retard provided by the controller would be 15°–20°. Severity and duration of detonation will determine the amount of retard. Twenty seconds after detonation ceases, the controller returns spark advance to normal. Vacuum and centrifugal advances operate in the normal manner on ESC HEI distributors Light-duty trucks have the controller mounted under the dash, and some turbocharged cars use a controller mounted in the engine compartment. ESC circuits may vary slightly, depending on the model year and application.

DIAGNOSIS OF ESC HEI SYSTEMS

Engine Fails To Start

The following steps should be taken to diagnose the ESC system when the engine fails to start:

1. Check all wiring connectors at the distributor and controller.

2. Disconnect the four-wire connector at the distributor. Connect a jumper wire across terminals A and C in the distributor connector. Failure of the engine to start indicates the defect is not in the ESC system. Diagnose the HEI system as outlined previously in this chapter.

3. If the engine starts when distributor terminals A and C are connected, reconnect the four-wire distributor connector. Connect a voltmeter from controller terminals F to K. A reading below 7 V indicates an open circuit or a resistance problem between the ignition switch and terminal F.

4. Check the brown, black, green, and white wires from the distributor connector to the controller by connecting an ohmmeter across each wire. An infinite reading indicates an open wire. When testing, disconnect both connectors. Connect an ohmmeter from each wire to ground. A low reading indicates a grounded condition.

5. If the tests in steps 3 and 4 are satisfactory, replace the controller.

Engine Detonates Excessively

Use the following procedure for diagnosing detonation complaints on an ESC system:

1. With the engine operating above 1,000, tap on the engine lightly near the detonation sensor while observing the timing marks with a timing light. If the timing retards, the ESC system is operating normally.

2. If timing does not retard in step 1, disconnect the controller connector. Connect an ohmmeter from controller terminal B to ground. A resistance reading between 175 and 375 Ω indicates a satisfactory detonation sensor. If the sensor resistance reading is incorrect, connect an ohmmeter from controller terminal B to the disconnected sensor wire. The sensor wire is open if the ohmmeter reading is infinite. The sen-

sor wire is grounded if a low reading is obtained when an ohmmeter is connected from terminal B to ground with the sensor wire disconnected.

3. Try to start the engine with the controller connector disconnected. If the engine starts, replace the ignition module.

4. If the engine fails to start with the controller connector disconnected, reconnect the controller connector. Connect a jumper wire from the disconnected detonation sensor wire, and position the other end of the jumper wire on top of the HEI coil. Observe the timing marks with a timing light while the engine is operating above 1,000 RPM. If timing retards, replace the sensor.

5. If timing does not retard in step 4, connect a voltmeter from controller terminals H to K. If the voltmeter registers more than 0.2 V with the ignition switch on, replace the controller. A reading below 0.2 V indicates an open circuit in the white wire from the ignition module to the controller.

Unsatisfactory Engine Performance

Whenever engine performance complaints are being diagnosed, the ESC system may be bypassed by disconnecting the distributor four-wire connector, and connecting a jumper wire across terminals A and C in the distributor connector. If the performance complaint is corrected, the problem is in the controller or wiring between the controller and the distributor. The voltage measured from controller terminal F to ground must exceed 11.6 V with the engine running. When the performance complaint is not corrected by connecting terminals A and C together, the ESC system is not the source of the complaint.

HIGH-ENERGY IGNITION USED WITH COMPUTER SYSTEMS

Design and Operation

A seven-wire ignition module is used in the computer command control (3C) HEI system. The additional three-module terminals are connected to the electronic control module (ECM). Conventional vacuum and centrifugal advance mechanisms have been discontinued, and the correct spark advance is supplied by the ECM in relation to the input sen-

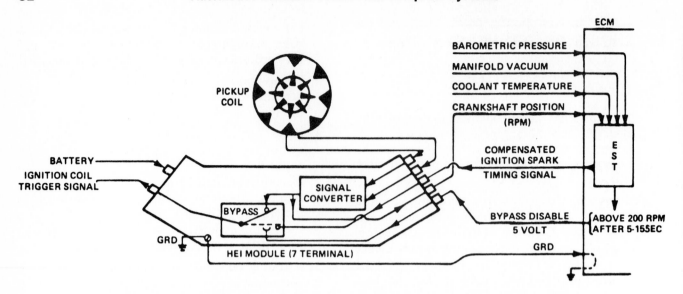

FIGURE 6-29. Computer Command Control High-Energy Ignition System, Start Mode. (Courtesy of GM Product Service Training, General Motors Corporation)

sor signals that it receives. Input sensor signals from the barometric pressure sensor, manifold absolute pressure sensor, coolant temperature sensor, and the crankshaft position sensor are used by the ECM to determine the correct spark advance, as shown in Figure 6-29.

While the engine is cranking, the pickup coil signal goes directly through the module signal converter and bypass circuit. The ECM does not affect spark advance when the engine is being started. Approximately 5 seconds after the engine is started,

a 5-V disable signal is sent from the ECM to the module. This signal opens the module bypass circuit and completes the compensated ignition spark timing circuit from the ECM to the module, as illustrated in Figure 6-30.

The pickup coil signal now travels through the module signal converter, ECM, and the compensated ignition spark timing circuit to the module. Ignition spark advance is controlled by the electronic spark-timing (EST) circuit in the ECM under this condition. The distributor pickup coil signal

FIGURE 6-30. Computer Command Control High-Energy System, Run Mode. (Courtesy of GM Product Service Training, General Motors Corporation)

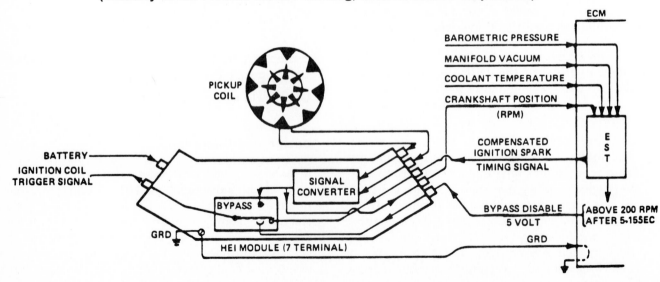

that is sent through the module to the ECM is referred to as a crankshaft position revolutions per minute (RPM) signal because it provides the ECM with an engine speed signal. This same HEI system is used with many General Motors throttle body and port fuel injection systems.

GENERAL MOTORS HIGH-ENERGY IGNITION USED WITH SOME FUEL INJECTION SYSTEMS

Design and Operation

Some throttle body injection systems have a conventional pickup coil and a Hall Effect switch in the distributor. The conventional pickup coil is used only when the engine is being cranked. A signal is sent from the Hall Effect switch through the ECM to the HEI module while the engine is running. The Hall Effect switch sends a revolutions per minute (RPM) signal to the ECM, and the R terminal on the HEI module that was used on other systems for this purpose is no longer connected to the ECM, as shown in Figure 6-31.

When the engine is being cranked, the conventional pickup coil signal is sent directly to the transistor in the HEI module. A 5-V disable signal is sent from the ECM through the tan-black wire and the winding in the HEI module to ground. This 5-V disable signal changes the HEI module circuit into the EST mode, as shown in the illustration in the lower left-hand side of Figure 6-31. The signal from the Hall Effect switch can now travel through the ECM and HEI module circuit to the transistor in the module.

The "time variable" circuit in the ECM varies the signal from the Hall Effect switch to provide the precise spark advance that is required by the engine. Each time a signal is received by the transistor in the HEI module, it opens the primary ignition circuit to provide magnetic collapse in the ignition coil and high induced voltage in the secondary winding to fire the spark plug. In the EST mode, the circuit is open from the conventional pickup coil to the transistor in the HEI module. The Hall Effect switch will provide a signal each time a reflector blade that is attached to the distributor shaft rotates past the switch. The HEI module circuit shown in Figure 6-31 is for illustration purposes; the actual module circuit would be much more complex.

When the basic timing is being adjusted on the ignition systems used with General Motors 3C ig-

FIGURE 6-31. High-Energy Ignition with Hall Effect Switch. (Courtesy of GM Product Service Training, General Motors Corporation)

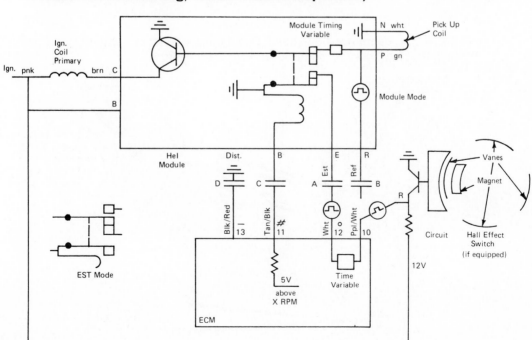

nition, or the ignition systems used with fuel injection systems, one of the following procedures will be necessary:

1. On some systems, a set timing connector must be grounded.

2. The four-wire connector near the distributor must be disconnected.

3. The tan wire in the four-wire distributor connector must be disconnected.

4. In other systems, the test terminal in the assembly line diagnostic link (ALDL) must be grounded. (The ALDL is explained in Chapter 9.)

The underhood emission label should provide the correct procedure for checking the basic timing on each vehicle.

COMPUTER-CONTROLLED COIL IGNITION SYSTEMS USED WITH PORT FUEL INJECTION

Design and Operation

The main components in the computer-controlled coil ignition (C^3I) system are the cam sensor, crank sensor, and the electronic coil module, as pictured in Figure 6-32.

The cam sensor and the crank sensor both contain Hall Effect switches. A conventional distrib-

utor is not required with the C^3I system, and the cam sensor is mounted in the engine block in place of the distributor. This sensor is driven by the camshaft. The crank sensor is mounted at the front of the crankshaft. This sensor is operated by an interruption ring attached to the crankshaft pulley. Signals from the crank sensor inform the electronic coil module and the ECM when each piston is at top dead center (TDC). This signal is used by the ECM and the coil module for correct timing and spark advance. A signal is also sent from the cam sensor to the coil module and the ECM. This sensor generates one signal for each sensor revolution. The ECM uses this signal to time the injector opening. For example, if the engine has sequential fuel injection (SFI), the ECM begins to energize the injectors in the correct sequence when it receives the cam sensor signal.

Three ignition coils are mounted in the electronic coil module assembly, and an electronic module is located underneath the coils. The C^3I system has two spark plug wires connected to each coil. On some V6 engines, plug wires 1 and 4, 5 and 2, and 6 and 3 are paired together. With this type of system, each coil fires both spark plugs at the same time. The wires are paired so that one spark plug is firing when the piston is on the compression stroke and the other piston is on the exhaust stroke. If a spark plug fires when the piston is on the exhaust stroke, it has no effect.

When the engine is being cranked, the crank sensor signal goes directly to the coil module, which opens the primary circuit of each coil. Once the en-

FIGURE 6-32. Computer-Controlled Coil Ignition System. (Courtesy of GM Product Service Training, General Motors Corporation)

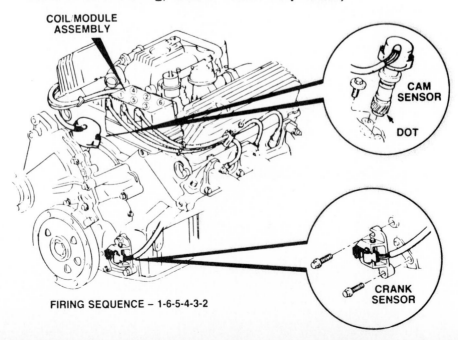

COIL MODULE ASSEMBLY

CAM SENSOR

DOT

CRANK SENSOR

FIRING SEQUENCE – 1-6-5-4-3-2

gine starts, the crank sensor signal goes through the reference low or reference high circuit to the ECM. This signal is sent from the ECM through the EST circuit to the coil module. On the basis of the input signals, the ECM varies the crank sensor signal to provide the precise spark advance required by the engine. An initial timing adjustment is not required on the C³I system. The wiring diagram for a C³I system is provided in Figure 6-33.

With the ignition switch on, voltage is supplied from terminal P on the coil module through the module circuit to the blue input wires on each coil primary winding. The coil module opens and closes the circuit from each primary winding to ground. When the ignition switch is turned on, voltage is supplied from terminal M on the coil module through terminal H to terminal A on the crank and cam sensor.

Some engines have a combined crank and cam sensor at the front of the crankshaft. The wiring diagram for this circuit is shown in Figure 6-34.

A tachometer lead is connected to the coil

module, and the terminal on this wire is usually located near the coil module. Only digital tachometers should be connected to the tachometer lead.

Later Model Distributorless Direct Ignition System (DIS)

On 1987 and later model DIS systems, a magnetic sensor-type pickup is mounted in the engine block. This pickup is triggered by a timing disc that is an integral part of the crankshaft. When the timing disc notches rotate past the pickup, a pickup signal is sent to the coil module and the electronic control module (ECM). The ECM and the coil module use this signal to open each primary ignition coil circuit at the correct instant, which supplies the precise spark advance required by the engine. When number 1 spark plug is at top dead center (TDC) on the compression stroke, a wider timing disc notch rotates past the pickup. This signal is also used by

FIGURE 6-33. Computer-Controlled Coil Ignition System Wiring Diagram. (Courtesy of GM Product Service Training, General Motors Corporation)

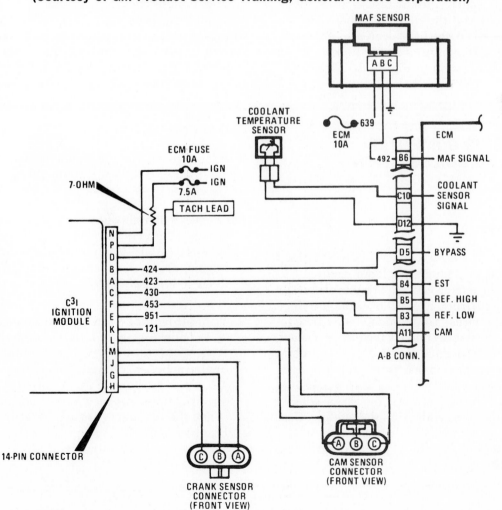

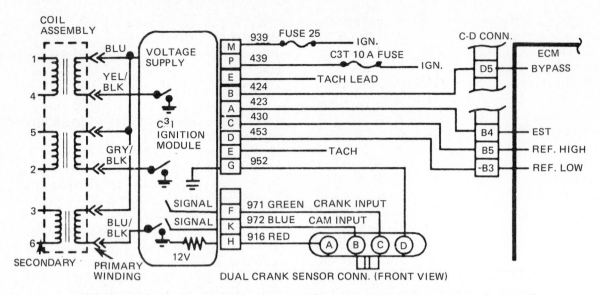

FIGURE 6-34. Computer-Controlled Coil Ignition System with Combined Crank and Cam Sensor. (Courtesy of GM Product Service Training, General Motors Corporation)

the ECM to sequence number 1 injector. Since the pickup is mounted in a fixed position in the engine block, timing adjustments are not possible.

An electronic spark control (ESC) sensor sends a signal to the ECM if the engine detonates. When this signal is received, the ECM decreases spark advance. The DIS system is used on 4-cylinder and V6 engines. The DIS system in Figure 6-35 is from a V6 engine.

DIAGNOSIS OF COMPUTER-CONTROLLED COIL IGNITION SYSTEMS

Engine Misfires

If the engine is misfiring, connect an ST 125 HEI test spark plug to a plug wire from each coil to ground while the engine is being cranked. If one or two coils are not firing, check those coil windings and spark plug wires. An ohmmeter may be connected across the two secondary terminals on each coil to test the secondary winding. This winding should have less than 15,000 Ω resistance.

If the coil secondary winding and spark plug wires are satisfactory, remove the six assembly screws and detach the coils from the module. Connect a 12-V test light across the primary coil terminals while the engine is being cranked. If the test

light blinks but the test spark plug did not fire in the previous test, replace the coil assembly. When the light remains on or fails to come on, the coil module is defective.

If the test spark plug fires on each plug wire while the engine is being cranked, the ignition system is satisfactory. Check the engine compression and the fuel system to locate the cause of the engine misfiring.

Engine Fails to Start

If the test spark plug does not fire when it is connected to any of the spark plug wires, proceed as follows:

1. Connect a digital voltmeter from coil module terminals M and P to ground with the ignition switch on. If 12 V is not available at these terminals, check the fuses and connecting wires. Refer to Figure 6-36 for terminal identification.

2. With the ignition switch off, disconnect the ECM A-B connector and connect a digital voltmeter from ECM terminal B5 to ground while the engine is being cranked. This reading should be 1–7 V and varying. If this reading is not within specifications, proceed to the sensor diagnosis. When the reading is normal at terminal B5, the crank and cam sensor signal is being generated; therefore, proceed to step 3.

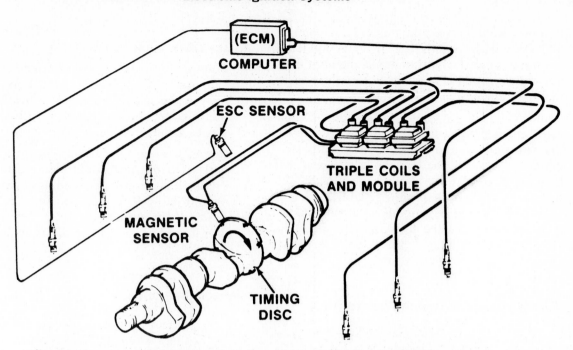

FIGURE 6-35. Later Model DIS Ignition System with Crankshaft Timing Disc. (Courtesy of GM Product Service Training, General Motors Corporation)

FIGURE 6-36. Computer-Controlled Coil Ignition System with Combined Cam and Crank Sensor. (Courtesy of GM Product Service Training, General Motors Corporation)

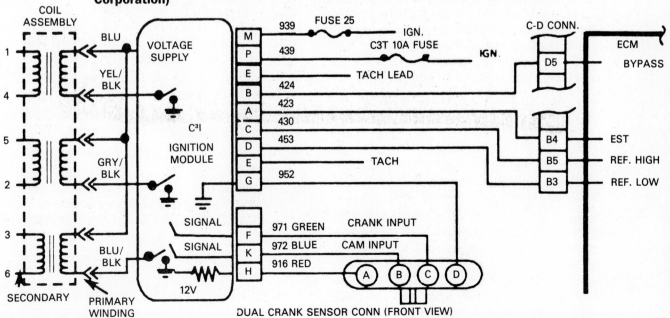

3. Remove the coils from the module and connect a 12-V test light from the blue primary coil leads to ground with the ignition on. If the light does not come on and the readings in step 1 are satisfactory, replace the coil module.

4. Connect a 12-V test light across the primary leads on each coil while the engine is cranked. The test light should blink. If the test light remains on, replace the coil module. If the test light blinks on all three coils, connect the ST 125 test spark plug to the plug wire from each coil and ground while the engine is being cranked. If no spark occurs, the coils are defective. It is very unlikely that all three coils will be unsatisfactory.

Sensor Diagnosis if Engine Fails to Start

This sensor diagnosis applies to the C³I system with combination cam and crank sensor that is illustrated in Figure 6-36. Diagnosis would be similar to the C³I system with separate cam and crank sensors. If the voltmeter reading is not satisfactory in step 2 of the "engine fails to start" diagnosis, proceed with the sensor diagnosis as follows:

1. Check voltage from coil module terminal M and P to ground, as in step 1 of the "engine fails to start" diagnosis.

2. Disconnect the crank and cam sensor connector and connect a digital voltmeter across terminals A and B. With the ignition switch on, the voltmeter should read 5–11 V. If the reading is not within specifications, check the wires from terminals A and B to coil module terminals H and K for open circuits and grounds. If the wires are satisfactory, replace the coil module.

3. Repeat the procedure in step 2 on terminals A and C in the sensor connector; the same voltmeter reading should be obtained. If the reading is not satisfactory, check the wires from terminals A and C to coil module terminal H and F for open circuits and grounds. If the wires are satisfactory, replace the coil module.

4. With the ignition switch on, connect a digital voltmeter across terminals A and D in the sensor connector. This voltmeter reading should be 5–11 V. When the reading is not within specifications, check the wire from terminal D to coil module terminal G for open circuits and grounds. If the wire is satisfactory but the volt-

meter reading is out of specifications, replace the coil module.

5. Connect four short jumper wires between the cam and crank sensor terminals and the terminals in the wiring harness connector to allow meter connections to these terminals with the sensor connected.

6. Connect a digital voltmeter from terminals B and C to ground while the engine is being cranked. The voltmeter reading at each terminal should be 7–9 V and varying. If voltmeter readings at either terminal are not within specifications, the cam and crank sensor is defective.

Sensor Clearance

The dual sensor bracket cannot be moved to adjust basic timing. However, the clearance should be 0.030 in (0.762 mm) between each side of the crankshaft pulley discs and the sensors. This clearance may be adjusted by loosening the clamp bolt and moving the sensor up or down in the bracket.

Cam Sensor Installation

If a separate cam sensor is mounted on the engine block and the sensor is removed, it must be reinstalled correctly. The cam sensor drive gear has a dot on one side that must face opposite the cam sensor disc window. When the cam sensor is installed, this dot must face away from the timing chain with number 1 piston TDC on the compression stroke. The cam sensor harness must face toward the timing chain. When the cam sensor is being installed, use the following procedure:

1. Rotate the engine until number 1 piston is at TDC on the compression stroke.

2. Rotate the crankshaft until the timing mark on the harmonic balancer is 25° after TDC.

3. Remove the spark plug wires from the coil assembly.

4. Disconnect the cam sensor connector and connect three short jumper wires between the sensor terminals and the wiring harness connector.

5. Connect a digital voltmeter from the center cam sensor terminal to ground. Three terminals are connected to the cam sensor.

6. With the ignition switch on, rotate the cam sensor counterclockwise until the sensor switch

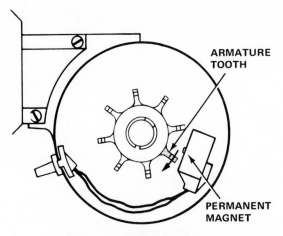

**TOOTH AWAY FROM
PERMANENT MAGNET**

FIGURE 6-37. Solid State Ignition Pickup Coil. (Courtesy of Ford Motor Co.)

closes. This will result in a decrease on the voltmeter reading.

7. Tighten the cam sensor clamp in this position. Reinstall the plug wires and reconnect the sensor connector.

The distributorless direct ignition system (DIS) can be diagnosed in basically the same way as the C³I system.

FORD SOLID STATE IGNITION SYSTEMS

Design

The solid state ignition system has an electronic module mounted on the fender apron. A pickup coil containing a permanent magnet is mounted in the distributor. When the distributor shaft rotates, the armature teeth move past the head on the pickup coil, as illustrated in Figure 6-37. The armature is

secured to the shaft by a roll pin. Armature teeth are matched to the number of cylinders. Clearance between the armature teeth and pickup coil head is nonadjustable because the pickup coil is riveted to the distributor plate.

Orange and purple wires are connected from the distributor pickup coil leads to the module. The module is grounded to the distributor pickup plate through the black wire. A 1.4-Ω resistance wire is connected in series between the ignition switch "run" contacts and the coil "bat" terminal, as outlined in Figure 6-38.

When the engine is being cranked, the resistor wire can be bypassed by the wire connected from the ignition switch "start" terminal to the coil "bat" terminal. A white wire is connected from the ignition switch "start" terminal to the module. The "DEC" terminal of the coil is connected to the module through the green wire. Voltage is supplied to the module from the coil side of the resistor wire by a blue wire.

Dura-Spark II Ignition Systems

A six-wire module is used on the Dura-Spark II system. The blue wire used on the previous seven-wire system is discontinued. Other circuit connections are similar in the six- and seven-wire systems, as outlined in Figure 6-39.

The Dura-Spark II system has a large distributor cap, 8-mm (0.0314 in) plug wires, and a wide rotor gap, as illustrated in Figure 6-40. Additional space between distributor cap terminals helps to prevent the leakage of high secondary voltage.

Dura-Spark I Ignition Systems

Dura-Spark I ignition systems are similar to Dura-Spark II ignition systems except that:

1. The primary circuit resistance wire is discontinued in Dura-Spark I systems, as shown in Figure 6-41.

FIGURE 6-38. Solid State Ignition System, Seven-Wire Module. (Courtesy of Ford Motor Co.)

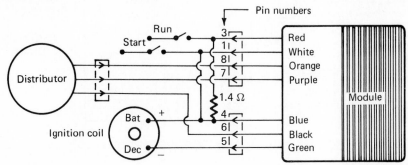

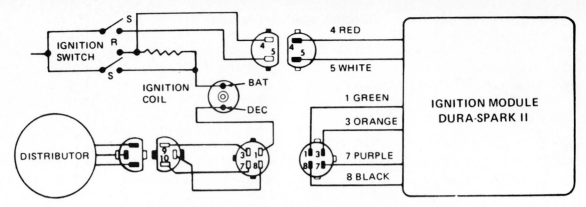

FIGURE 6-39. Dura-Spark II Ignition System.
(Courtesy of Ford Motor Co.)

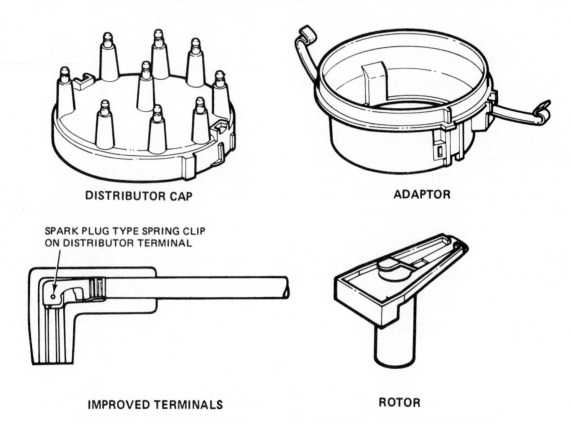

DISTRIBUTOR CAP

ADAPTOR

SPARK PLUG TYPE SPRING CLIP
ON DISTRIBUTOR TERMINAL

IMPROVED TERMINALS

ROTOR

FIGURE 6-40. Dura-Spark II Distributor Cap. (Courtesy of Ford Motor Co.)

FIGURE 6-41. Dura-Spark I Ignition System. (Courtesy of Ford Motor Co.)

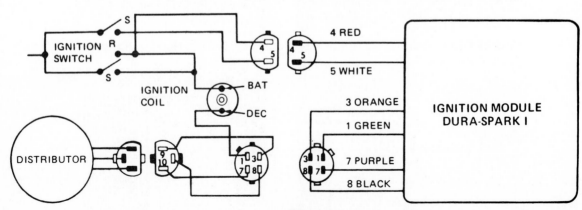

2. Primary coil winding resistance is lowered from 1–2 Ω to 0.75 Ω.

3. A variable dwell circuit is used in the module to increase dwell in relation to engine speed. The module is not interchangeable with Dura-Spark II modules.

Dual-Mode Ignition Timing Systems

Three extra wires are used on dual-mode ignition modules. For economy calibration, the extra wires are connected from the module to a vacuum switch. For altitude calibration, the additional three wires are connected from the module to a barometric pressure switch. Except for the extra three module wires, dual-mode ignition timing systems (see Figure 6-42) are the same as Dura-Spark II ignition systems.

The barometric pressure switch connected to the module retards the ignition timing 3°–6° below 2,400 ft (731 m) elevations. At elevations between 2,400 ft (731 m) and 4,300 ft (1,310 m), the barometric pressure switch may affect ignition timing. Above 4,300 ft (1,310 m), the barometric pressure switch has no effect on ignition timing. The barometric pressure switch signal provides correct ignition timing at high elevations and prevents spark knock at lower altitudes.

For economy calibration, the vacuum switch signal to the module retards ignition timing 3°–6°

when manifold vacuum is below 6 in Hg (20 kPa). When manifold vacuum is above 10 in Hg (33 kPa), the vacuum switch signal does not affect ignition timing. Between 6 in Hg (20 kPa) and 10 in Hg (33 kPa) of manifold vacuum, the vacuum switch signal to the module may affect ignition timing. The vacuum switch signal prevents spark knock under heavy engine load conditions. When checking basic ignition timing, remember that the wires connected to the barometric pressure switch, or vacuum switch, should be disconnected.

Thick Film Integrated (TFI) Ignition Systems

TFI ignition systems are used on some models of Ford Escort and Lynx. The main differences in TFI systems are:

1. A TFI module is attached to the distributor housing, as pictured in Figure 6-43. Module circuitry increases dwell time as engine speed increases.

2. TFI ignition coil windings are set in epoxy. An iron frame surrounds the coil windings.

3. The primary circuit resistance wire normally connected between the ignition switch and the coil primary winding is eliminated on TFI systems.

4. Pickup coil leads are connected to the TFI module inside the distributor housing, as illustrated in Figure 6-44.

FIGURE 6-42. Dual-Mode Ignition Timing System. (Courtesy of Ford Motor Co.)

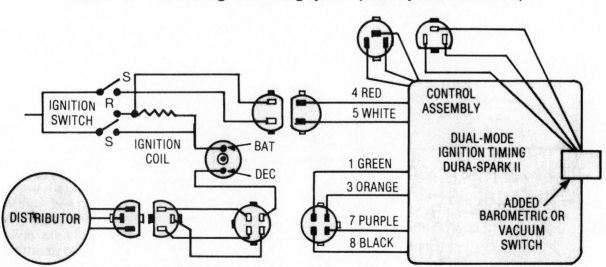

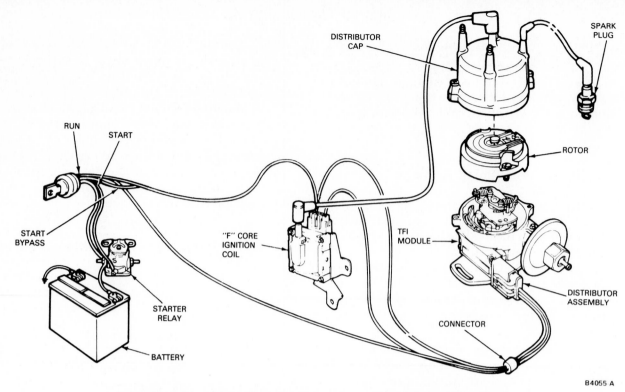

FIGURE 6-43. TFI Ignition System. (Courtesy of Ford Motor Co.)

FIGURE 6-44. TFI Distributor. (Courtesy of Ford Motor Co.)

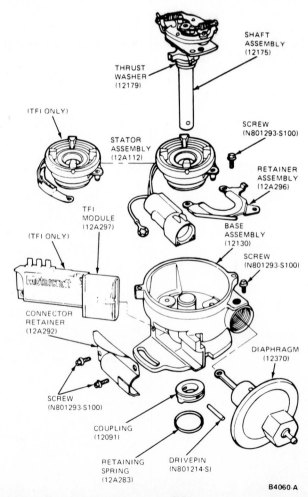

TABLE 6-1. Diagnosis of Ford Electronic Ignition Systems

	Test Voltage Between	Specifications	Testing
Key on	Red wire and ground	over 11.5 V	voltage to module
	Blue wire and ground omit 76 and up	over 11.5 V	ignition wire
	Green wire and ground	over 11.5 V	resistor wire and primary winding
Key on	Battery terminal of coil and ground	5–7 V	ignition
	Ground Tach terminal of coil when making above test		resistor wire
Key on cranking	Battery terminal of coil and ground	over 9 V	bypass wire
	Ground Tach terminal of coil when making above test		

	Test Resistance Between	Specifications	Testing
Key off	Purple and orange wire	400–800 Ω	pickup coil open
	Purple or orange wire and ground	inf. × 100 scale	pickup coil grounds
	Black wire and ground	0 Ω	ground wire
	Primary coil terminals	1–2 Ω	primary winding
	Primary coil terminal to coil tower	7000-13000 Ω	secondary winding

DIAGNOSIS OF FORD ELECTRONIC IGNITION SYSTEMS

Dura-Spark II and Seven-Wire Module Systems

Dura-Spark II systems and seven-wire module versions of Ford solid state ignition systems may be diagnosed using Table 6-1. Most test connections are made at the wiring harness side of the module connectors, with the module connectors disconnected. Some test connections are made directly to the coil terminals, as indicated in Table 6-1.

Dura-Spark I ignition systems may be diagnosed using the chart in Table 6-1, except for the following differences:

1. In the second "key on" testing step, the voltage reading would be 12 V or greater.

2. Step 3, "key on cranking" should be deleted because Dura-Spark I does not have a resistor bypass circuit.

3. Primary coil winding resistance would be 0.75 Ω.

TFI Test Lamp Method

Connect a 12-V test lamp to the ignition coil negative terminal, as pictured in Figure 6-45. Failure of the test lamp to flash while the engine is being cranked indicates a defective pickup coil or module. If the test lamp flashes while the engine is being cranked, the pickup coil and module are satisfactory. If no spark occurs at the spark plugs and the test lamp flashes during cranking, the secondary circuit is defective.

Connect the spark tester from the coil secondary lead to ground as indicated in Figure 6-46. If the spark tester fails to fire during engine cranking, the ignition coil is defective. Connect the spark tester from a spark plug wire to ground. A defective distributor cap, rotor, or plug wire is indicated if the tester does not fire now but did fire when connected to the secondary coil wire.

TFI Voltmeter, Ohmmeter Method

Connect a voltmeter from the positive primary coil terminal to ground, as shown in Figure 6-47. With the ignition switch on, the voltmeter reading should exceed 12 V. A zero voltmeter reading indicates an

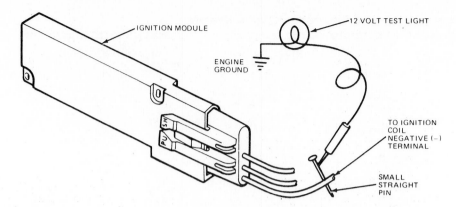

FIGURE 6-45. TFI Test Lamp Diagnosis. (Courtesy of Ford Motor Co.)

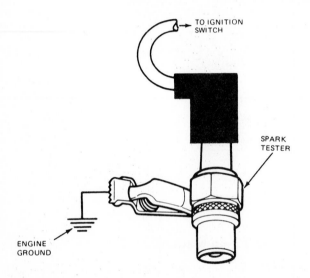

FIGURE 6-46. TFI Spark Tester. (Courtesy of Ford Motor Co.)

FIGURE 6-47. TFI Voltmeter Diagnosis. (Courtesy of Ford Motor Co.)

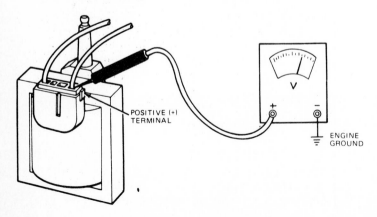

open circuit between the ignition switch and the positive coil terminal.

Disconnect the module connector, and connect a voltmeter from terminals 1 and 2 to ground, as pictured in Figure 6-48. When the ignition switch is on, the voltmeter should register 12 V or more at both terminals. A zero reading at terminal 1 indicates an open circuit in the coil primary winding or in the wire from the coil negative terminal to terminal 1. If the voltmeter reading is zero at terminal 2, an open circuit exists in the wire from the coil positive terminal to terminal 2.

With the ignition switch in the start position, connect a voltmeter from module connector terminal 3 to ground. The voltmeter should register over 9 V during engine cranking. A zero voltmeter reading indicates an open circuit in the ignition switch start terminal, or in the wire from the start terminal to module terminal 3.

Connect the ohmmeter leads across the primary coil terminals as shown in Figure 6-49. A resistance reading between 0.3 and 1 Ω indicates a satisfactory primary winding. The primary winding is shorted if the resistance is below 0.3 Ω. A reading above 1 Ω indicates excessive primary resistance.

Connect the ohmmeter from the secondary coil terminal to one of the primary terminals, as pictured in Figure 6-50. A resistance reading below 9,000 Ω indicates a shorted winding. Secondary resistance is excessive if the resistance reading exceeds 11,500 Ω.

Connect the ohmmeter to the pickup coil leads as illustrated in Figure 6-51. A resistance below 800 Ω indicates a shorted pickup coil. Pickup coil resistance is excessive if the ohm reading exceeds 975. If the voltmeter and ohmmeter tests are satisfactory and there is no spark at the spark plugs while the engine is cranking, the module or coil is defective.

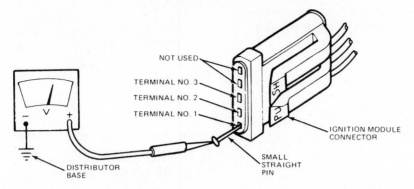

FIGURE 6-48. TFI Voltmeter Diagnosis. (Courtesy of Ford Motor Co.)

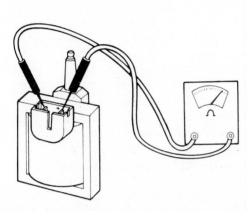

FIGURE 6-49. Testing Primary Winding of TFI Coil. (Courtesy of Ford Motor Co.)

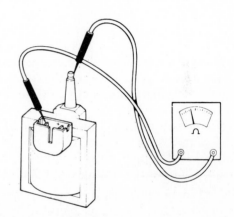

FIGURE 6-50. Testing Secondary Winding of TFI Coil. (Courtesy of Ford Motor Co.)

FIGURE 6-51. Testing TFI Pickup Coil. (Courtesy of Ford Motor Co.)

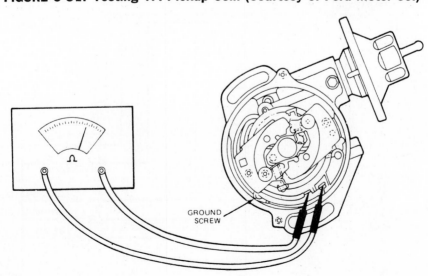

DURA-SPARK III IGNITION SYSTEMS

Design

Dura-Spark III ignition systems are used in the EEC III system. The same wiring connections for the primary ballast resistor, coil, and the ignition module were used in the EEC II system. This ignition system has a five-wire ignition module, as pictured in Figure 6-52.

The distributor is held in the engine block with a hold-down bolt and clamp. A slot in the distributor housing fits over the hold-down bolt and prevents the distributor from being rotated. The distributor and the hold-down clamp are illustrated in Figure 6-53.

Internal distributor components are shown in Figure 6-54. The only purpose of the distributor is to distribute the high secondary voltage from the coil to the spark plugs. Rotor alignment on the Dura-Spark III distributor can be adjusted as follows:

1. Position number one piston at top dead center of the compression stroke. The timing mark on the crankshaft pulley must be aligned with the zero position on the timing mark indicator.

2. The alignment tool must fit between the slots in the sleeve and adapter, as pictured in Figure 6-55.

Spark advance is determined by the electronic control assembly (ECA), which eliminates the need for vacuum and centrifugal advance mechanisms. The crankshaft position sensor signal is sent through the ECA to the Dura-Spark III module. Signals from the crankshaft position signals are sent from ECA terminal 17 to the ignition module through circuit 144, as pictured in Figure 6-56.

IGNITION DIAGNOSIS OF EEC II AND EEC III SYSTEMS

When Engine Fails To Start

Step 1. Check for spark at several spark plug wires with a spark tester while cranking the engine. If spark occurs, the ignition system is capable of starting the engine. The fuel system, or engine condition, must be the cause of the no-start situation.

Step 2. Continue with the Dura-Spark III diagnosis if the spark tester did not fire in step 1. Connect the spark tester to the coil secondary lead. Check for spark at the tester while cranking the engine. If the spark tester fires while the engine is cranking but no spark was available in step 1, the distributor cap or rotor is defective.

Step 3. If the spark tester will not fire when connected to the coil secondary lead, connect the diagnostic test adapter, as outlined in Figure 6-57. With the ignition switch on, momentarily connect the test adapter lead to the battery positive terminal. If

FIGURE 6-52. Dura-Spark III Ignition System. (Courtesy of Ford Motor Co.)

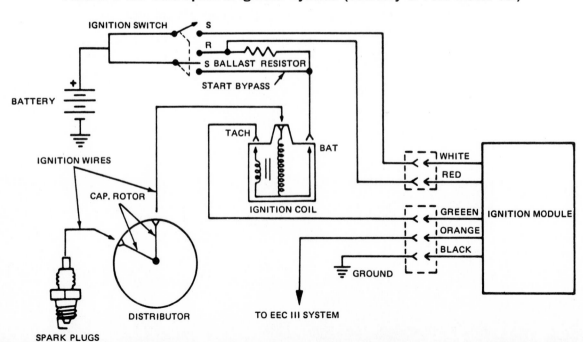

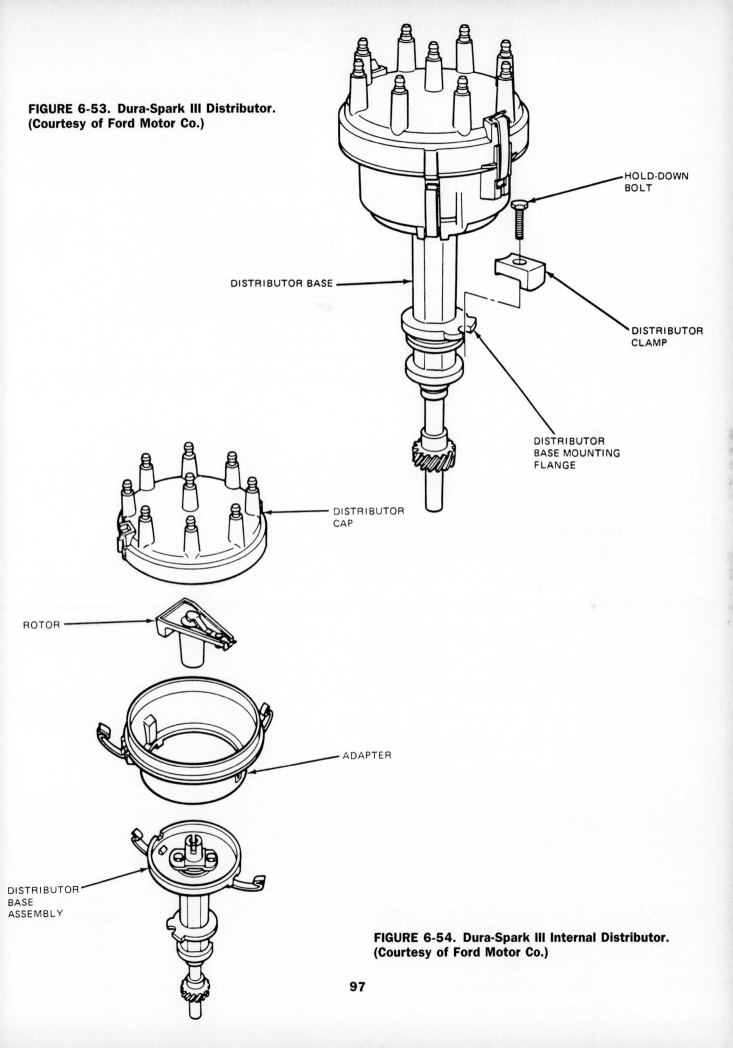

FIGURE 6-53. Dura-Spark III Distributor.
(Courtesy of Ford Motor Co.)

HOLD-DOWN
BOLT

DISTRIBUTOR BASE

DISTRIBUTOR
CLAMP

DISTRIBUTOR
BASE MOUNTING
FLANGE

DISTRIBUTOR
CAP

ROTOR

ADAPTER

DISTRIBUTOR
BASE
ASSEMBLY

FIGURE 6-54. Dura-Spark III Internal Distributor.
(Courtesy of Ford Motor Co.)

97

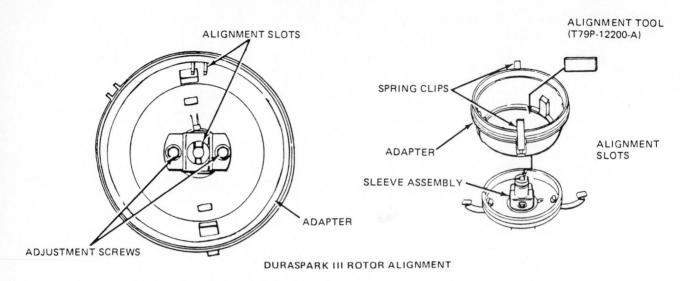

FIGURE 6-55. Dura-Spark III Rotor Alignment. (Courtesy of Ford Motor Co.)

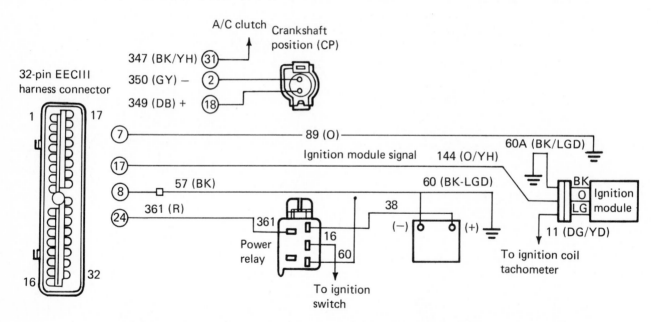

FIGURE 6-56. Crankshaft Position Sensor Signal. (Courtesy of Ford Motor Co.)

spark occurs at the spark tester, the module and ignition coil are satisfactory, and the crankshaft position sensor, or ECA, is defective. Lack of spark at the tester indicates a defective ignition module, ignition coil, or connecting wires.

Step 4. Disconnect the crankshaft position (CP) sensor and connect an alternating current (AC) voltmeter to the sensor terminals while the engine is

being cranked. The lowest AC voltage scale must be used for these tests. If there is no AC voltage signal, the CP sensor is defective. Reconnect the CP sensor connector and disconnect the ECA connector. While the engine is being cranked, connect an AC voltmeter to terminals 2 and 18 in the ECA connector. If an AC voltage signal was available at the CP sensor terminals but there is no AC voltage signal at terminals 2 and 18, the CP sensor wires

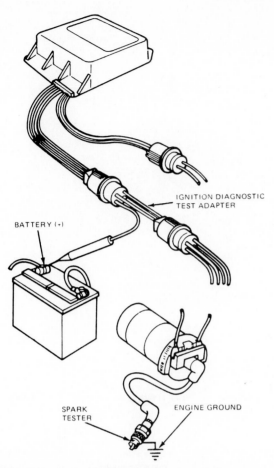

BATTERY (+)

IGNITION DIAGNOSTIC
TEST ADAPTER

SPARK
TESTER

ENGINE GROUND

FIGURE 6-57. Dura-Spark III Diagnosis. (Courtesy of Ford Motor Co.)

are defective. (See Figure 6-56 for terminal identification.)

Step 5. With the ignition switch on, connect a direct current (DC) voltmeter from terminal 24 in the ECA connector to ground. If the voltage is below 12 V, check the power relay circuit.

Step 6. Reconnect the ECA connector and probe the orange wire on ECA terminal 17 with a steel pin. While the engine is being cranked, connect an AC voltmeter from the pin to ground. If the voltage readings in steps 4 and 5 were satisfactory and there is no AC voltage signal at the orange wire, the ECA is defective.

Step 7. Disconnect the three-wire ignition module connector and connect an AC voltmeter from the orange wire in the wiring harness to ground while the engine is being cranked. If the voltage signal in step 6 is satisfactory but there is no voltage signal in step 7, the orange wire from the ECA to the module is defective. If there was no spark in step 3, the tests in Table 6-2 should be used to check the primary ignition circuit wiring and the ignition coil.

Disconnect the module connectors. Test connections are made to the wiring harness side of the module connectors except those test connections that are completed at the coil terminals. If no defects are found in the diagnostic procedure, the module or ignition coil is defective.

TABLE 6-2. Dura-Spark III Diagnosis

	Test Voltage Between	Specifications	Testing
Key on	Red wire and ground	12 V	voltage to module
	Green wire and ground	12 V	ballast resistor coil primary
Key on	Positive primary coil terminal and ground	6–8 V	ballast resistor
	Ground coil tach terminal		
Key on, cranking	Positive primary coil terminal and ground	9 V	resistor bypass
	Ground coil tach terminal		
	Test Resistance Between	**Specifications**	**Testing**
Key off	Black wire and ground	0 Ω	ground wire
	Coil primary terminals	0.8-1.6 Ω	coil primary
	Coil primary terminal to coil tower	7,700 to 10,500 Ω	coil secondary

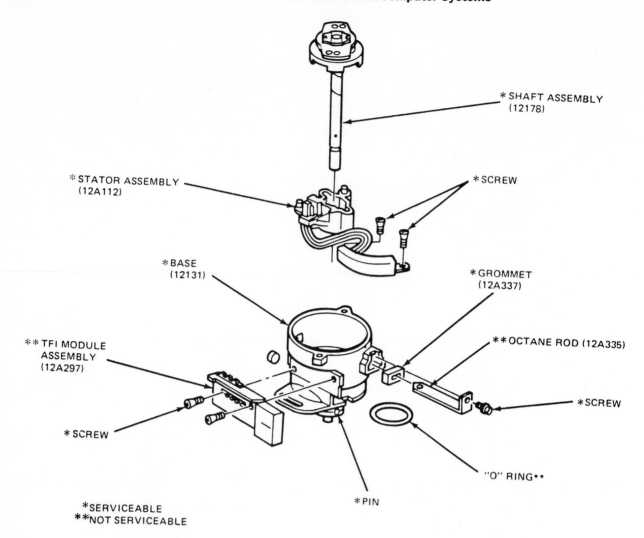

FIGURE 6-58. EEC IV TFI Distributor. (Courtesy of Ford Motor Co.)

EEC IV THICK FILM INTEGRATED (TFI) IGNITION SYSTEMS

Design and Operation

TFI ignition systems are used in all EEC IV systems. The distributor in the EEC IV TFI ignition system is similar to that in the conventional TFI ignition system. A Hall Effect device referred to as a profile ignition pickup (PIP) is used in place of the conventional pickup coil in the distributor. The EEC IV TFI distributor is illustrated in Figure 6-58.

EEC IV TFI ignition systems use a six-wire TFI module. The primary ignition circuit is connected to the module in the same way as in the conventional TFI ignition systems. However, the EEC IV TFI ignition system has three extra wires that are

connected from the module to the electronic control assembly (ECA), as shown in Figure 6-59.

The electrical ground circuit between the ECA and the TFI module is provided by circuit 16. A PIP signal is sent from the Hall Effect device to the ECA through circuit 56. This signal informs the ECA when each piston is at 10° before top dead center (BTDC). The ECA provides a spark output (SPOUT) signal through circuit 36 to the TFI module, which provides the precise spark advance that is required by the engine under all operating conditions. This eliminates the need for conventional advance mechanisms on the EEC IV TFI distributor.

Diagnosis of Electronic Engine Control IV (EEC IV) TFI Ignition Systems

When checking the initial ignition timing, the spark output (SPOUT) circuit number 36 must be discon-

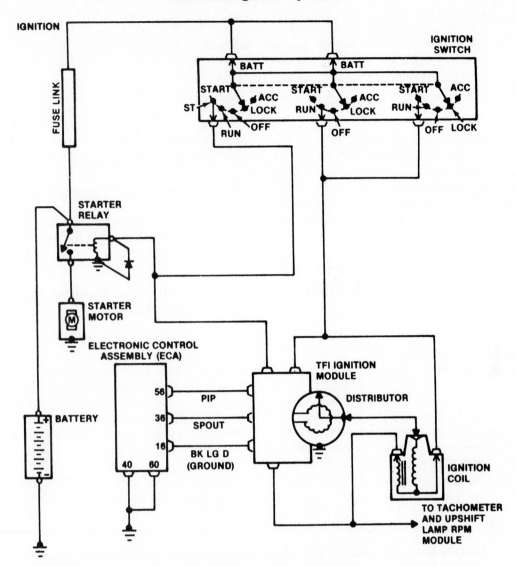

FIGURE 6-59. EEC IV TFI Ignition Wiring. (Courtesy of Ford Motor Co.)

nected at the distributor connector. A separate in-line timing connector is provided for this wire. Refer to Figure 6-59 for circuit identification.

When the in-line timing connector is disconnected, the timing is fixed by the TFI module in a limited operation strategy (LOS) mode. In this mode the distributor may be rotated to adjust the initial timing. If the engine fails to start, the following diagnostic procedures should be followed:

1. Perform the voltmeter and ohmmeter tests on TFI ignition as outlined previously in this chapter. The profile ignition pickup (PIP) cannot be tested with an ohmmeter.

2. Connect a 12 V test lamp from the coil tach terminal to ground while the engine is being cranked. A flashing test lamp indicates that the

PIP and the module are turning the primary circuit on and off. If the test lamp fails to flash, the PIP module or electronic control assembly (ECA) is defective.

3. If the test lamp flashes in step 2, connect a test spark plug from the coil secondary wire to ground while the engine is being cranked. If the spark plug fails to fire, the coil is defective.

4. If the test spark plug fires in step 3, connect the test spark plug to several plug wires while the engine is being cranked. When the test spark plug fires in step 3 but fails to fire on the spark plug wires, the distributor cap or rotor is defective.

5. When the test lamp does not flash in step 2, test wires 36 and 56 between the module and the ECA

for open circuits and grounds with an ohm-meter. If these wires are satisfactory, probe wires 36 and 56 at the ECA with the terminals connected to the ECA. Connect a voltmeter from each wire to ground while the engine is being cranked. Each voltmeter reading should be 3 V to 6 V. If the reading on the number 56 PIP wire is not within specifications, the PIP or the module is defective. Since these components cannot be tested individually, replacement is necessary. When the voltmeter reading on the 56 PIP wire is satisfactory but the reading on the number 36 SPOUT wire is not within specifications, the ECA is defective.

CHRYSLER ELECTRONIC IGNITION SYSTEMS

Conventional Systems with Five-Terminal Module

The distributor pickup coil leads are connected to terminals 4 and 5 on the module. A dual ballast resistor contains the conventional ignition resistor and an auxiliary resistor. Terminal 3 on the module is connected to the auxiliary resistor as illustrated in Figure 6-60. The ignition resistor is connected in series between the ignition switch and the positive primary coil terminal. Module terminal 1 is con-

nected to the ignition switch side of the dual ballast resistor. Terminal 2 on the module is connected to the negative primary coil terminal.

A roll pin secures the reluctor to the distributor shaft, and the number of reluctor tips matches the number of engine cylinders. Clearance between the reluctor tips and the pickup coil head should be adjusted to 0.006 in (0.152 mm) using a non-magnetic feeler gauge, as shown in Figure 6-61.

Systems with Four-Terminal Module

The auxiliary resistor and the number 3 module terminal are discontinued in the four-terminal module system, as pictured in Figure 6-62.

Systems with Four-Terminal Module and Dual Pickup Coils

The dual pickup coils are referred to as "start" and "run" pickups. A common wire from both pickup coils is connected to module terminal 5. The dual relay contacts connect either the "run" or the "start" pickup lead to module terminal 4, as illustrated in Figure 6-63. While the engine is running, the run pickup is connected to the module terminal 4. While the engine is being cranked, the relay winding is energized and the relay coil magnetism moves the relay points to the start position to connect the start pickup to module terminal 4. When the start pickup coil signal is sent to the module, the basic

FIGURE 6-60. Chrysler Five-Terminal Module System. (Courtesy of Chrysler Canada Ltd.)

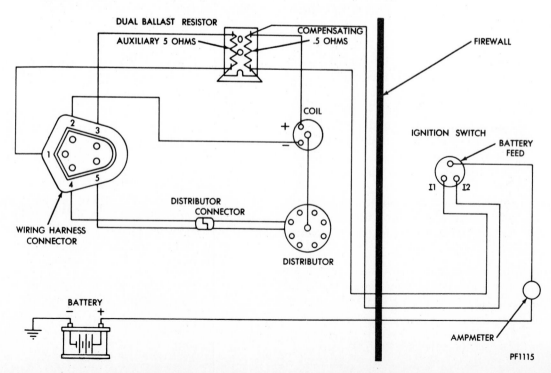

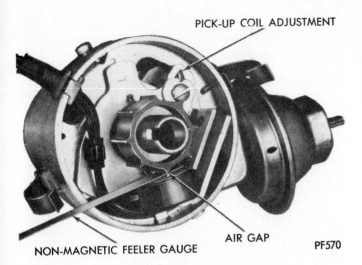

FIGURE 6-61. Adjusting Pickup Coil Clearance. (Courtesy of Chrysler Canada Ltd.)

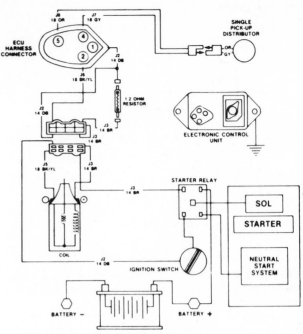

FIGURE 6-62. Four-Terminal Module System. (Courtesy of Chrysler Canada Ltd.)

ignition timing is retarded to provide easier starting.

Systems with Five-Terminal Modules and Hall Effect Pickup Assemblies

Hall Effect pickup assemblies contain a permanent magnet and a semiconductor material. The high points of the reflector blade rotate past the semiconductor material. When the reflector blade tip moves into alignment with the permanent magnet, magnetic buildup occurs across the semiconductor material. When the reflector blade begins moving out of alignment with the permanent magnet, a magnetic collapse occurs across the semiconductor material and signals the module to open the primary ignition circuit, as illustrated in Figure 6-64.

FIGURE 6-63. Electronic Ignition System with Four-Terminal Module and Dual Pickup Coils. (Courtesy of Chrysler Canada Ltd.)

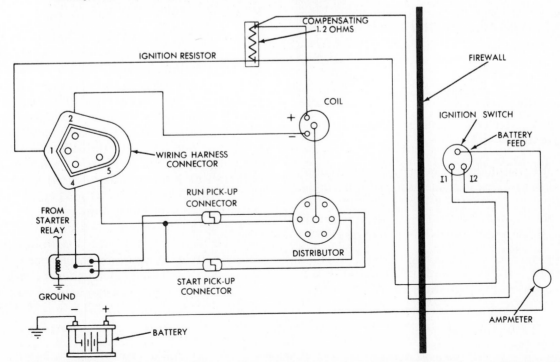

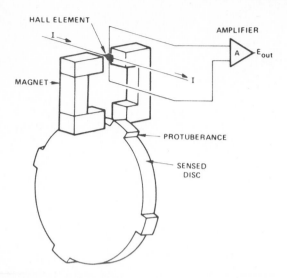

FIGURE 6-64. Hall Effect Pickup Assembly. (Reprinted with permission © 1980 Society of Automotive Engineers)

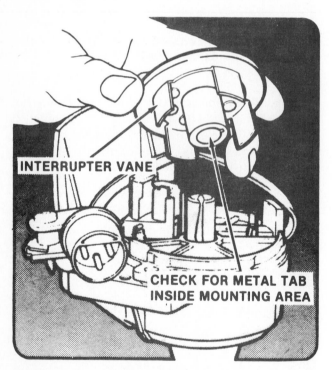

FIGURE 6-65. Hall Effect Reflector Blade. (Courtesy of Sun Electric Corp.)

The reflector blade is attached to the distributor rotor, and a ground tab connects the reflector blade to the distributor shaft, as shown in Figure 6-65. When there is no electrical contact between the reflector blade and the distributor shaft, the engine will fail to start.

Three leads are connected from the Hall Effect pickup assembly to the module. The "ign" terminal

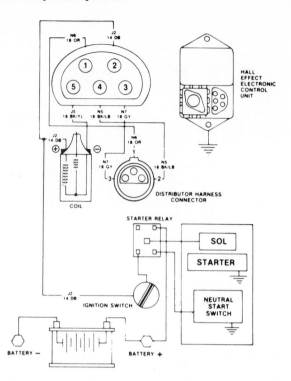

FIGURE 6-66. Electronic Ignition System with Hall Effect Pickup Assembly. (Courtesy of Chrysler Canada Ltd.)

on the ignition switch is connected to the coil positive primary terminal and the module terminal 2. The primary circuit resistor is discontinued in the Hall Effect system. The negative primary coil terminal is connected to the module terminal 5, as shown in Figure 6-66.

Distributor Used with Electronic Fuel Injection (EFI) Systems

The distributors used with single-point (throttle body) injection systems are very similar to the Hall Effect distributors used with electronic ignition systems. Chrysler 2.2L turbocharged engines are equipped with multi-point (port) fuel injection. The distributors used with these engines contain a Hall Effect switch above the pickup plate for ignition triggering and a second Hall Effect switch under the pickup plate for injector sequencing. This second Hall Effect switch is referred to as a "sync" pickup. (Refer to Chapter 9 for a complete description of Chrysler EFI systems.)

A labyrinth distributor is used in 1986 and later model 2.2L engines. In these distributors, the Hall Effect switches for ignition triggering and injector sequencing are both located under the pickup plate, which reduces electromagnetic interference (EMI).

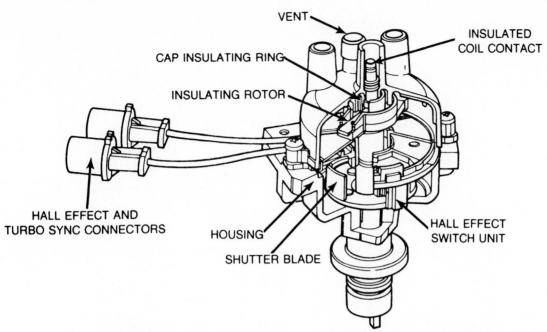

FIGURE 6-67. Labyrinth Distributor. (Courtesy of Chrysler Canada Ltd.)

A plate with four shutter blades is attached to the distributor shaft. These shutter blades rotate through both Hall Effect switches. The shutter blade for number 1 cylinder has a hole drilled through it. When this shutter blade rotates through the "sync" pickup, the computer energizes injectors 1 and 2. Injectors 3 and 4 are energized when the next crankshaft revolution is completed. The labyrinth distributor has improved insulating qualities around the center distributor cap terminal, as shown in Figure 6-67.

A photo-optical-type distributor is used on the 3.0L V6 engine. This engine is manufactured by Mitsubishi and is installed in some Chrysler products. In the photo-optical-type distributor, a disc is attached to the distributor shaft. A set of slots is located around the outer edge of this disc, and a second set of slots is positioned on the inner area of the disc. Both rows of slots are chemically etched into the disc. A light-emitting diode is located above each set of slots, and a photo diode is installed under the disc directly below the light-emitting diode, as illustrated in Figure 6-68.

When the light-emitting diodes shine light through a slot in the disc, the photo diodes send a current to the integrated circuit in the engine controller. As the disc rotates, the blank between the slots blocks the light from the light-emitting diode, which shuts off the current flow in the integrated circuit. The outer set of slots provides a signal every two degrees of crankshaft rotation. This set of slots is used to provide precise timing and spark advance

at speeds up to 1,200 RPM. At slow engine speeds, the engine RPM tends to decrease between cylinder firings, which causes pulsations in engine speed. The outer row of slots instantaneously triggers the ignition system at the correct crankshaft position regardless of these engine pulsations, which results

FIGURE 6-68. Photo-Optical-Type Distributor. (Courtesy of Chrysler Canada Ltd.)

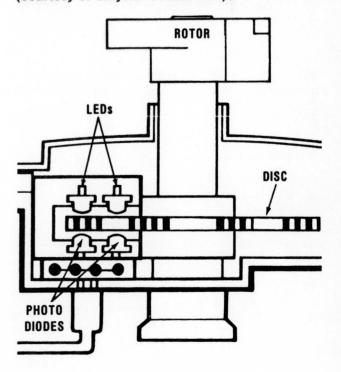

in a smoother running engine at idle and low speeds. The inner row of slots is used to sequence the injectors in the port injection system, and this row of slots also triggers the ignition system at engine speeds above 1,200 RPM.

DIAGNOSIS OF CHRYSLER ELECTRONIC IGNITION SYSTEMS

Engine Fails To Start, Four- and Five-Wire Modules

The diagnostic chart in Table 6-3 may be used when diagnosing Chrysler electronic ignition systems. For systems with four-wire modules, the auxiliary resistor test and the voltage test at module cavity 3 are eliminated. Test connections referring to cavities are completed at the disconnected module connector. Replacement ignition resistors have a resistance specification of 1.25 Ω regardless of temperature. (The specification for the original ignition resistor is provided in Table 6-3.)

A zero voltmeter reading at cavity 1 in the module connector indicates an open circuit in the circuit from the ignition switch to the resistors, or in the circuit from the resistors to cavity 1, as illustrated in Figure 6-69.

Figure 6-70 pictures a voltmeter connected from cavity 2 in the module connector to ground. A zero volt reading with the ignition switch on would be caused by an open circuit in the ignition resistor,

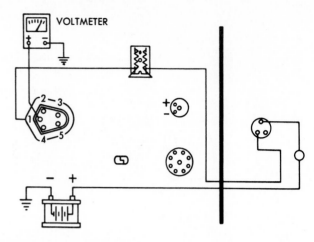

FIGURE 6-69. Voltmeter Test at Cavity 1. (Courtesy of Chrysler Canada Ltd.)

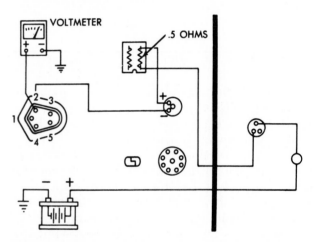

FIGURE 6-70. Voltmeter Test at Cavity 2. (Courtesy of Chrysler Canada Ltd.)

TABLE 6-3. Chrysler Electronic Ignition Diagnosis

	Test Voltage Between	Specifications	Testing
Key on	Cavity 1, 2, 3, and ground	11.5 V or over	Open circuits
	Ground negative primary coil when making the next test.		
Key on, cranking	Positive primary terminal and ground	9 V or over	Resistor bypass circuit
	Test Resistance Between	Specifications	Testing
Key off	Cavity 4 and 5	350–550 Ω	Pickup coil
	Cavity 4 or 5 and ground	Infinity	
	Primary coil terminals	1.4–1.79 Ω at 70°	Primary winding
	Primary terminal to coil tower	8000–11700 Ω at 70°	Secondary winding
	Ignition resistor terminals	0.5–0.6 Ω at 70°	Ballast resistor
	Auxiliary resistor terminals	4.75–5.75 Ω at 70°	Ballast resistor

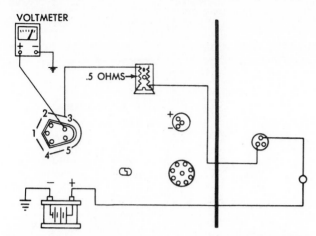

FIGURE 6-71. Voltmeter Test at Cavity 3. (Courtesy of Chrysler Canada Ltd.)

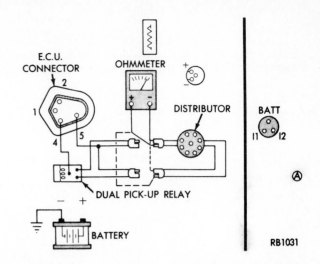

FIGURE 6-72. Testing Dual Pickup Coils. (Courtesy of Chrysler Canada Ltd.)

coil primary winding, or connecting wires, as outlined in Figure 6-70.

An open circuit exists in the auxiliary resistor or connecting wire if a zero volt reading is obtained at cavity 3, as pictured in Figure 6-71.

Four-Wire Module Systems with Dual Pickup Coils

The diagnostic chart in Table 6-3 should be used to diagnose systems with four-wire modules and dual pickup coils. If an infinite ohmmeter reading is obtained when the ohmmeter leads are connected across cavities 4 and 5, an open circuit is located in the run pickup coil, relay points, or connecting wires. Individual pickup coils can be tested by connecting an ohmmeter across each pair of pickup leads at the disconnected distributor connectors, as outlined in Figure 6-72. Resistance readings below 150 Ω or above 900 Ω indicate a defective pickup coil.

Connect the ohmmeter leads across relay terminals 4 and 5 to test the relay winding. If the resistance reading is below 20 Ω or above 30 Ω, replace the relay. The relay contacts are connected from terminals 1 to 3 and 2 to 3. When the relay winding is not energized, one set of contacts should indicate zero ohms resistance on an ohmmeter. The other set of relay contacts should have zero ohms resistance if the relay winding is energized from a 12 V source. (See Figure 6-73 for the location of the relay terminal.)

Five-Wire Systems with Hall Effect Pickup Assembly

Systems with Hall Effect pickups may be diagnosed using the chart in Table 6-4. Disconnect the module

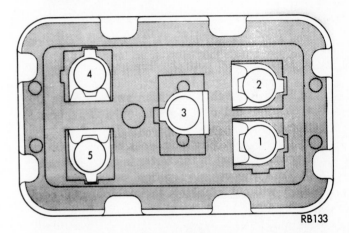

FIGURE 6-73. Relay Terminals, Dual Pickup System. (Courtesy of Chrysler Canada Ltd.)

and distributor connectors before proceeding with the tests.

On Chrysler oxygen (O_2) feedback systems with Hall Effect distributors, the ignition system may be tested as shown in Figure 6-74. Defective pickups on Chrysler electronic fuel injection (EFI) systems will set fault codes in the logic module memory. (Refer to Chapter 8 for O_2 feedback diagnosis and Chapter 9 for EFI diagnosis.)

SCOPE DIAGNOSIS OF MODULES AND PICKUP COILS

Pickup Coil Testing

Various scopes have pickup coil testing capabilities. The scope accessory leads must be connected to the pickup coil leads as pictured in Figure 6-75. A module adapter is installed on the end of the red

107

TABLE 6-4. Diagnosis of Chrysler Electronic Ignition with Hall Effect Pickup

	Test Voltage Between	Specifications	Testing
Key on	Cavity 2 to ground	12 V	Module power supply
	Cavity 5 to ground	12 V	Coil primary

	Test Resistance Between	Specifications	Testing
Key off	Distributor terminal 3 to cavity 3	0 Ω	wire 3
	Distributor terminal 2 to cavity 4	0 Ω	wire 2
	Distributor terminal 1 to cavity 1	0 Ω	wire 1

Connect Module Connector

Key on	Momentarily connect terminal 2 to 3 in distributor wiring harness connector, Figure 6-74. Hold coil secondary lead ½″ away from engine block. If spark occurs at coil, lead, replace pickup. Module or ignition coil is defective if no spark occurs at coil lead.

HALL EFFECT DISTRIBUTOR WIRING
HARNESS CONNECTOR

JUMPER WIRE

FIGURE 6-74. Testing Systems with Hall Effect Pickup. (Courtesy of Chrysler Canada Ltd.)

accessory lead. Sharp pinpoints on the accessory lead clips puncture the pickup wire insulation. Silicone sealer may be used to seal the puncture marks.

During the testing of pickup coils, the program test selector switch must be in position 6 and the special test selector set in the distributor pickup coil position. Selector switches are illustrated in Figure 6-76.

Normal pickup coil waveforms, as illustrated in Figure 6-77, should appear on the scope screen while the engine is cranking. The pickup coil is defective if no waveform is available. Hall Effect pickup assemblies cannot be tested with the scope.

FIGURE 6-75. Pickup Coil Test Connections. (Courtesy of Bear Automotive Inc.)

STEP 1. Attach RED ACCESSORY lead clip to MODULE ADAPTER. Attach the other Adapter lead to the P.U. coil wire Piercing clip.

Module Adapter

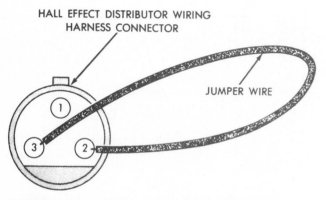

P.U. Coil

DIST.

OR.

BLK.

PUR.

STEP 2. Attach Piercing clip to each P.U. Coil wire.

NOTE: Connections can be made to this connector side. Color code may differ - trace Pick-up coil wires across connector to connecting wires.

STEP 3. Attach BLACK ACCESSORY lead clip to the P.U. coil wire Piercing clip.

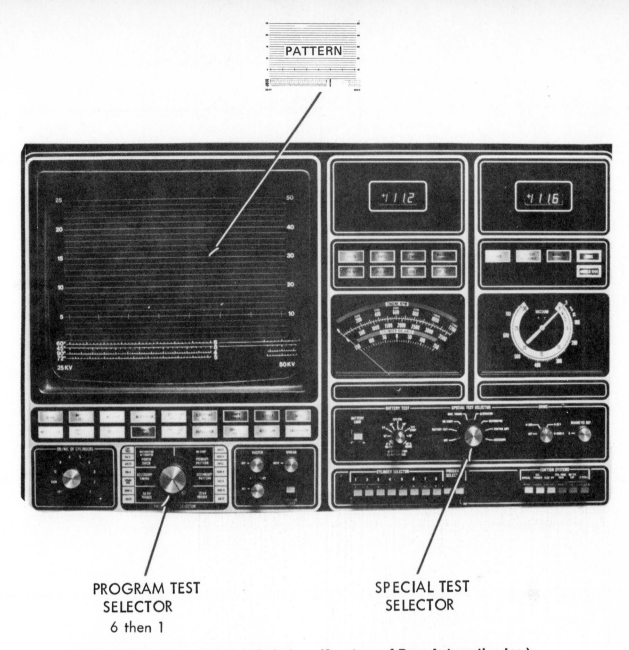

PATTERN

PROGRAM TEST
SELECTOR
6 then 1

SPECIAL TEST
SELECTOR

FIGURE 6-76. Scope Selector Switches. (Courtesy of Bear Automotive Inc.)

FIGURE 6-77. Normal Waveform of Pickup Coil. (Courtesy of Bear Automotive Inc.)

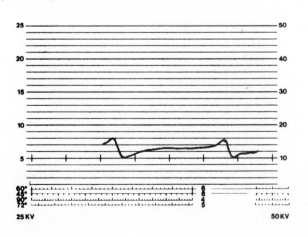

Module Testing

The scope accessory leads are connected in the same way for pickup coil and module testing. Connect the other scope leads to the ignition system in the normal manner. On HEI systems, with ignition coils mounted in the distributor cap, ground the center cap terminal. Position the program test selector in position 1 and the special test selector in the control unit position. The ignition switch must be off before moving the special test selector switch. When the ignition switch is on, a waveform should appear on the scope, as pictured in Figure 6-78. Module waveforms may vary, depending on the ignition system. In the control unit test position,

109

FIGURE 6-78. Normal Module Waveform. (Courtesy of Bear Automotive Inc.)

the scope accessory leads are used to trigger the module, and the scope ignition leads pick up the waveform as the module opens and closes the primary circuit. The module is defective if no waveform appears in the control unit test position.

Questions

1. In most electronic ignition systems the dwell is determined by the _____.

2. When the ignition switch is turned on in a high energy ignition (HEI) system, the primary current flow would be 6 amperes. T F

3. An intermittent open circuit in the HEI pickup coil leads could result in _____ _____ .

4. If a test light connected from the negative primary coil terminal to ground flashes during engine cranking when a no-start complaint is being diagnosed, the module or pickup coil is defective. T F

5. If, while the engine is being cranked, a spark tester fires when connected from the coil secondary wire to ground but fails to fire when connected from the plug wires to ground, the _____ or _____ _____ is defective.

6. The electronic module retard (EMR) system reduces emission levels by providing faster _____ warm-up.

7. A defective electronic spark control (ESC) controller could cause a no-start complaint. T F

8. The ESC system retards the spark advance when the engine _____ .

9. Dura-Spark I ignition systems have a primary circuit resistance wire connected in series between the ignition switch and the positive primary coil terminal. T F

10. Dura-Spark I and Dura-Spark II ignition modules are interchangeable. T F

11. The vacuum switch in a dual-mode ignition timing system retards the spark advance when the manifold vacuum is below 6 in Hg (20 kPa). T F

12. In a thick film integrated (TFI) ignition system, the ignition module is mounted externally from the distributor. T F

13. In a Chrysler-type Hall Effect pickup assembly, a no-start complaint could be caused by high resistance between the reflector blade and the distributor shaft. T F

14. Cavity 1 in a conventional Chrysler electronic ignition system is connected to the _____ .

15. The Hall Effect pickup assembly in a Chrysler electronic ignition system may be tested with an ohmmeter. T F

7

COMPUTER-CONTROLLED DISTRIBUTOR ADVANCE SYSTEMS

CHRYSLER ELECTRONIC LEAN BURN (ELB) SYSTEMS

Design

In the electronic lean burn (ELB) system, a spark control computer is used to determine the spark advance. The ignition module is mounted in the spark control computer. Conventional centrifugal and vacuum advance mechanisms are discontinued in the ELB system, except in some early model systems having a centrifugal advance mechanism. The spark control computer provides more accurate spark advance in relation to engine operating conditions, and thus reduces exhaust emissions. Dual distributor pickup coils are connected to the spark control computer. These pickup coils are referred to as the "start" pickup and the "run" pickup. The start pickup coil has a larger connector than the run pickup, as illustrated in Figure 7-1.

The dual ballast resistor and the ignition coil used in the ELB system are similar to the coil and resistor used in the Chrysler electronic ignition system. A vacuum transducer is mounted in the computer and connected to the intake manifold vacuum. In some later model systems, the vacuum transducer is connected to ported manifold vacuum above the throttles. The vacuum transducer contains a mov-

able metallic core surrounded by a coil of wire, as illustrated in Figure 7-2.

A diaphragm in the vacuum transducer is linked to the metallic core. Therefore, the movement of the metallic core is controlled by the vacuum applied to the diaphragm.

Two terminals on the vacuum transducer are used to connect the vacuum transducer winding to the computer. The throttle transducer is mounted on a bracket near the carburetor. A metallic throttle transducer core is connected to the throttle linkage. The winding surrounding the metallic core is connected to the two transducer terminals, and external wires connect the transducer terminals to the computer. Figure 7-3 illustrates a throttle transducer with the core removed.

A carburetor switch is mounted on the end of the idle stop solenoid stem, but insulation separates the switch electrically from the stem. When the throttle is in the idle position, the carburetor switch is grounded by the idle speed screw, as pictured in Figure 7-4. In some systems, the idle stop solenoid is not used, and the carburetor switch is mounted on a post attached to the carburetor.

The coolant switch contains a bimetal strip and a set of contact points. When the engine coolant temperature is below 150°F (66°C), the coolant switch contacts are closed. The coolant switch and the carburetor switch share a common electrical connection to the spark control computer.

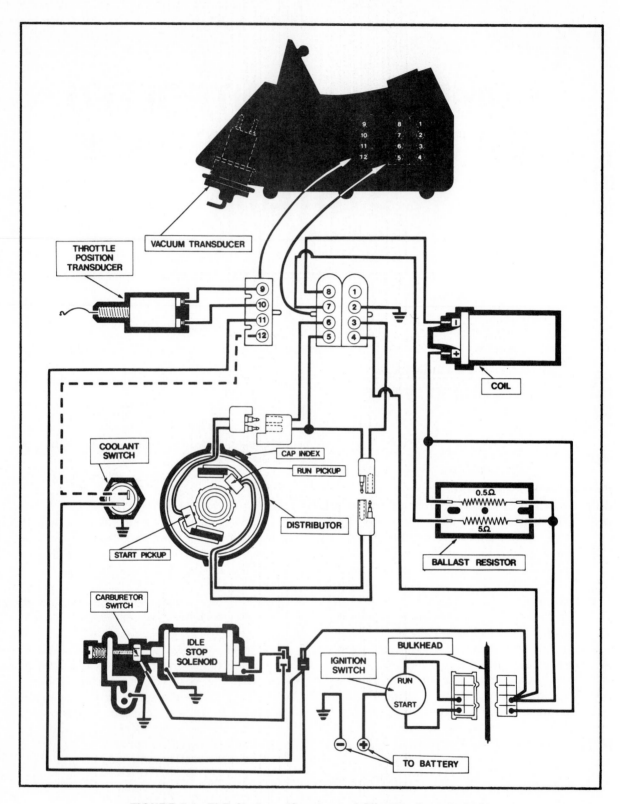

FIGURE 7-1. ELB System. (Courtesy of Chrysler Canada Ltd.)

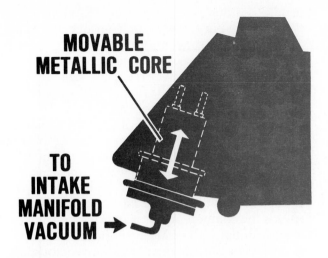

FIGURE 7-2. Vacuum Transducer. (Courtesy of Chrysler Canada Ltd.)

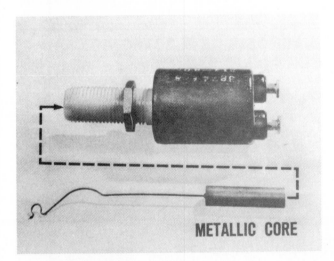

FIGURE 7-3. Throttle Transducer. (Courtesy of Chrysler Canada Ltd.)

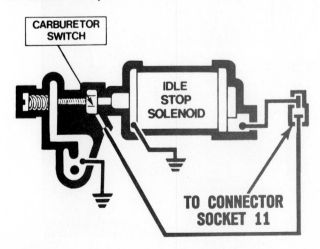

FIGURE 7-4. Carburetor Switch. (Courtesy of Chrysler Canada Ltd.)

Operation

When the engine is being cranked, the start pickup signal is delivered to the ignition control unit in the spark control computer. When the engine starts, the run pickup signal is sent through the spark control computer to the ignition control unit. The run pickup signal is approximately 45° advanced in relation to the start pickup signal. Electrical input signals from the throttle transducer and vacuum transducer are used by the computer to determine the precise spark advance required by the engine.

The throttle transducer sends a signal to the spark control computer in relation to engine speed. At wide-open throttle, the spark control computer will provide approximately 10° advance from the throttle transducer signal. An air temperature sensor in the computer is connected in the throttle transducer circuit. Figure 7-5 illustrates the air temperature sensor.

The computer is mounted on the side of the air cleaner in order to allow intake air to circulate through the computer. As the intake air temperature is increased, the spark advance provided by the throttle transducer signal is reduced to prevent detonation in the engine. If intake air temperature is above 150°F (66°C), the air temperature sensor will reduce the throttle transducer advance to zero.

Spark advance in relation to engine load is provided by the vacuum transducer signal to the spark control computer. This signal is cancelled when the carburetor switch is grounded by the idle speed screw, or when the coolant switch contacts are closed at coolant temperatures below 150°F

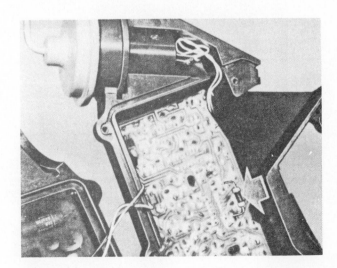

FIGURE 7-5. Air Temperature Sensor. (Courtesy of Chrysler Canada Ltd.)

(66°C). When the throttle is opened with the engine at normal operating temperature, the spark control computer will provide a gradual spark advance in response to the vacuum transducer signal. After the throttle has been open for 7 minutes, the spark control computer will provide 35° of spark advance if the manifold is 16 in Hg (54 kPa), as shown in Figure 7-6.

If the vacuum applied to the vacuum transducer is 12 in Hg (40 kPa), the spark control computer will provide 25° of spark advance after the throttle has been open for 7 minutes. When the throttle is moved from the part throttle to the full throttle position, the spark advance supplied from the vacuum transducer signal will be cancelled by the computer because of the extremely low manifold vacuum at full throttle. However, the computer will remember the former spark advance at part throttle and will immediately restore the spark advance when the throttle is returned from full- to part-throttle position.

When the throttle is moved from the part-throttle to the idle position, the carburetor switch will be grounded by the idle speed screw. When the carburetor switch is grounded, any spark advance that was supplied by the computer in response to the vacuum transducer signal is immediately cancelled. The computer will immediately restore part of the vacuum transducer spark advance if the throttle is reopened within 3½ minutes. If the throttle is left in the idle position for more than 3½ minutes, the computer memory of vacuum transducer advance is erased, as illustrated in Figure 7-7.

The exact amount of spark advance supplied by the computer varies with engine size, type of transmission, and rear axle ratio. Each size of engine will have a computer that requires a different amount of manifold vacuum to obtain maximum vacuum transducer spark advance. Therefore, individual computer advance specifications must be used for each computer. Each time the engine is started, the computer supplies 7–9 degrees of spark advance in addition to the basic timing. After 90 seconds the computer returns the spark advance to the basic timing setting. Most computers also supply some spark advance in relation to the reluctor speed. This spark advance is referred to as speed advance.

DIAGNOSIS OF ELECTRONIC LEAN BURN (ELB) SYSTEMS

Engine Fails To Start

Computer terminals in the diagnostic procedure can be identified by referring to Figure 7-1. The ignition switch must be in the "off" position before the

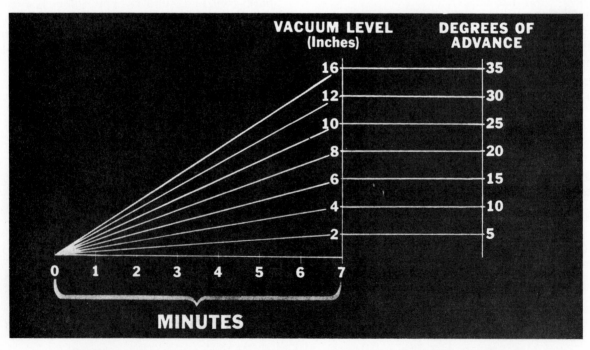

FIGURE 7-6. Vacuum Transducer Advance Operation. (Courtesy of Chrysler Canada Ltd.)

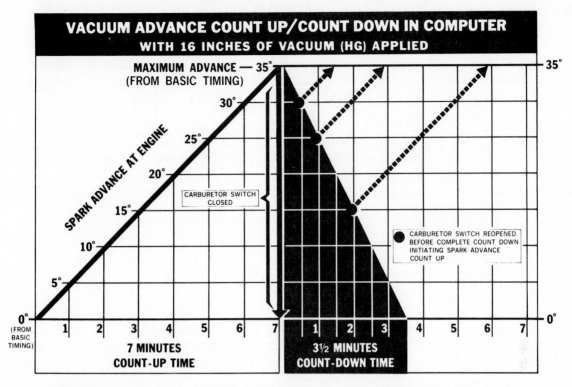

FIGURE 7-7. Vacuum Transducer Advance Operation. (Courtesy of Chrysler Canada Ltd.)

computer connectors are disconnected. The original temperature sensitive ignition resistor had a specified resistance of 0.5 Ω at 70°F (21°C). Replacement ignition resistors have 1.25 Ω resistance regardless of temperature. When the engine fails to start, the ELB system may be diagnosed by using the voltmeter and ohmmeter tests provided in Table 7-1.

Unsatisfactory Performance or Economy

The test procedure shown in Table 7-2 should be used to diagnose performance and economy complaints in the ELB system.

TABLE 7-1. ELB Diagnosis When Engine Fails To Start.

	Test Voltage Between	Specifications	Testing
Key on; coolant switch disconnected; carb switch screw insulated	Carburetor switch and ground	5 V or over	Open circuits
Key on; dual connector disconnected at computer	Terminal 4 and ground	12 V	Open circuit 4 wire
	Terminal 7 and ground	12 V	Open circuit 5 ohm resistor
	Terminal 8 and ground	12 V	Open circuit ignition resistor or coil primary

TABLE 7-1 (continued).

	Test Resistance Between	Specifications	Testing
Key off; single and dual computer connectors disconnected	Terminal 11 and carburetor switch	Low reading	Open circuit 11 wire
	Terminal 2 and ground	Low reading	Computer ground
	Terminal 5 and 6	150–900 Ω	Start pickup coil
	Terminal 5 or 6 and ground	Infinity	Start pickup coil
	Coil primary terminals	1–2 Ω	Primary winding
	Primary terminal to coil tower	8,000–12,000 Ω	Secondary winding
	Ignition resistor terminals	1.25 Ω	Ignition resistor
	Auxiliary resistor terminals	5 Ω	Auxiliary resistor

Ignition coil should be tested using coil tester. If all tests are satisfactory, replace spark control computer.

TABLE 7-2. ELB Diagnosis, Unsatisfactory Performance or Economy Complaints.

	Test Resistance Between	Specifications	Testing
Key off; single and dual computer connectors disconnected	Terminal 3 and 5	150–900 Ω	Run pickup coil
	Terminal 3 or 5 and ground	Infinity	Run pickup coil
	Terminal 9 and 10	50–90 Ω	Throttle transducer
	Terminal 9 or 10 and ground	Infinity	Throttle transducer
Coolant switch wire disconnected	Coolant switch terminal and ground	Below 150° coolant temperature 0 Ω	Temperature switch
		Above 150° coolant temperature infinity	
Carb switch insulated from idle screw, and coolant switch wire disconnected	Carb switch and ground	Infinity	

	Test Vacuum Transducer	Specifications
Engine warm; carb switch insulated from screw	Specified vacuum applied to transducer	Manufacturer's specifications
	Connect timing light, wait 7 minutes	
	Check advance using advance meter on timing light	

Timing should immediately return to original setting when carburetor switch is grounded.

	Test Throttle Transducer	Specifications
Engine running at idle carb switch grounded	Wait 90 seconds	Manufacturer's specifications
	Observe timing mark with timing light	
	Move throttle transducer out 1″	
	Check advance using advance meter on timing light	

Timing mark should immediately return to basic setting as throttle transducer is pushed inward to idle position.

CHRYSLER ELECTRONIC SPARK ADVANCE (ESA) SYSTEMS

Design and Operation

The electronic spark advance (ESA) system has a redesigned computer with one dual connector. Other changes in this system include a single distributor pickup coil and a single ignition resistor. The 5-Ω resistor that was connected between the ignition switch and the computer in the electronic lean burn (ELB) system is not used in this system (see Figure 7-8).

The ESA system and the ELB system operate in basically the same way. A start mode in the ESA computer prevents any possibility of spark advance when the engine is being cranked. The throttle transducer is discontinued on some ESA systems.

Some spark control computers provide instant spark advance in response to the vacuum transducer signal in the ESA system.

DIAGNOSIS OF ELECTRONIC SPARK ADVANCE (ESA) SYSTEMS

Engine Fails To Start

The computer terminals in the diagnostic procedure may be identified by referring to Figure 7-8. The ignition switch must be in the "off" position before the computer connector is disconnected. If the engine will not start, the ESA system may be diagnosed by using the voltmeter and ohmmeter tests in Table 7-3.

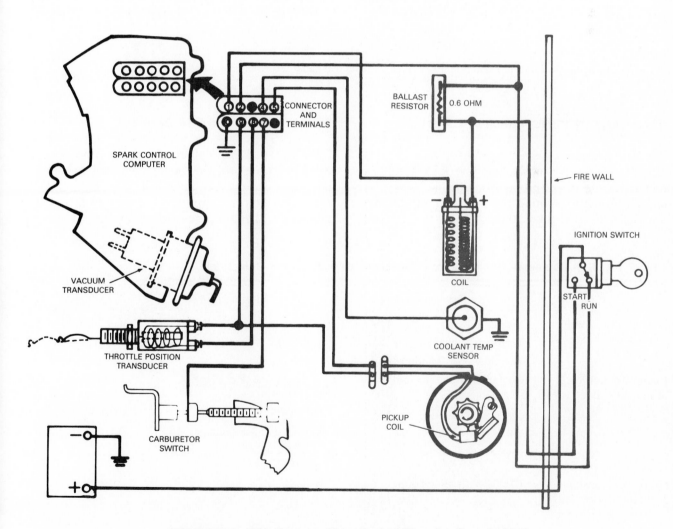

FIGURE 7-8. ESA System. (Courtesy of Chrysler Canada Ltd.)

TABLE 7-3. ESA Diagnosis When Engine Fails To Start.

	Test Voltage Between	Specifications	Testing
Key on; dual connector disconnected at computer	Terminal 1 and ground	12 V	Open circuit ignition resistor or coil primary
	Terminal 2 and ground	12 V	Open circuit ignition switch or wire 2
Key on; cranking; dual connector disconnected at computer	Coil positive terminal and ground	9 V or over	Bypass circuit
	Connect jumper wire from negative coil terminal to ground while making this test.		
	Test Resistance Between	**Specifications**	**Testing**
Key off; dual connector disconnected at computer	Terminal 10 and ground	Low reading	Ground wire
	Terminal 5 and 9	150–900 Ω	Pickup coil
	Terminal 5 or 9 and ground	Infinity	Pickup coil
	Coil primary terminals	1–2 Ω	Primary winding
	Primary terminal to coil tower	8,000–12,000 Ω	Secondary winding
	Ignition resistor terminals	1.25 Ω	Ignition resistor

Ignition coil should be tested using coil tester. If all tests are satisfactory, replace spark control computer.

Unsatisfactory Performance or Economy

The manufacturer's specifications for each computer must be followed when the ESA system is being diagnosed. The procedure in Table 7-4 may be used to diagnose an economy or performance complaint in the ESA system.

Effects of ELB and ESA Defects

If the start pickup is defective in the ELB system with dual pickup coils, the engine will fail to start. A defective run pickup signal will cause the engine to run on the start pickup signal. There will be no spark advance if the run pickup is not available, because

TABLE 7-4. ESA Diagnosis, Unsatisfactory Performance or Economy Complaint.

	Test Resistance Between	Specifications	Testing
Key off; dual connector disconnected at computer	Terminal 7 and carb switch	Low reading	Open circuit 7 wire
	Terminal 8 and 9	50–90 Ω	Throttle transducer
	Terminal 8 or 9 and ground	Infinity	Throttle transducer
Coolant switch wire disconnected	Coolant switch terminal and ground	Below 150°F coolant temperature 0 Ω	Temperature switch
		Above 150°F coolant temperature infinity	
Carb switch insulated from idle screw, and coolant switch wire disconnected	Carb switch and ground		

TABLE 7-4 (continued).

	Test Vacuum Transducer	Specifications
Engine warm; carb switch insulated from screw	Specified vacuum applied to transducer	Manufacturer's specifications
	Connect timing light, wait 8 minutes; check advance using advance meter on timing light	

Timing should immediately return to original setting when carb switch is grounded.

	Test Throttle Transducer	Specifications
Engine running at idle; carb switch grounded	Wait 90 seconds after engine is started	Manufacturer's specifications
	Observe timing mark with timing light	
	Move throttle transducer out 1″	
	Check advance using advance meter on timing light	

Timing mark should immediately return to basic setting as throttle transducer is pushed inward to idle position.

the computer normally controls the run pickup signal to determine the spark advance required by the engine. The term "limp-in mode" is used when the engine runs on the start pickup signal. Fuel economy and engine performance will decrease significantly in the limp-in mode. A quick check of the run pickup may be performed by disconnecting the start pickup while the engine is idling. If the engine stalls when the start pickup is disconnected, this means that the engine was running on the start pickup and that the run pickup is defective.

The spark advance supplied by the throttle transducer will be lost if the throttle transducer is defective. Adjustment of the throttle transducer may be checked by measuring the specified distance between the transducer body and the bracket, as shown in Figure 7-9.

A carburetor switch that is grounded continuously, or a coolant temperature switch that is closed at all times, will cause a complete loss of vacuum transducer advance.

FORD ELECTRONIC ENGINE CONTROL (EEC I) SYSTEMS

Electronic Control Assembly (ECA)

The EEC I system uses an electronic control assembly (ECA) to control spark advance, thermactor pump operation, and exhaust gas recirculation (EGR). Seven sensors are used to monitor various engine conditions and functions. The information is relayed from the sensors to the ECA, where it is processed and converted to output signals that are used in the control functions. A calibration assembly and a processor assembly are contained in the ECA. The ECA is mounted under the instrument panel on the driver's side of the vehicle, as shown in Figure 7-10.

The calibration assembly is a solid state memory bank programmed with pertinent information about the engine's specific calibration characteristics. The processor compares the input sensor signals with the

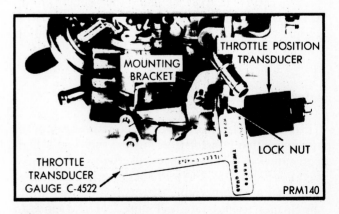

FIGURE 7-9. Throttle Transducer Adjustment. (Courtesy of Chrysler Canada Ltd.)

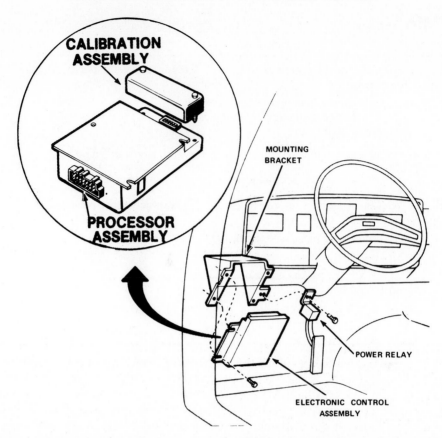

FIGURE 7-10. Electronic Control Assembly (ECA). (Courtesy of Sun Electric Corp.)

information stored in the calibration assembly and delivers the specific output signals required for precise control of spark advance, thermactor pump, and EGR. A power relay mounted near the ECA supplies battery voltage to the ECA when the ignition switch is on.

Inlet Air Temperature (IAT) Sensor

The inlet air temperature sensor contains a temperature sensitive resistor (thermistor) that has a high resistance at low temperatures. Sensor resistance decreases as the temperature of the sensor increases. The inlet air temperature sensor is mounted in the air cleaner, as illustrated in Figure 7-11, and sends an electrical signal to the ECA in relation to air cleaner temperature.

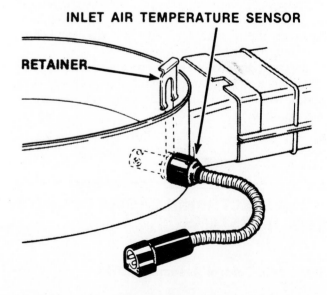

FIGURE 7-11. Inlet Air Temperature (IAT) Sensor. (Courtesy of Sun Electric Corp.)

Throttle Position (TP) Sensor

An exact throttle position signal is sent from the throttle position sensor to the ECA. This sensor is a variable resistor connected to the throttle shaft, as shown in Figure 7-12.

Engine Coolant Temperature (ECT) Sensor

A thermistor in the engine coolant temperature sensor supplies a high resistance at low temperatures. As the temperature of the sensor increases, its resistance becomes lower. If the engine coolant temperature is less than 70°F (21°C), or greater than 230°F (110°C), the coolant temperature sensor signals the ECA to stop all EGR flow. Periods of prolonged idle may cause the engine to overheat. Under this condition, the coolant temperature sensor signals the ECA to advance the ignition timing and thereby reduce engine coolant temperature. The ECT sensor is located near the rear of the intake manifold, as illustrated in Figure 7-13.

Manifold Absolute Pressure (MAP) Sensor

Manifold absolute pressure is defined as atmospheric pressure minus manifold vacuum. A capsule in the MAP sensor is sensitive to manifold vacuum and atmospheric pressure. The capsule converts the vacuum and atmospheric pressure changes to an output voltage that is relayed to the ECA. The MAP sensor signal is used by the ECA to determine the spark advance and EGR flow required by the engine. A hose is connected from the intake manifold to the MAP sensor, which is mounted on the left rocker arm cover, as shown in Figure 7-14.

Exhaust Gas Recirculation (EGR) Valve and Position Sensor

The top part of the EGR valve houses a pintle position sensor. The ECA supplies a reference voltage to the sensor and monitors the resultant output voltage developed by the sensor as it follows the pintle position. The ECA is able to monitor the EGR valve opening continuously because the output voltage from the sensor is directly proportional to pintle position. The EGR valve and the built-in pintle position sensor are illustrated in Figure 7-15.

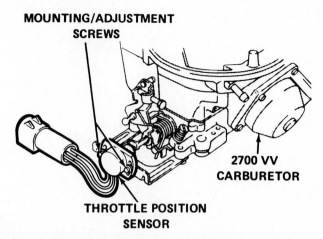

FIGURE 7-12. Throttle Position (TP) Sensor. (Courtesy of Sun Electric Corp.)

Crankshaft Position (CP) Sensor

The CP sensor is mounted in the rear of the engine block. In order to control ignition timing, the electronic control assembly (ECA) must know when each piston reaches top dead center (TDC). It identifies TDC by sensing crankshaft position as it

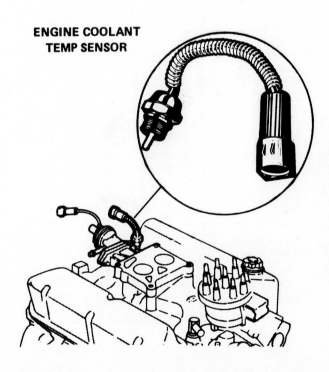

FIGURE 7-13. Engine Coolant Temperature (ECT) Sensor. (Courtesy of Sun Electric Corp.)

MANIFOLD ABSOLUTE
PRESSURE SENSOR

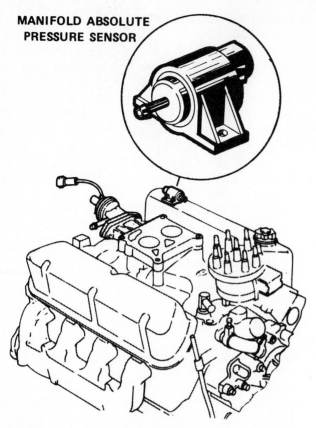

FIGURE 7-14. Manifold Absolute Pressure (MAP) Sensor. (Courtesy of Sun Electric Corp.)

relates to piston position. A thin steel pulse ring with four equally spaced (90° apart) lobes is permanently pressed on the flywheel end of the crankshaft, as illustrated in Figure 7-16.

The crankshaft position (CP) sensor is an electromagnetic device mounted near the pulse ring. As the pulse wheel turns, the sensor develops a signal each time a pulse ring lobe passes the sensor. The ECA calculates engine revolutions per minute (RPM) and establishes the basic ignition timing from the CP sensor signals. Basic ignition timing is not adjustable because the CP sensor is in a fixed position in the engine block, as illustrated in Figure 7-17.

Barometric Pressure (BP) Sensor

Atmospheric pressure changes with weather conditions and increasing or decreasing altitudes. Less exhaust gas recirculation (EGR) flow is required by a vehicle driven at higher altitudes compared with a vehicle driven at sea level. For this reason, some means of determining atmospheric pressure is needed to permit altitude compensation of EGR flow. The BP sensor monitors engine compartment atmospheric pressure. The electronic control assembly (ECA) applies a reference voltage to the BP

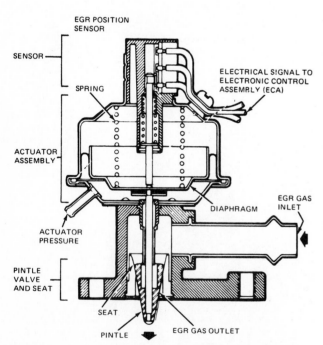

FIGURE 7-15. Exhaust Gas Recirculation (EGR) Valve with Sensor. (Courtesy of Ford Motor Co.)

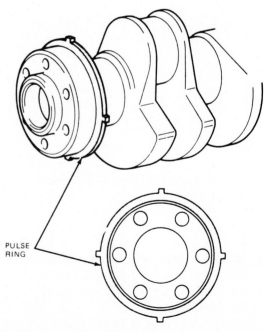

FIGURE 7-16. Pulse Ring of Crankshaft Position (CP) Sensor. (Courtesy of Ford Motor Co.)

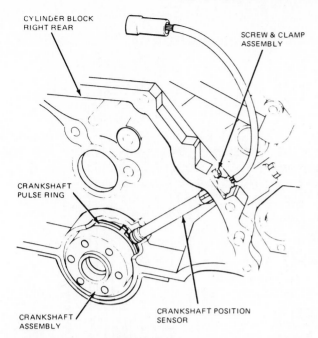

FIGURE 7-17. Crankshaft Position (CP) Sensor. (Courtesy of Ford Motor Co.)

sensor and monitors the resultant sensor output voltage which is proportional to atmospheric pressure. Figure 7-18 pictures a BP sensor.

OPERATION OF ELECTRONIC ENGINE CONTROL (EEC I) SYSTEMS

Ignition Spark Advance Control

When the ignition switch is turned on, the power relay will be energized, and battery voltage will be supplied through the power relay contacts and the red wire to terminal F on the electronic control assembly (ECA). The second red wire connected to the power relay supplies voltage to the thermactor air bypass solenoid and exhaust gas recirculation (EGR) pressure and vent solenoids, as illustrated in Figure 7-19.

The signal from the crankshaft position sensor is sent to the ECA, and an output signal is sent from terminal G on the ECA through the orange and yellow wire to the ignition module. The ECA processes the information that is received from the input sensors and delivers an output signal from terminal G that provides the precise ignition spark advance required by the engine.

Exhaust Gas Recirculation (EGR) Control

The EGR valve is opened by thermactor pump pressure. Thermactor pump pressure is applied to the EGR vent and pressure solenoids that are mounted on the right-hand rocker arm cover, as shown in Figure 7-20.

Thermactor pump pressure is available from the air outlet on the side of the thermactor air bypass valve to the EGR vent solenoid and pressure solenoid, as shown in Figure 7-21.

Thermactor pump pressure applied to the EGR valve is controlled by the EGR vent and pressure solenoids, which are operated by the ECA. In this way, the ECA controls EGR flow on the basis of the data received from the input sensors. When the input sensor information demands an increase in EGR flow, the ECA will energize both the vent and pressure solenoid windings. Under this condition, the vent solenoid plunger closes the vent solenoid port, and the pressure solenoid plunger is lifted to supply thermactor pump pressure to the EGR valve, as pictured in Figure 7-22.

When the input sensor signals to the ECA call for a constant EGR flow, the ECA will enter the maintain-EGR-flow mode. During this mode, the ECA deenergizes the pressure solenoid and maintans current flow through the vent solenoid winding. The pressure solenoid plunger moves downward to trap the thermactor pump pressure in the hose to the EGR valve. Upward movement of the vent solenoid plunger closes the vent on the plunger to maintain

FIGURE 7-18. Barometric Pressure (BP) Sensor. (Courtesy of Sun Electric Corp.)

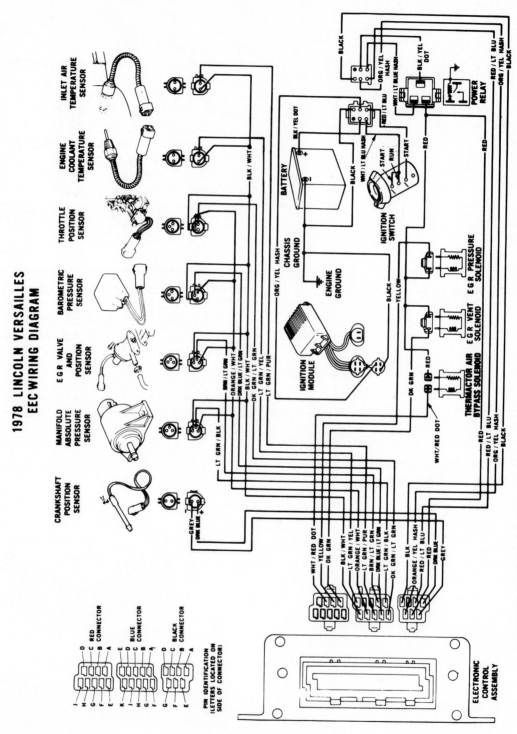

1978 LINCOLN VERSAILLES EEC WIRING DIAGRAM

INLET AIR TEMPERATURE SENSOR

ENGINE COOLANT TEMPERATURE SENSOR

THROTTLE POSITION SENSOR

BAROMETRIC PRESSURE SENSOR

E G R VALVE AND POSITION SENSOR

MANIFOLD ABSOLUTE PRESSURE SENSOR

CRANKSHAFT POSITION SENSOR

BATTERY

CHASSIS GROUND

ENGINE GROUND

IGNITION MODULE

IGNITION SWITCH

POWER RELAY

E G R PRESSURE SOLENOID

E G R VENT SOLENOID

THERMACTOR AIR BYPASS SOLENOID

ELECTRONIC CONTROL ASSEMBLY

PIN IDENTIFICATION
(LETTERS LOCATED ON SIDE OF CONNECTOR)

RED CONNECTOR

BLUE CONNECTOR

BLACK CONNECTOR

FIGURE 7-19. Electronic Engine Control (EEC I) System. (Courtesy of Ford Motor Co.)

124

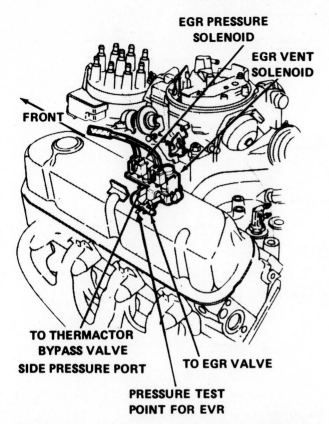

EGR PRESSURE
SOLENOID

EGR VENT
SOLENOID

FRONT

TO THERMACTOR
BYPASS VALVE
SIDE PRESSURE PORT

TO EGR VALVE

PRESSURE TEST
POINT FOR EVR

FIGURE 7-20. Exhaust Gas Recirculation (EGR) Vent and Pressure Solenoids. (Courtesy of Sun Electric Corp.)

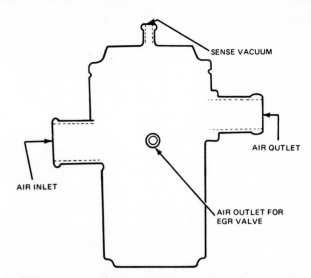

SENSE VACUUM

AIR INLET

AIR OUTLET

AIR OUTLET FOR
EGR VALVE

FIGURE 7-21. EGR Air Outlet on Air Bypass Valve. (Courtesy of Ford Motor Co.)

the thermactor pump pressure applied to the EGR valve, as shown in Figure 7-23.

Under certain conditions, such as idle or wide open throttle operation, the input sensors will signal the ECA to enter the decrease-EGR-flow mode.

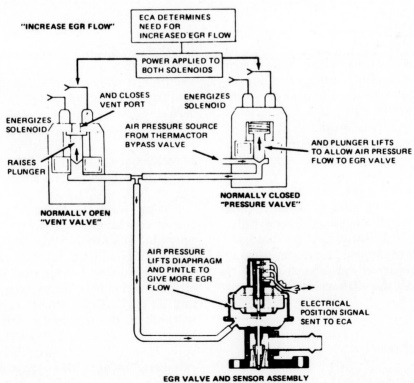

"INCREASE EGR FLOW"

ECA DETERMINES
NEED FOR
INCREASED EGR FLOW

POWER APPLIED TO
BOTH SOLENOIDS

AND CLOSES
VENT PORT

ENERGIZES
SOLENOID

ENERGIZES
SOLENOID

AIR PRESSURE SOURCE
FROM THERMACTOR
BYPASS VALVE

AND PLUNGER LIFTS
TO ALLOW AIR PRESSURE
FLOW TO EGR VALVE

RAISES
PLUNGER

NORMALLY CLOSED
"PRESSURE VALVE"

NORMALLY OPEN
"VENT VALVE"

AIR PRESSURE
LIFTS DIAPHRAGM
AND PINTLE TO
GIVE MORE EGR
FLOW

ELECTRICAL
POSITION SIGNAL
SENT TO ECA

EGR VALVE AND SENSOR ASSEMBLY

FIGURE 7-22. Increase EGR Flow Mode. (Courtesy of Ford Motor Co.)

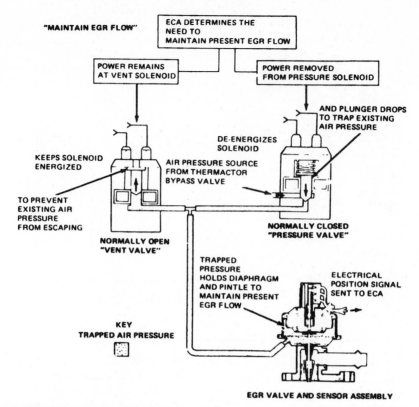

FIGURE 7-23. Maintain-EGR-Flow Mode. (Courtesy of Ford Motor Co.)

Under this condition, the ECA deenergizes the vent solenoid and the pressure solenoid windings. When the vent solenoid winding is deenergized, the plunger moves downward to vent the thermactor pump pressure applied to the EGR valve. Additional thermactor pump air applied to the EGR valve is shut off by the downward movement of the pressure solenoid plunger when the solenoid is deenergized, as illustrated in Figure 7-24.

An external EGR cooler is used to reduce exhaust gas temperature and thus improve exhaust flow characteristics, engine operation, and EGR valve life. The cooler assembly is a heat exchanger that uses engine coolant to lower exhaust gas temperature as the exhaust gas flows from the exhaust manifold through the cooler to the EGR valve (see Figure 7-25).

Thermactor Pump Control

The electronic control assembly (ECA) operates the thermactor air bypass solenoid in order to control the thermactor pump output. When the ECA energizes the thermactor air bypass solenoid manifold, vacuum is available through the solenoid to the thermactor air bypass valve, as illustrated in Figure 7-26.

When manifold vacuum is applied to the air bypass valve, the upward movement of the air bypass poppet valve assembly will allow thermactor pump air to flow to the exhaust ports, as shown in Figure 7-27.

Airflow from the thermactor pump will be directed to the exhaust ports when the engine is operating at normal cruise conditions. Under certain operating conditions such as cold engine coolant, prolonged idle, or deceleration, the ECA will deenergize the thermactor air bypass solenoid. The resultant downward movement of the solenoid plunger will shut off the manifold vacuum applied to the air bypass valve. Since this allows the air bypass poppet valve to move downward, thermactor air will be exhausted through the bypass valve to the atmosphere.

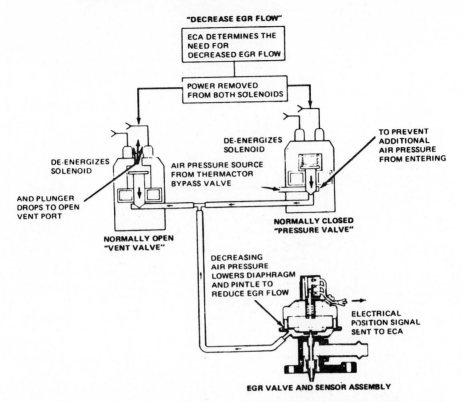

FIGURE 7-24. Decrease-EGR-Flow Mode. (Courtesy of Ford Motor Co.)

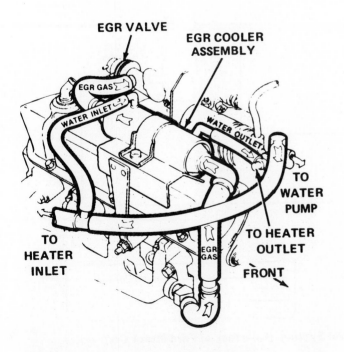

FIGURE 7-25. Exhaust Gas Recirculation (EGR) Cooler. (Courtesy of Sun Electric Corp.)

ELECTRONIC ENGINE CONTROL (EEC I) DIAGNOSIS

Engine Fails To Start

A spark tester, as outlined in Chapter 6, may be used to test for spark at the spark plugs and the coil secondary wire while the engine is being cranked. If spark occurs at the coil secondary wire but no spark is available at the spark plugs, the distributor cap or rotor is defective. If no spark is available at the coil secondary wire, the Dura-Spark II ignition system, electronic control assembly (ECA), crankshaft position (CP) sensor, or power relay is defective.

With the ignition switch on, connect a voltmeter from ECA terminal F to ground, and the power relay red wire to ground. Battery voltage should be available in both tests. If battery voltage is present at the power relay red wire, but no voltage is available at ECA terminal F, the red wire from the relay to ECA terminal F has an open circuit. (See Figure 7-28 for terminal and wire identification.)

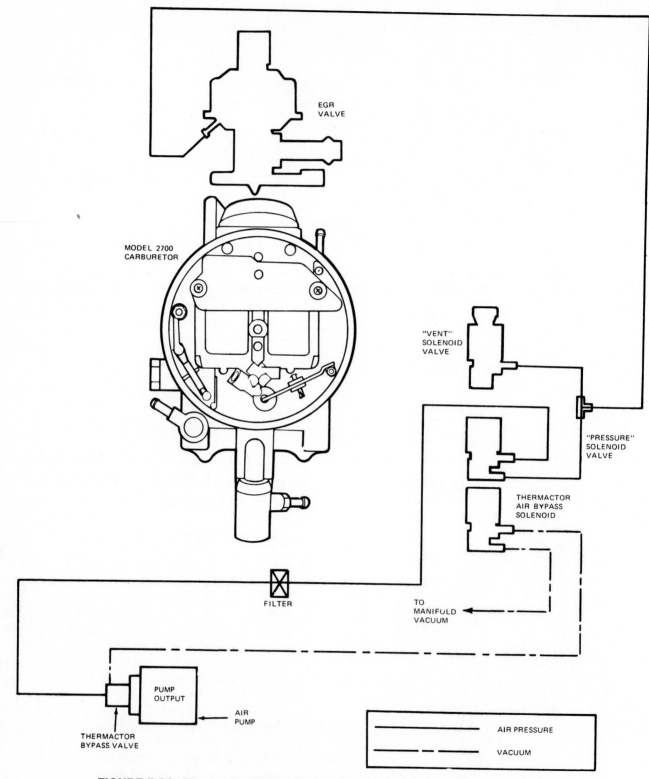

FIGURE 7-26. Thermactor Pump Control System. (Courtesy of Ford Motor Co.)

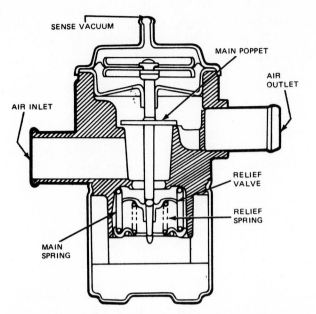

FIGURE 7-27. Thermactor Air Bypass Valve. (Courtesy of Ford Motor Co.)

If no power is available at the power relay red wire with the ignition switch on, connect the voltmeter from the power relay red and yellow wire to ground and the black and yellow wire to ground. Battery voltage should be available at the red and yellow wire with the ignition switch on; therefore if no voltage is available, the wire, or the ignition switch, has an open circuit. If battery voltage is not available at all times at the black and yellow wire terminal on the power relay, the black and yellow wire has an open circuit. When battery voltage is available at the power relay red and yellow wire and the black and yellow wire, but there is no voltage available at the red wire, the power relay is defective. Before replacing the relay, be sure the relay ground is satisfactory.

Probe the module orange and yellow wire and connect an AC voltmeter from the orange and yellow wire to ground. Use the lowest scale on the AC voltmeter. A pulsating voltmeter reading while the engine is being cranked indicates the crankshaft

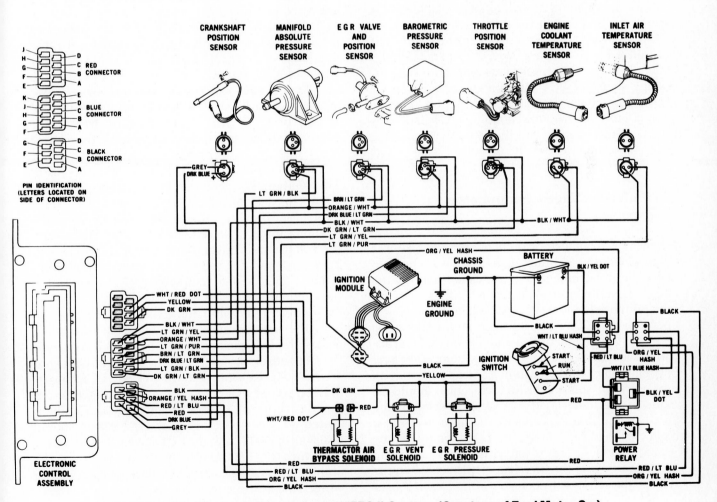

FIGURE 7-28. Electronic Engine Control (EEC I) System. (Courtesy of Ford Motor Co.)

position (CP) sensor and the electronic control assembly (ECA) are satisfactory. When an AC voltage signal is available at the orange and yellow wire connected to the module, but there is no spark at the coil secondary wire while the engine is being cranked, the module or the ignition coil is defective.

If there is no AC voltage signal at the module orange and yellow wire while the engine is being cranked, check the AC voltage signal from ECA terminal G to ground while the engine is being cranked. When an AC voltage signal is available at ECA terminal G, but there is no voltage signal available at the module orange and yellow wire, the orange and yellow wire has an open circuit or a grounded condition.

If an AC voltage signal is not available at ECA terminal G while the engine is being cranked, check the AC voltage signal at ECA terminals A and E. The ECA is defective if a voltage signal is present at terminals A and E while the engine is being cranked but there is no voltage signal from terminal G to ground.

When an AC voltage signal is not available at ECA terminals A and E while the engine is being cranked, connect the AC voltmeter leads to the crankshaft position (CP) sensor leads. If an AC voltage signal is available at the CP sensor leads, but there is no AC voltage signal at ECA terminals A and E, the CP sensor leads from the sensor to the ECA have an open circuit or a grounded condition. When there is no AC voltage signal from the CP sensor while the engine is being cranked, the CP sensor is defective.

Unsatisfactory Engine Performance or Economy

When a performance or an economy complaint is being diagnosed, a timing light should be used to check the initial timing and the advance timing. If the engine is running at the specified idle speed, the initial timing should be 28°–32°. When the engine is operating at 1,900 RPM, advance timing should be 48°–52°. If the initial timing and advance timing are within the specifications, the EEC I system is working normally. Since the initial timing and advance timing may vary on some vehicles, reference should be made to the underhood specification decal. If the initial timing and the advance timing remain at a fixed 10° the EEC I system is operating in the "default mode." This is an emergency operating mode that allows the vehicle to be driven when the ECA or the sensors are defective. Since each sensor

contributes to the total timing advance, the timing should not be in the default mode when only one sensor is defective.

The electronic control assembly (ECA) reference voltage that is applied to the barometric pressure (BP) sensor, EGR valve position (EVP) sensor, and the manifold absolute pressure (MAP) sensor may be tested by connecting a voltmeter from the orange and white wire at each sensor to ground. With the ignition switch on, the reference voltage at each sensor should be 8.5–9.5 V. When the reference voltage is not within specifications at any of the four sensors, the ECA is defective. If the reference voltage is not within specifications at some of the four sensors, a resistance problem exists in the orange and white wire, or the sensor providing the incorrect reading is defective.

Test the modified voltage output from the four sensors by connecting a voltmeter from the green wire at each sensor to ground. The green wire is a dual-colored wire at each sensor. Specifications for the modified voltage output from the four sensors are listed in Table 7-5.

The specifications in Table 7-1 are based on normal engine operating temperature with the ignition switch on. If the modified voltage output from the MAP sensor or BP sensor is not within the specified range, the sensor should be replaced. When the modified voltage output from the TP sensor or EVP sensor is incorrect, the sensor diagnosis may be verified by testing the sensor resistance with an ohmmeter connected from the green wire to black and white wire at each sensor. The ignition switch must be off and the sensor connector disconnected during the ohmmeter tests. Table 7-6 provides the sensor resistance specifications. If the TP sensor or EVP sensor is not within the voltage and resistance specification provided in Tables 7-5 and 7-6, the sensor should be replaced.

The engine coolant temperature (ECT) sensor and the inlet air temperature (IAT) sensor may be tested by connecting a voltmeter across the sensor leads with the ignition switch on. The voltage reading across the ECT sensor should be 1.9–3.7 volts, and the IAT sensor should provide a voltage reading of 4.1–6.3 volts. Resistance in the IAT and ECT sensors may be tested by connecting an ohmmeter across the disconnected sensor leads with the ignition switch off. The resistance of the IAT sensor should be 6,500–45,000 Ω, and the ECT sensor should have a resistance of 1,500–6,000 Ω. If the voltmeter or ohmmeter readings are not within specifications on either sensor, the sensor is defective.

TABLE 7-5. Sensor Voltage Specifications.

	Elevation	Volts
Manifold absolute pressure (MAP) sensor, and barometric pressure (BP) sensor	0–1,000 ft (0–312 m)	6.7–7.8
	1,000–2,000 ft (312–625 m)	6.5–7.4
	2,000–3,000 ft (625–937 m)	6.2–7.3
	3,000–4,000 ft (937–1,249 m)	6.0–7.0
	4,000–5,000 ft (1,249–1,561 m)	5.8–6.8
	5,000–6,000 ft (1,561–1,873 m)	5.5–6.6
	6,000–7,000 ft (1,873–2,185 m)	5.3–6.4
Throttle position (TP) sensor		1.7–2.1
EGR valve position (EVP) sensor		1.1–1.6

TABLE 7-6. Sensor Resistance Specifications.

	Ohms
Throttle position (TP) sensor: orange and white wire to black and white wire;	3,000–5,000
brown and green wire to black and white wire	580–1,1000
EGR valve position (EVP) sensor: orange and white wire to black and white wire;	2,800–53,000
orange and white wire to brown and green wire	350–940

In order to test the EGR system, a "T" fitting should be used to connect a vacuum gauge to the hose joining the vent and pressure solenoids to the EGR valve. When the engine is operating at idle speed, the pressure gauge should register zero. At 1,800 RPM and normal engine temperature, the pressure gauge should indicate ¾ to 1 PSI. If the pressure reading is not within specifications, the vent and pressure solenoids are defective. All vacuum hoses and wiring connections should be checked for proper installation before the solenoids are replaced.

Throttle Position (TP) Sensor Adjustment

When checking the TP sensor adjustment, connect a voltmeter from the sensor green wire to ground. With the ignition switch on, the engine not running, and the throttle in the idle position, the voltmeter should read 1.75–2.00 V. If the voltage reading is not within specifications, loosen the two sensor mounting screws and rotate the sensor until the voltage reading is correct. Tighten the mounting screws while holding the sensor to maintain the specified voltage reading.

Rotor Alignment Adjustment

A special bilevel rotor and distributor cap are used on the electronic engine control (EECI) distributor, as illustrated in Figure 7-29.

The bilevel distributor cap requires special installation of the spark plug wires in the distributor cap. For example, the rotor distributes a spark to number 1 spark plug wire, and then passes wires 2 and 7 before delivering a spark to wire 5, which is next in the firing order on a 302 V8 engine, as shown in Figure 7-30.

An alignment adjustment is required on the bilevel rotor. When piston 1 is at top dead center (TDC) of the compression stroke and the timing mark on the crankshaft pulley is aligned with the zero degree position on the timing indicator, the rotor alignment tool should fit between the rotor and cap adapter alignment slots, as indicated in Figure 7-31.

If the alignment tool will not fit in the rotor and cap adapter slots, loosen the rotor retaining screws and turn the rotor until the alignment tool fits into the slots. The crankshaft should not be turned when the rotor has been removed from the distributor, or when the distributor has been removed from the engine.

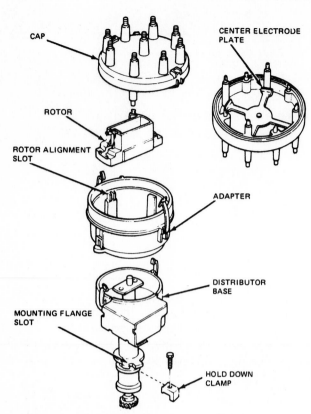

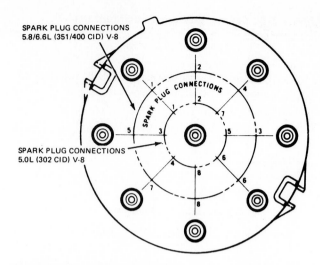

FIGURE 7-30. Bilevel Distributor Cap Wire Installation. (Courtesy of Ford Motor Co.)

FIGURE 7-29. Bilevel Distributor Cap and Rotor. (Courtesy of Ford Motor Co.)

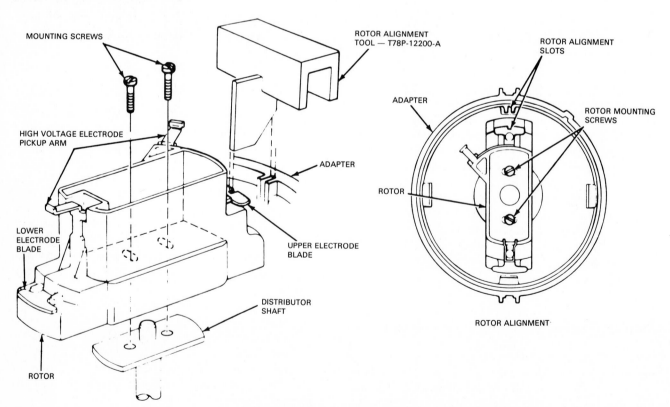

FIGURE 7-31. Bilevel Rotor Alignment. (Courtesy of Ford Motor Co.)

Questions

Electronic Lean Burn (ELB) and Electronic Spark Advance (ESA) Systems

1. Two requirements for vacuum transducer advance are:
   a. _____
   b. _____

2. The air temperature sensor is connected in the _____ circuit.

3. Engine detonation is reduced by the air temperature sensor when the air intake temperature is cold.   T   F

4. The basic timing should be checked as soon as the engine is started.   T   F

5. Many ELB computers supply full advance from the vacuum transducer signal _____ _____ minutes after the throttle is opened.

6. In a dual pickup ELB system, a defective run pickup coil would cause a loss of spark advance.   T   F

7. A coolant temperature switch that is closed at all temperatures would cause a complete loss of _____ transducer advance.

8. The run pickup coil has a larger connector than the start pickup coil.   T   F

9. The carburetor switch is grounded by the _____ _____.

Electronic Engine Control (EEC I) Systems

1. The power relay supplies battery voltage to the _____ when the ignition switch is turned on.

2. The exhaust gas recirculation (EGR) valve is opened by _____ _____ _____ _____.

3. Adjustment of the initial timing is accomplished by rotating the distributor.   T   F

4. Output signals from the barometric pressure (BP) sensor are used by the electronic control assembly (ECA) to control the _____ _____ flow.

5. A defective ECA could prevent the engine from starting.   T   F

6. If the circuit in the power relay contacts is open, a "no-start" complaint can be expected.   T   F

7. In the "default mode," the initial timing and the advance timing will be fixed at _____ _____ degrees.

8. If the manifold absolute pressure (MAP) sensor is defective, the system will operate in the "default mode."   T   F

9. The modified voltage output signal from the throttle position (TP) sensor is adjustable.   T   F

8

COMPUTER-CONTROLLED CARBURETOR SYSTEMS

CATALYTIC CONVERTERS

Operation

Many vehicles are equipped with three-way catalytic converters that operate in two stages. The first converter stage contains rodium, which is used to reduce nitrous oxide (NOx) to nitrogen and oxygen. Platinum or palladium in the second stage acts as an oxidization catalyst to change carbon monoxide (CO) and hydrocarbons (HC) to carbon dioxide (CO_2) and water (H_2O). Oxygen must be injected into the second stage to provide the necessary oxidization; therefore a thermactor pump is connected to the converter, as illustrated in Figure 8-1.

To control all three pollutants, a three-way catalytic converter requires an exact air-fuel ratio of 14.7:1. Since lean air-fuel ratios will provide low HC and CO levels, the converter can be very efficient in reducing these emission levels below the required emission standards. However, these ratios will generate high cylinder temperatures and increased nitrous oxide (NOx) emissions, and converter efficiency in correcting the NOx emissions is low, as shown in Figure 8-2.

On the other hand, rich air-fuel ratios provide low levels of NOx and high converter efficiency in correcting NOx emissions, but they also cause high emissions of HC and CO and converter efficiency in correcting these emission levels is extremely low. Therefore, if the air-fuel ratio is kept in a narrow range near 14.7:1, the two-stage converter is able to operate at approximately 90 percent efficiency in correcting all three pollutants. Computer systems can provide the necessary rigid control of air-fuel ratios.

GENERAL MOTORS COMPUTER COMMAND CONTROL (3C) SYSTEM

Electronic Control Module (ECM)

The computer used with the 3C system is referred to as an electronic control module (ECM). A removable programmable read-only memory (PROM) is located in the ECM, as illustrated in Figure 8-3.

The ECM is the control center of the 3C system. It constantly monitors information it receives from the various sensors and provides commands in order to perform the correct functions. Specific information about each vehicle is programmed into the PROM so the ECM can be tailored to meet the requirements of the specific engine, transmission, and differential combination of a particular vehicle. The ECM receives signals from the following devices:

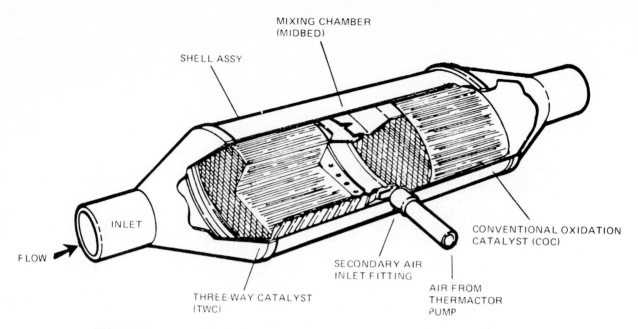

FIGURE 8-1. Two-Stage Catalytic Converter. (Courtesy of Ford Motor Co.)

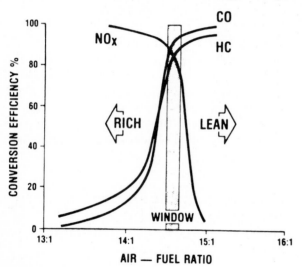

FIGURE 8-2. Catalytic Converter Efficiency. (Courtesy of Sun Electric Corp.)

FIGURE 8-3. Electronic Control Module (ECM). (Courtesy of General Motors of Canada Ltd.)

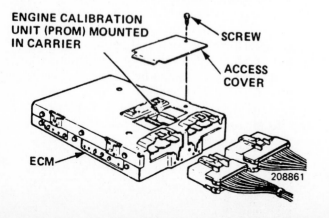

1. Oxygen (O_2) sensor
2. Throttle position sensor (TPS)
3. Coolant temperature sensor
4. Manifold absolute pressure (MAP) and barometric pressure (BP) sensors
5. Vehicle speed sensor (VSS)
6. Distributor pickup coil, crankshaft position, revolutions per minute (RPM)

The ECM, in turn, controls the following:

1. Mixture control (MC) solenoid
2. Electronic spark-timing (EST) system
3. Exhaust gas recirculation (EGR) valve
4. Air injection reactor (AIR) pump
5. Torque converter clutch (TCC) lockup
6. Early fuel evaporation (EFE) system
7. Idle speed control (ISC) motor

Sources of ECM Input

Oxygen (O_2) Sensor. An exhaust gas oxygen (O_2) sensor is located in one of the exhaust manifolds. Rich air-fuel ratios supply fuel to mix with all the oxygen entering the engine. Rich air-fuel ratios create low oxygen levels in the exhaust because excess fuel mixes with all the oxygen entering the engine. When lean mixtures are used, oxygen levels in the exhaust are high because of the lack of fuel entering the cylinders. Exhaust gas is applied to the outside of the oxygen-sensing element and atmospheric pressure is supplied to the inside of the sensing element, as shown in Figure 8-4.

Rich mixtures and low levels of oxygen in the exhaust cause the sensing element to generate higher voltage, as shown in Figure 8-5. High levels of oxygen in the exhaust and lean mixtures result in low oxygen sensor voltage because high oxygen levels are present on both sides of the sensing element. The oxygen sensor signal to the ECM is used to control the air-fuel ratio in the carburetor. Unleaded gasoline must be used to obtain satisfactory oxygen sensor operation.

Throttle Position Sensor (TPS). The throttle position sensor is a variable resistor operated by the accelerator pump linkage, as illustrated in Figure 8-6.

A reference voltage is supplied to the TPS from the ECM. The TPS output varies in relation to throttle opening. At wide-open throttle and sudden acceleration, the TPS signal calls for mixture enrichment.

Coolant Temperature Sensor. The resistance of the coolant sensor varies in relation to coolant temperature. The ECM operating mode is selected from the coolant sensor signal. Open-loop operation occurs during engine warm-up, when a richer mixture is necessary. In the open-loop mode, the ECM maintains the carburetor mixture. Closed-loop operation occurs when the coolant sensor reaches a predetermined temperature. The ECM controls the air-fuel ratio in response to an oxygen sensor signal in the closed-loop mode. The ECM may also use the coolant sensor signal to assist in regulating ignition, air injection, and exhaust gas recirculation. Figure 8-7 illustrates a coolant temperature sensor.

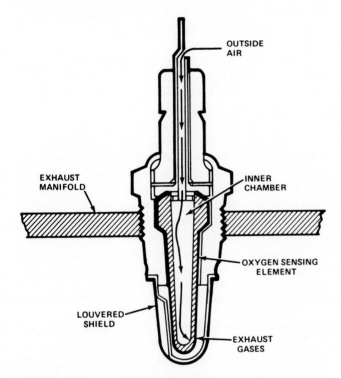

FIGURE 8-4. Oxygen (O_2) Sensor. (Courtesy of General Motors of Canada Ltd.)

FIGURE 8-5. Oxygen Sensor Voltage Output. (Courtesy of Sun Electric Corp.)

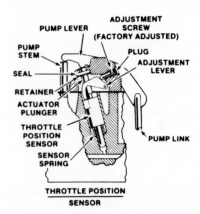

FIGURE 8-6. Throttle Position Sensor (TPS). Courtesy of General Motors of Canada Ltd.)

FIGURE 8-7. Coolant Temperature Sensor. (Courtesy of General Motors of Canada Ltd.)

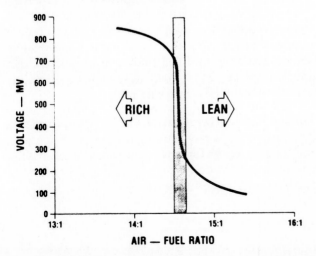

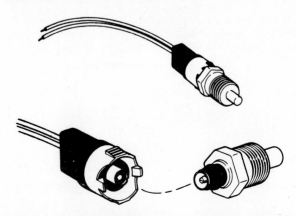

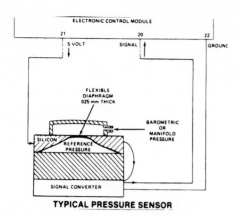

FIGURE 8-8. Manifold Absolute Pressure Sensor. (Courtesy of General Motors of Canada Ltd.)

Manifold Absolute Pressure Sensor. A reference voltage is supplied to the manifold absolute pressure (MAP) sensor from the ECM, as illustrated in Figure 8-8.

The voltage output signal from the MAP sensor to the ECM varies in relation to manifold vacuum. Spark control is managed by the ECM in response to the MAP sensor signal. Other ECM functions may be affected by the MAP sensor signal. A barometric pressure (BP) sensor may be used with the MAP sensor on some applications.

Vehicle Speed Sensor (VSS). The speedometer cable rotates a reflector blade past a light-emitting diode and photo cell in the VSS, as shown in Figure 8-9.

Reflector blade speed and VSS output signal are directly proportional to vehicle speed. The ECM controls torque converter clutch lockup from the VSS signal.

FIGURE 8-9. Vehicle Speed Sensor. (Courtesy of General Motors of Canada Ltd.)

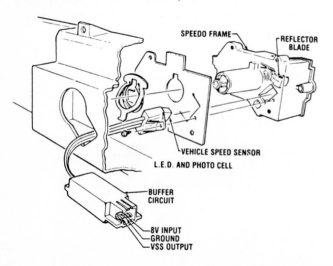

ECM Control Functions

Mixture Control (MC) Solenoid Management. The 3C carburetor contains a mixture control (MC) solenoid that is controlled by the ECM. Movement of the solenoid plunger controls the metering rods in the main jets and the air bleed pin in the idle circuit. The mixture control solenoid plunger is spring loaded in the upward position. The solenoid plunger moves down when the solenoid winding is energized. When the oxygen sensor provides a rich signal, the ECM energizes the mixture control solenoid winding, the metering rods move down, and the air-fuel mixture becomes leaner. The ECM deenergizes the mixture control solenoid winding when the oxygen sensor indicates a lean air-fuel ratio.

Deenergizing the solenoid winding allows the mixture control solenoid plunger and metering rods to move upward and provide a richer air-fuel mixture. The ECM energizes the mixture control solenoid winding ten times per second to provide an air-fuel ratio of 14.7:1. When the mixture control solenoid plunger moves upward, it forces the idle air bleed pin upward; as a result, airflow into the idle circuit is reduced and idle air-fuel mixture is richer. Most defects in the oxygen sensor, mixture control solenoid winding, or ECM will result in upward movement of the solenoid plunger and metering rods, which will cause an extremely rich air-fuel ratio and excessive fuel consumption. (A 3C carburetor with a mixture control solenoid is illustrated in Figure 8-10.)

Electronic Spark Timing (EST) Management. A seven-wire ignition module is used in the computer command control (3C) high-energy ignition HEI system. The additional three-module terminals are connected to the electronic control module (ECM). Conventional vacuum and centrifugal advance mechanisms are discontinued, and the correct spark advance is supplied by the ECM in relation to the input sensor signals that it receives. Input sensor signals from the barometric pressure sensor, manifold absolute pressure sensor, coolant temperature sensor, and the throttle position sensor are used by the ECM to determine the correct spark advance. (Refer to Chapter 6 for a complete explanation of HEI systems.)

Torque Converter Clutch (TCC) Management. The torque converter clutch eliminates slippage in the torque converter under normal cruise driving conditions. A friction disc is splined to the converter turbine, as shown in Figure 8-11.

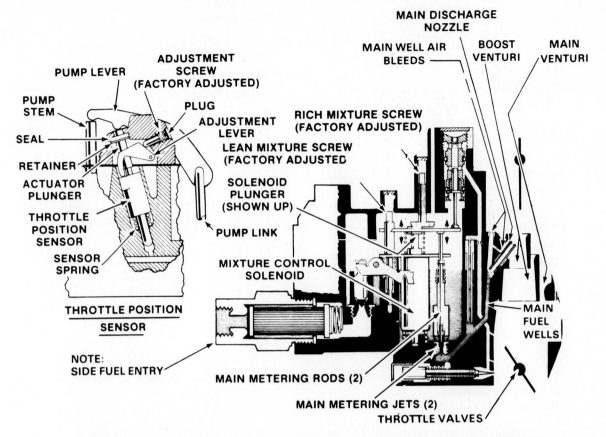

FIGURE 8-10. Carburetor with Mixture Control (MC) Solenoid. (Courtesy of General Motors of Canada Ltd.)

Converter lockup takes place when oil enters the converter hub and forces the friction disc against the front of the converter. In the lockup mode, the flywheel is connected through the front of the converter to the friction disc, turbine, and transmission input shaft. The converter becomes unlocked when oil is directed through the hollow input shaft and the friction disc is forced away from the front of the converter. Oil flow to the converter is controlled by an apply valve in the transmission. The apply valve is controlled by the TCC solenoid.

When the vehicle reaches a predetermined speed in high gear, the computer grounds the TCC solenoid winding. Once the TCC solenoid is energized, a hydraulic bleed port in the direct clutch circuit closes, and the direct clutch oil pressure rises. The increase in oil pressure moves the converter apply valve and directs oil in the converter hub to lock

FIGURE 8-11. Lockup Torque Converter. (Courtesy of General Motors of Canada Ltd.)

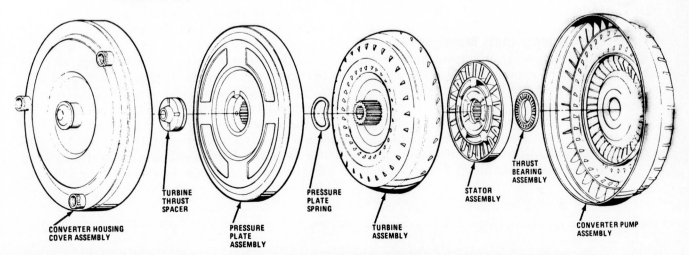

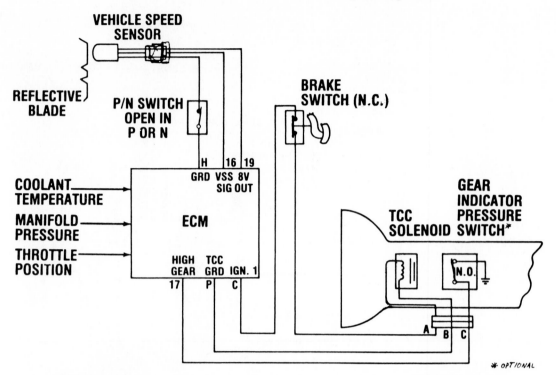

FIGURE 8-12. Torque Converter Clutch Circuit. (Courtesy of General Motors of Canada Ltd.)

up the converter. Other input signals to the computer may affect TCC lockup, as indicated in Figure 8-12.

Early Fuel Evaporation (EFE) Management. The heat riser valve in the EFE system is operated by a vacuum diaphragm, as shown in Figure 8-13. A cold coolant sensor signal causes the ECM to energize the EFE solenoid and apply vacuum to the power actuator. If the heat riser valve is closed, the intake manifold will become heated as exhaust gas is forced through the intake manifold crossover passage. The ECM will deactivate the EFE solenoid at a specific coolant temperature and allow the heat riser valve to open. The EFE vacuum is also supplied to the air injection system, as illustrated in Figure 8-14.

Air Injection Reactor (AIR) Management. The AIR system supplies air from the air pump to the exhaust ports or to the catalytic converter, as illustrated in Figure 8-15. Airflow to the exhaust ports lowers emission levels during warm-up and reduces oxygen sensor and converter warm-up time. Oxygen is necessary for converter operation once the converter has reached operating temperature.

A divert valve and a switching valve are used to control air flow from the AIR pump. During engine warm-up, the ECM deenergizes the divert valve solenoid winding and enables AIR pump pressure to move the divert valve diaphragm and spool valve upward. Air from the pump is diverted to the air cleaner when the spool valve is in the upward position, as indicated in Figure 8-16.

When the ECM energizes the divert solenoid winding, the solenoid valve shuts off AIR pump pressure applied to the divert valve diaphragm, and the diaphragm and spool valve are able to move to the downward position. As a result, air is directed from the AIR pump to the air-switching valve. High manifold vacuum on deceleration can lift the divert valve diaphragm and cause air from the pump to exhaust to the air cleaner.

When the engine coolant is cold, vacuum from the EFE system is applied to the lower air-switching valve vacuum port and shuts off the flow of air from the AIR pump to the air-switching valve diaphragm. As a result, the diaphragm and spool valve remain in the downward position, and air from the AIR pump is directed to the exhaust ports. Vacuum to the lower air-switching valve port is shut off when the coolant is warm and AIR pump pressure can force the diaphragm and spool valve upward. The flow of air through the air-switching valve will

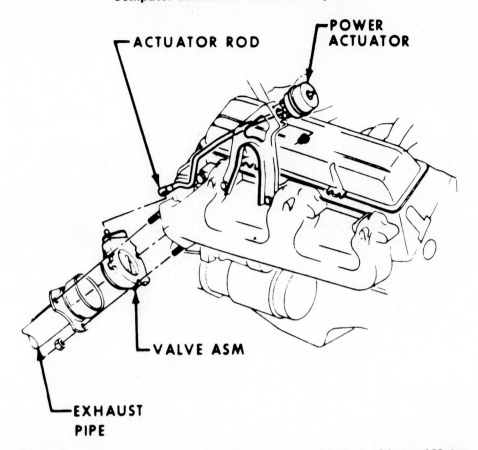

FIGURE 8-13. Early Fuel Evaporation (EFE) System. (Courtesy of General Motors of Canada Ltd.)

FIGURE 8-14. Early Fuel Evaporation Vacuum Circuit. (Courtesy of General Motors of Canada Ltd.)

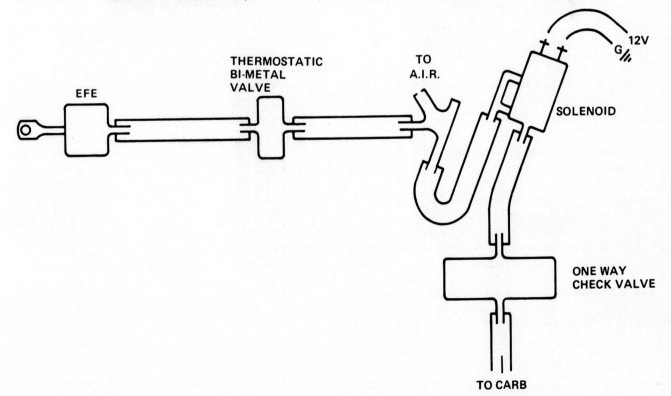

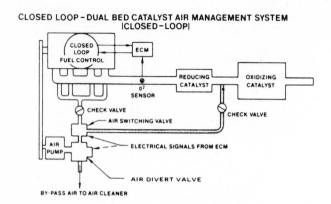

FIGURE 8-15. Air Injection Reactor System. (Courtesy of General Motors of Canada Ltd.)

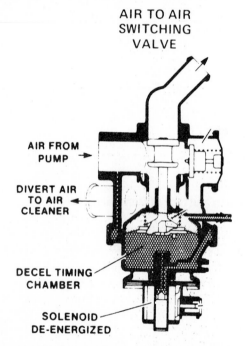

FIGURE 8-16. Air Injection System Reactor (AIR) Divert Valve. (Courtesy of General Motors of Canada Ltd.)

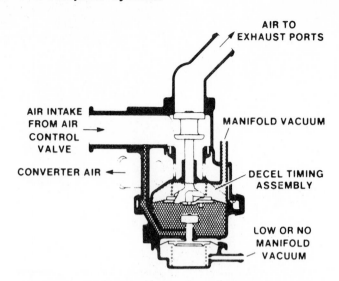

FIGURE 8-17. Air-Switching Valve. (Courtesy of General Motors of Canada Ltd.)

be directed to the converter when the spool valve is in the upward position, as indicated in Figure 8-17.

Manifold vacuum applied to the upper air-switching valve port could lift the diaphragm and spool valve on deceleration with a cold engine. Air will be directed momentarily to the converters to prevent manifold backfiring.

Exhaust Gas Recirculation (EGR) Management. The EGR valve directs exhaust from the exhaust system to the intake manifold and lowers NOx emissions. Vacuum to the EGR valve is controlled by a

solenoid operated by the ECM. The ECM shuts off EGR vacuum when the coolant is cold by energizing the solenoid. Warm engine coolant conditions signal the ECM to deactivate the solenoid and thus allow vacuum to the EGR valve. The vacuum port connected to the EGR system is always above the throttles. Vacuum should never be applied to the EGR valve until the throttles are opened. An EGR valve that is open at idle speed will cause rough idling. Figure 8-18 illustrates the EGR valve and the vacuum control circuit.

Idle Speed Control (ISC) Management. The idle speed control (ISC) motor is a small reversible motor that is operated by the ECM to control idle speed. When the engine coolant is cold, the fast idle cam will determine the fast idle speed. The ISC motor will extend slightly more than normal when a cold signal is sent from the coolant temperature sensor to the ECM. The air conditioner clutch switch signals the ECM to maintain engine idle speed when the air-conditioning compressor clutch is engaged. If the transmission selector is moved from drive to park, the park-neutral switch signals the ECM to reduce idle speed to specifications. The reference signal from the distributor will send a revolutions-per-minute (RPM) signal to the ECM as long as the engine is running. A "low battery voltage" signal to terminal R on the ECM will cause the ECM to increase idle speed to assist in recharging the battery. When a "hot" signal is received from the coolant temperature sensor, the ECM extends the ESC motor plunger and increases the idle speed so the engine temperature will decrease.

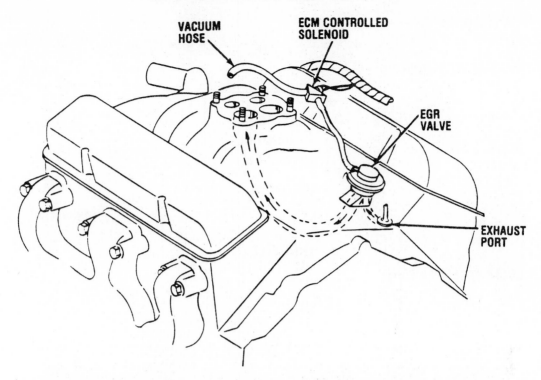

FIGURE 8-18. Exhaust Gas Recirculation (EGR) Valve. (Courtesy of General Motors of Canada Ltd.)

The ISC motor does not require adjusting because the ECM determines the correct idle position. A tang on the throttle lever contacts the ISC motor plunger and closes a throttle switch in the ISC motor. The throttle switch is connected to the ECM; the ECM will not control the ISC motor unless the throttle switch is closed. The system must be in the closed-loop mode before the ECM will control the ISC motor. During a cylinder output test, the coolant sensor must be disconnected to put the system in open-loop operation; otherwise the ISC motor will try to correct the drop in speed as each cylinder is shorted out. (See Figure 8-19 for the ISC motor and related ECM circuit.)

DIAGNOSIS OF COMPUTER COMMAND CONTROL (3C) SYSTEMS

Self-Diagnostics

A check-engine light on the instrument panel is energized by the ECM whenever certain system defects occur. Service personnel can ground a diagnostic test terminal in a connector under the dash, as indicated in Figure 8-20.

Various diagnostic connectors can be used, depending on the year and make of the vehicle. The check-engine light should be on for a few seconds each time the engine is started. Part-throttle operation for 5 minutes is sometimes required before the check-engine light will indicate a system defect.

When the diagnostic test terminal is grounded with the ignition switch on, the check-engine light begins flashing a code 12 and all other trouble codes that are stored in the computer memory. The codes are given in numerical order and repeated three times. One flash followed by a pause and two more flashes in quick succession indicate code 12, as shown in Figure 8-21.

The initial code 12 that is provided when the diagnostic test lead is grounded with the ignition switch on indicates that the ECM is capable of diagnosing the system. The ECM of most 3C systems manufactured before 1982 will perform the following functions when the diagnostic test lead is grounded with the ignition switch on:

1. Flash a code 12 and any other codes stored in the computer memory.

2. Send a 30° dwell signal to the mixture control (MC) solenoid.

3. Energize all solenoids controlled by the ECM.

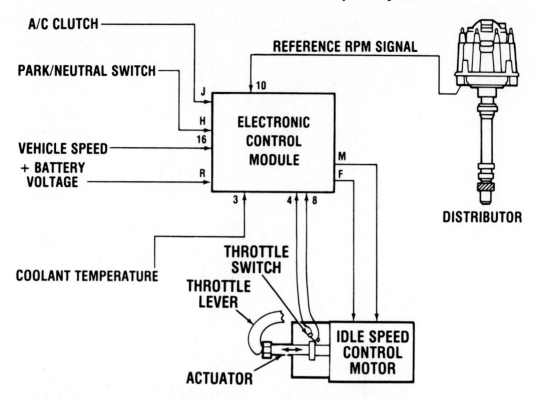

FIGURE 8-19. Idle Speed Control Motor. (Courtesy of General Motors of Canada Ltd.)

FIGURE 8-20. Assembly Line Diagnostic Link (ALDL). (Courtesy of GM Product Service Training, General Motors Corporation)

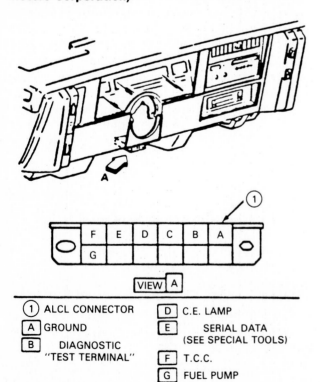

① ALCL CONNECTOR
Ⓐ GROUND
Ⓑ DIAGNOSTIC "TEST TERMINAL"
Ⓓ C.E. LAMP
Ⓔ SERIAL DATA (SEE SPECIAL TOOLS)
Ⓕ T.C.C.
Ⓖ FUEL PUMP

4. Remove the open-loop mode time when the engine coolant is cold.

5. Remove start-up enrichment and blended enrichment.

If the diagnostic test terminal is grounded with the engine running, the ECM will function normally except that the system will go into closed loop immediately when a warm engine is started and the oxygen sensor is hot.

When the diagnostic test lead of most 1982 and later 3C systems is grounded with the ignition switch on, the ECM will perform the following functions:

1. Flash code 12 and any other trouble codes stored in the ECM memory that would indicate system defects.

2. Send a 30° dwell signal to the mixture control (MC) solenoid.

3. Energize all solenoids controlled by the ECM.

4. Pulse the idle speed control motor in and out.

When the diagnostic test lead is grounded with the engine running, the ECM will follow another routine:

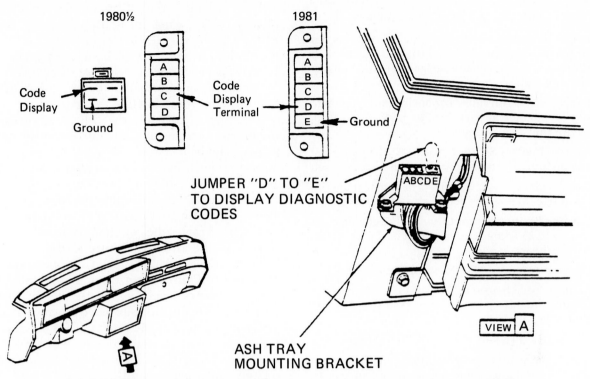

FIGURE 8-21. Check-Engine Light. (Courtesy of General Motors of Canada Ltd.)

1. No trouble codes will be set.

2. The open-loop timer, start-up enrichment, and blended enrichment functions of the ECM will be taken out.

3. The electronic spark timing (EST) circuit in the ECM will provide a fixed spark advance that may be used to set initial timing.

The check-engine light will be on as long as a system defect exists. When an intermittent defect occurs, the check-engine light will go out once the defect disappears. However, the trouble code caused by the defect will be stored in the computer memory. Figure 8-22 indicates the possible trouble codes.

Code 44 would be caused by an excessively lean exhaust gas oxygen (O_2) sensor signal. The actual defect may be in the O_2 sensor or the wires from the sensor to the ECM. A code 44 might also be displayed because the ECM is defective and therefore unable to receive the O_2 sensor signal. If the carburetor delivers an extremely lean mixture, it could cause a code 44 to be set in the computer memory. Most trouble codes indicate a defect in a specific circuit and not necessarily in the sensor that is affected.

In some systems, the check-engine light is con-nected to terminal G on the ECM, and the light is operated by the ECM. In other models, a lamp driver module is connected between the check-engine light and ECM terminal G, as shown in Figure 8-23.

The lamp driver module is taped to the wiring harness. The check-engine light may be tested by grounding terminal D in the diagnostic connector, which may be referred to as an assembly line diagnostic link (ALDL). On "J" cars, terminal B must be grounded to test the check-engine light. If the check-engine light does not come on when ter-minal D or B is grounded, the gauge fuse, check-engine light bulb, or connecting wires are defective.

The trouble codes may be cleared from the com-puter memory by disconnecting the ECM fuse in the fuse panel for 10 seconds. This fuse supplies power to terminal R on the ECM, as illustrated in Figure 8-24.

Dwellmeter

A dwellmeter may be connected from the mixture control (MC) solenoid connector to ground, as pic-tured in Figure 8-25. The dwellmeter control switch should be located in the 6-cylinder position regard-less of the number of cylinders in the engine being

The trouble codes indicate problems as follows:

12 - No reference pulses to the ECM. This code is not stored in memory and will only flash while the fault is present. Normal code with ignition "ON," engine not running.

13 - Oxygen Sensor Circuit – The engine must run up to five (5) minutes at part throttle, under road load, before this code will set.

14 - Shorted Coolant Sensor Circuit – The engine must run up to two (2) minutes before this code will set.

15 - Open Coolant Sensor Circuit – The engine must run up to five (5) minutes before this code will set.

21 - Throttle Position Sensor Circuit – The engine must run up to 25 seconds, at specified curb idle speed, before this code will set.

23 - Open or grounded M/C Solenoid Circuit.

24 - Vehicle Speed Sensor (VSS) Circuit – The car must operate up to five (5) minutes at road speed before this code will set.

32 - Barometric Pressure Sensor (BARO) Circuit low.

34 - Manifold Absolute Pressure (MAP) or Vacuum Sensor Circuit – The engine must run up to five (5) minutes, at specified curb idle speed, before this code will set.

35 - Idle Speed Control (ISC) Switch Circuit shorted. (Over 1/2 throttle for over two (2) sec.)

41 - No distributor reference pulses at specified engine vacuum. This code will store.

42 - Electronic Spark Timing (EST) Bypass Circuit grounded or open.

43 - ESC retard signal for too long; causes a retard in EST signal.

44 - Lean Oxygen Sensor indication – The engine must run up to five (5) minutes, in closed loop, at part throttle before this code will set.

44 & 55 - (At same time) – Faulty Oxygen Sensor Circuit.

45 - Rich System indication – The engine must run up to five (5) minutes, in closed loop, at part throttle before this code will set.

51 - Faulty calibration unit (PROM) or installation. It takes up to 30 seconds before this code will set.

54 - Shorted M/C Solenoid Circuit and/or faulty ECM.

55 - Grounded V ref (term. 21), faulty Oxygen Sensor or ECM.

FIGURE 8-22. 3C Trouble Codes. (Courtesy of General Motors of Canada Ltd.)

tested. One end of the MC solenoid winding is supplied with 12 V when the ignition switch is on. The ECM controls the air-fuel mixture by completing the circuit from the MC solenoid winding to ground 10 times per second. A dwellmeter is normally used to measure the number of degrees that the distributor cam rotates while the ignition points are closed. Specifically, the dwellmeter is measuring the length of time that current flows through the primary ignition circuit. During 3C system diagnosis the dwellmeter is connected from the ECM side of the MC solenoid winding to ground. The dwellmeter will measure the "on time" of the MC solenoid winding, as illustrated in Figure 8-26.

The ECM varies the "on time" of the MC solenoid to control the air-fuel mixture in the carburetor. Upon receiving a "rich" signal from the oxygen sensor, the ECM will energize the MC solenoid for a longer period of time. As a result, the MC solenoid plunger and metering rods move downward. In the downward position, the thick part of the metering rods is in the main jets and a leaner air-fuel ratio is provided. The ECM cycles the MC solenoid on and off ten times per second. If the ECM calls for a leaner air-fuel ratio, it may energize the MC solenoid for 90 percent of the time on each cycle. The metering rods will then stay down longer and a leaner air-fuel ratio is provided. The 6-cylinder scale on the dwellmeter reads from 0 to 60°. When the ECM energizes the MC solenoid winding for 90 percent of the time in each cycle, the dwellmeter will register 54° (90 percent of 60 is equal to 54), as indicated in Figure 8-27.

A lean oxygen sensor signal to the ECM will cause the ECM to decrease the "on time" of the MC solenoid winding. If the ECM provides an "on time" of 10 percent on each cycle of the MC solenoid, the solenoid plunger and metering rods will remain in the upward position for 90 percent of the time on each cycle. The upward metering rod position allows the thinner part of the metering rods to be in the main jets and thus provides a richer air-fuel ratio. When the engine is at normal operating temperature and the 3C system is in closed loop, the dwellmeter reading should vary continuously between 6° and 54°. This variation indicates that the ECM and the sensors are controlling the air-fuel mixture in the normal manner.

Under certain operating conditions such as a cold engine, the system will be in the open-loop mode and a fixed reading will be provided on the dwellmeter because the ECM is maintaining a slightly richer air-fuel ratio. In open-loop operation, the ECM is no longer controlling the MC solenoid from

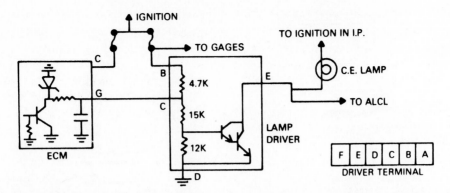

FIGURE 8-23. Check-Engine Light with Lamp Driver Module. (Courtesy of General Motors of Canada Ltd.)

FIGURE 8-24. Electronic Control Module (ECM) Terminals. (Courtesy of General Motors of Canada Ltd.)

ECM TERMINAL IDENTIFICATION

NOT ALL TERMINALS ARE USED ON ALL ENGINE APPLICATIONS

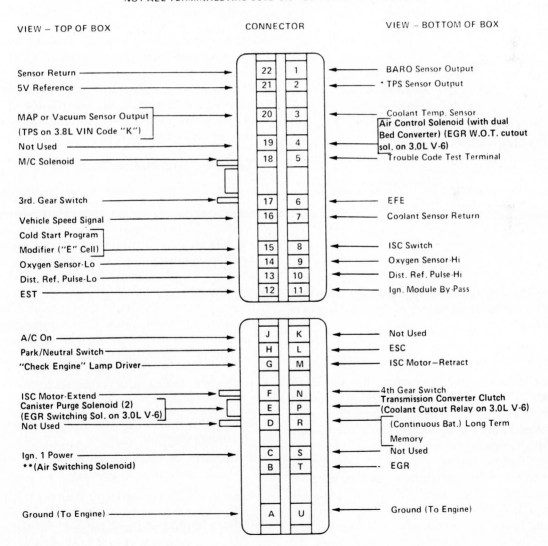

**Divert valve on applications that do not use a switching valve.

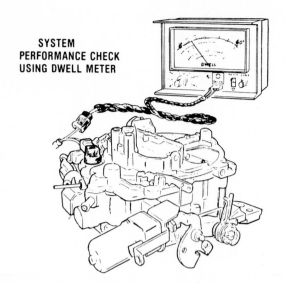

FIGURE 8-25. Dwellmeter Test Connection to the Mixture Control (MC) Solenoid. (Courtesy of General Motors of Canada Ltd.)

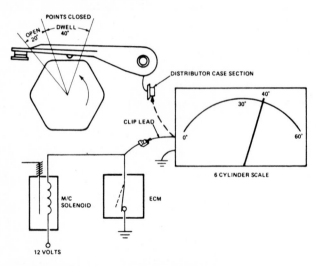

FIGURE 8–26. Dwellmeter Readings with Lean and Rich Air-Fuel Mixtures. (Courtesy of General Motors of Canada Ltd.)

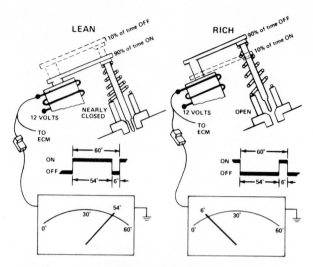

FIGURE 8–27. Dwellmeter Diagnosis of 3C System. (Courtesy of General Motors of Canada Ltd.)

relationship between the dwellmeter readings and MC solenoid cycling is illustrated in Figure 8-28.

If the carburetor is partly choked when the engine is idling, the dwellmeter pointer should move upscale. This indicates that the system is in closed loop and it will react immediately to correct the rich air-fuel mixture.

System Performance Test

Before proceeding with the diagnosis of the computer command control (3C) system, check all other sources of performance or economy complaints such as ignition components, engine compression, and the exhaust gas recirculation (EGR) valve. The absence of trouble codes in the ECM does not necessarily mean that the system is working normally. Certain defects in the system will not be indicated by trouble codes. The following system performance test may be performed to determine if the system is working normally:

1. Run the engine until it reaches normal operating temperature.
2. Ground the diagnostic test terminal with the engine running.
3. Connect a tachometer from the distributor "tach" terminal to ground.
4. Disconnect the mixture control (MC) solenoid and ground the solenoid dwell terminal. Once the solenoid is disconnected, the metering rods will move upward and provide a very rich air-fuel ratio.

the oxygen sensor signal. Wide-open throttle (WOT) operation will also cause a fixed reading on the dwellmeter. The oxygen sensor may cool down during prolonged idle operation and cause the system to go into the open-loop mode.

A high reading on the dwellmeter always indicates that the system is trying to provide a leaner air-fuel ratio; therefore the oxygen sensor signal to the ECM will indicate a rich mixture. A lean mixture signal from the oxygen sensor to the ECM is indicated by a low dwellmeter reading as the system tries to provide a richer air-fuel ratio. The

RELATIONSHIP OF DWELLMETER READINGS TO MIXTURE CONTROL SOLENOID CYCLING

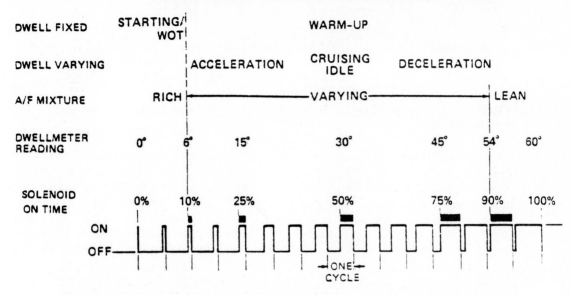

FIGURE 8-28. MC Solenoid Dwell Reading Interpretation. (Courtesy of General Motors of Canada Ltd.)

5. Operate the engine at 3,000 RPM and reconnect the MC solenoid. The grounded MC solenoid dwell lead will energize the solenoid winding continuously. The metering rods should then move downward and the air-fuel mixture should become leaner. If the engine slows down 100 RPM or more, the MC solenoid is working normally. A decrease of less than 100 RPM in engine speed indicates that the MC solenoid requires servicing, as outlined in the MC solenoid adjustments that are provided later in this chapter.

6. Disconnect the grounded MC solenoid dwell lead before slowing the engine to idle speed. Reconnect the MC solenoid and observe the dwellmeter. If the ECM and the sensors are working normally, the dwellmeter will vary continously between 6° and 54°. If the MC solenoid dwell reading is in a fixed position or out of the specified range, the system is defective.

When MC Solenoid Dwell Reading Is Fixed at 5°-10°

When the MC solenoid dwell reading is very low, the metering rods will be in the upward position for a longer period of time on each cycle, and the system will try to provide a richer air-fuel mixture. Thus the exhaust gas oxygen sensor signal will be excessively lean. The following diagnostic procedure may be used when the MC solenoid dwell reading is fixed at 5°-10°:

1. With the engine operating at normal temperature and idle speed, close the choke half way. If the MC solenoid dwell reading increases, check all possible sources of air leaks into the intake manifold. If no air leaks are found, perform the MC solenoid adjustments outlined later in this chapter, and check the carburetor for all possible causes of a lean air-fuel ratio, such as a low float level. Check the AIR pump to make sure it is pumping air into the converter in the closed-loop mode.

2. If the MC solenoid dwell reading does not increase when the choke is closed half way, disconnect the oxygen sensor wires. On dual-wire oxygen sensors, connect a 1.5-V flashlight battery to the tan and purple oxygen sensor wires that are connected to the ECM. Connect the positive battery terminal to the purple wire and the negative battery terminal to the tan wire. On single-wire oxygen sensors, connect a digital voltmeter from the battery positive terminal to the purple oxygen sensor wire connected to the ECM. The voltmeter must be on the 20-V scale. The voltmeter or the flashlight battery simulates the oxygen sensor signal.

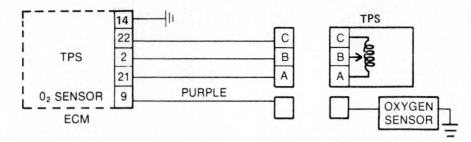

FIGURE 8-29. ECM and Oxygen Sensor Terminals. (Courtesy of General Motors of Canada Ltd.)

In systems with dual-wire oxygen sensors, if the MC solenoid dwell reading increases when the flashlight battery is connected, the tan wire connected from the oxygen sensor to terminal 14 on the ECM has an open circuit, or the oxygen sensor is defective. If the MC solenoid dwell reading does not increase, the purple wire connected from ECM terminal 9 to the oxygen sensor has an open circuit or a grounded condition, or the ECM is defective.

In systems with single-wire oxygen sensors, if the MC solenoid dwell reading increases when the voltmeter is connected, the wire connected from ECM terminal 14 to ground has an open circuit, or the oxygen sensor is defective. If the MC solenoid dwell reading does not increase, the purple wire from ECM terminal 9 to the oxygen sensor has an open circuit or a grounded condition, or the ECM is defective. (See Figure 8-29 for ECM and oxygen sensor terminal identification on single-wire oxygen sensor systems.)

When MC Solenoid Dwell Reading Is Fixed Between 10° and 50°

When the 3C system continues to operate in the open-loop mode with the engine at normal operating temperature, the MC solenoid dwell reading will be fixed between 10° and 50°. This problem could be caused by defects in the coolant temperature sensor, oxygen sensor, sensor wires connected to the ECM, or the ECM. If the engine coolant is at a low level and the engine is operating at a normal temperature, open-loop operation may occur. The following diagnostic steps may be used to locate the cause of the fixed dwell reading:

1. With the engine operating at 1,000 RPM, disconnect the coolant sensor terminal and connect a jumper wire to the sensor leads to the ECM. If the dwell reading changes, the coolant sensor is defective, or there is a high resistance connection at the sensor terminal. When the coolant

sensor wires are connected together and there is no change in the MC solenoid dwell reading, connect a jumper wire across coolant sensor terminals 3 and 7 at the ECM. If a dwell change occurs when terminals 3 and 7 are connected together, the coolant sensor wires have an open circuit or a grounded condition. The ECM is defective if a dwell change does not occur when terminals 3 and 7 are connected together.

2. Operate the engine at 1,000 RPM and disconnect the oxygen sensor. For dual-wire sensors, connect the two sensor wires together. If a single-wire sensor is used, connect the sensor wire from the ECM to ground. A decrease in the dwell reading indicates a defective oxygen sensor. If the dwell reading does not change, connect a jumper wire across terminals 9 and 14 at the ECM. If a decrease in the dwell reading occurs, the oxygen sensor wires have an open circuit. The ECM is defective if there is no dwell change when terminals 9 and 14 are connected together.

When MC Dwell Reading Is Fixed Above 50°

When the MC solenoid dwell reading is continuously high, the system is trying to provide a leaner air-fuel mixture. Therefore the oxygen sensor signal must be indicating an extremely rich air-fuel ratio. To diagnose this condition, proceed as follows:

1. Run the engine at 1,000 RPM and normal operating temperature.

2. Disconnect the positive crankcase ventilation (PCV) valve hose from the PCV valve. Place your thumb over the PCV hose and slowly uncover the hose to create a large vacuum leak. If the dwell reading decreases, check the mixture control solenoid adjustments and any carburetor defects

that could result in a rich air-fuel ratio. Check the vapor canister system for proper operation.

3. On systems with dual-wire oxygen sensors, disconnect the oxygen sensor and connect a jumper wire across the sensor leads connected to the ECM. If the dwell reading decreases, the oxygen sensor is defective. The ECM is defective if the dwell reading does not decrease when the sensor leads are connected together.

4. For systems with single wire oxygen sensors, disconnect the oxygen sensor and ground the sensor wire connected to the ECM. If the MC solenoid dwell reading does not change, the ECM is defective.

5. If the oxygen sensor wire connected to the ECM is grounded and the MC solenoid dwell reading drops to less than 10°, disconnect the ground wire from the sensor lead. Connect a digital voltmeter from the oxygen sensor wire to the ECM and ground with the ignition switch on. A voltmeter reading below 0.55 indicates a defective oxygen sensor. When the voltmeter reading exceeds 0.55 V, the ECM is defective, or the sensor wire is shorted to a wire that is connected to a 12-V source.

Mixture Control (MC) Solenoid Adjustments

Two adjustments are necessary on the MC solenoid. With the carburetor air horn removed, turn screw A in Figure 8-30 until it bottoms lightly. Kent Moore tool J28696, or the equivalent, must be used to turn screw A. Turn screw A two turns counterclockwise from the bottomed position. With the carburetor air horn installed, position the modified float gauge on the MC solenoid plunger. Push down on the gauge until the plunger bottoms. The MC solenoid plunger should travel 2/32–4/32 in (1.59 mm–3.17 mm), as indicated on the gauge. If plunger travel is not correct, remove the air horn plug above screw B and adjust screw B until the MC solenoid plunger travels 3/32 in (2.36 mm). Screw B can be adjusted with tool J28696.

3C Systems with Electronic Spark Control (ESC)

Some engines have a combined 3C and ESC system. The ESC system was explained in Chapter 6. The EST and bypass disable signals go from the ECM through the ESC controller to the ignition module, as illustrated in Figure 8-31.

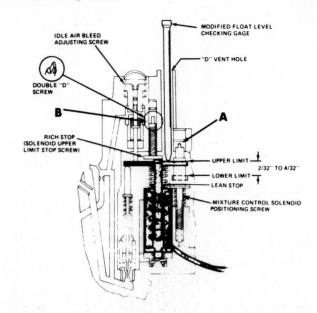

FIGURE 8-30. Mixture Control (MC) Solenoid Adjustments. (Courtesy of General Motors of Canada Ltd.)

A detonation signal from the detonation sensor to the ESC controller will cause the controller to retard the spark advance. The amount of spark retard is determined by the severity and duration of detonation. Maximum retard would be 28°.

If the timing does not change when the four-wire distributor connector is disconnected, the system could be in the start mode with the engine running. An open bypass disable circuit in the ESC controller could cause the system to remain in the start mode. Connect a jumper wire across ESC controller terminals D and C, as pictured in Figure 8-32. Recheck for timing change when the four-wire distributor connector is disconnected. If timing change now occurs, the ESC controller is defective.

A complaint of engine stalling 5 seconds after a restart could be the result of an open EST circuit in the ESC controller. The voltage available between controller terminals F and K must exceed 10 V when the ignition switch is on. Connect a jumper wire between controller terminals G and J. If the stalling problem is eliminated, the ESC controller is defective.

Engine detonation complaints may be diagnosed as follows:

1. With the engine operating at 1,500 RPM, tap the engine block near the detonation sensor. If RPM does not change, the ESC system is defective.

2. Connect an ohmmeter from controller terminal K to ground. An infinite reading indicates an

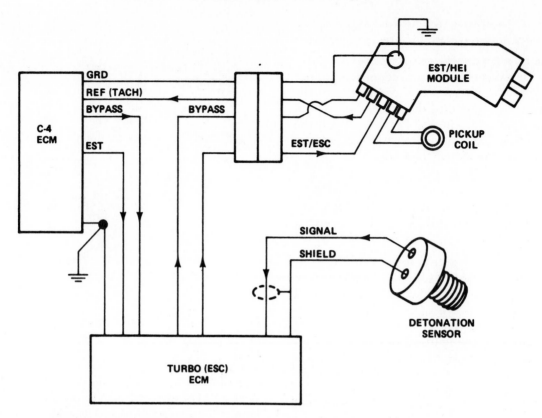

FIGURE 8-31. HEI Systems with C3 and ESC. (Courtesy of General Motors of Canada Ltd.)

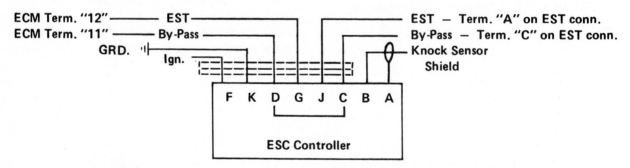

FIGURE 8-32. Diagnosis of C3 ESC Controller. (Courtesy of General Motors of Canada Ltd.)

open ground wire. If the ohmmeter reads zero, the ground wire is satisfactory.

3. Check the knock sensor wire, from controller terminal B to the sensor, for open circuits and grounds with an ohmmeter.

4. With the knock sensor wire connected, attach an ohmmeter from controller terminal B to terminal K. If the ohmmeter registers less than 175 Ω, or more than 375 Ω, replace the sensor.

5. If no defects have been located in steps 2, 3, and 4, replace the controller. (See Figure 8-33 for controller terminal identification.)

FIGURE 8-33. C3 ESC Controller Terminals. (Courtesy of General Motors of Canada Ltd.)

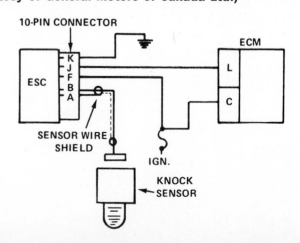

FORD ELECTRONIC ENGINE CONTROL II (EEC II) SYSTEMS

Sensors

The electronic engine control II (EEC II) systems are quite similar to the EEC I systems described in Chapter 7, except that an exhaust gas oxygen (EGO) sensor is installed in the EEC II system, as shown in Figure 8-34. In addition, the inlet air temperature sensor has been discontinued in the EEC II system, and the barometric (Baro) sensor and the manifold absolute pressure (MAP) sensor are combined in one unit known as the (BMAP) sensor. Figure 8-35 illustrates the combined BMAP sensor. Finally, the crankshaft position (CP) sensor has been relocated from the rear of the crankshaft to the front of the crankshaft, as pictured in Figure 8-36.

A four-lobe pulse ring is attached to the crankshaft vibration damper. Each time one of the pulse ring lobes rotates past the CP sensor, the sensor sends a signal to the electronic control assembly (ECA). Because the CP sensor is located in a fixed position, the basic timing is not adjustable. The engine coolant temperature (ECT) sensor, throttle position (TP) sensor, and the exhaust gas recirculation valve position (EVP) sensor are basically the same as in the EEC I system.

FIGURE 8-34. Exhaust Gas Oxygen (EGO) Sensor. (Courtesy of Ford Motor Co.)

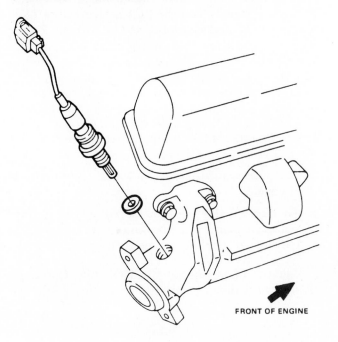

FRONT OF ENGINE

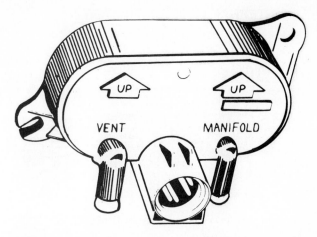

FIGURE 8-35. Barometric and Manifold Absolute Pressure (BMAP) Sensor. (Courtesy of Sun Electric Corp.)

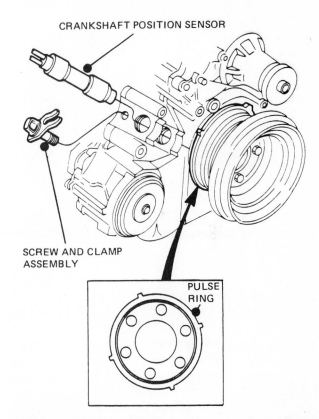

FIGURE 8-36. Crankshaft Position (CP) Sensor. (Courtesy of Ford Motor Co.)

Control Functions

A feedback carburetor actuator (FBCA) is mounted on the carburetor in the EEC II system. The FBCA is a small electric motor that is used to control a vacuum bleed pintle valve attached to the motor shaft, as illustrated in Figure 8-37.

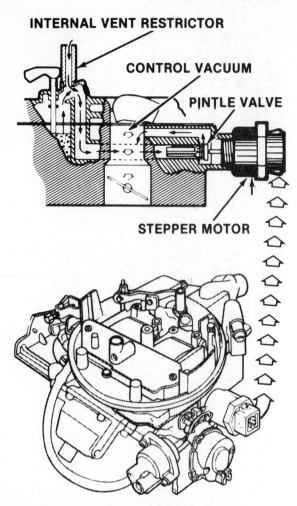

FIGURE 8-37. Feedback Carburetor Actuator (FBCA). (Courtesy of Sun Electric Corp.)

When the electronic control assembly (ECA) receives a signal from the exhaust gas oxygen (EGO) sensor that indicates an excessively rich mixture, the ECA energizes the FBCA and opens the vacuum bleed pintle valve a specific amount. Once the vacuum bleed pintle valve is open, a certain amount of control vacuum enters the float bowl. The normal atmospheric pressure in the float bowl is reduced by the control vacuum, and the air-fuel mixture becomes leaner because of the reduced pressure on the fuel in the float bowl. If the ECA receives an EGO sensor signal that indicates an excessively lean air-fuel ratio, the ECA energizes the FBCA so that it moves the vacuum bleed pintle valve toward the closed position. As a result, float bowl pressure increases and the air-fuel ratio becomes richer. The ECA operates the FBCA to provide an air-fuel ratio of 14.7:1.

Canister Purge Control

The electronic control assembly (ECA) operates a canister purge (CANP) solenoid valve that is connected between the intake manifold purge fitting and the carbon canister, as shown in Figure 8-38. When the CANP solenoid valve is activated, fuel vapors move from the carbon canister through the solenoid valve into the intake manifold. The ECA will activate the CANP solenoid valve under the following conditions:

1. Engine coolant temperature must be in the normal operating range.
2. Engine revolutions per minute (RPM) must be above a calibration value.
3. The time since the engine was started must be above a calibration value.
4. The engine is not in the closed-throttle mode.

When the engine is shut off, the CANP solenoid valve is closed and the electric bowl vent is open. Under this condition, fuel vapors from the fuel tank and the carburetor float bowl collect in the carbon canister. When the engine is idling, the electric bowl vent solenoid and the CANP solenoid are closed, and therefore only the fuel vapors from the fuel tank collect in the carbon canister.

Throttle Control

The electronic engine control II (EEC II) system uses a throttle kicker actuator and a throttle kicker solenoid (TKS) to control engine speed under certain conditions. When the ECA activates the TKS, manifold vacuum is applied to the throttle kicker actuator and the actuator provides a faster idle speed. The throttle kicker actuator is pictured in Figure 8-39 and the TKS in Figure 8-40.

The ECA will activate the TKS valve under the following conditions:

1. The engine coolant temperature is below a cold value or at an overheat value.
2. The air-conditioning compressor clutch is engaged.
3. The TKS valve will be activated for a few seconds each time the engine is started regardless of coolant temperature.

Vacuum diagrams for the TKS valve in the activated and deactivated positions are shown in Figure 8-41.

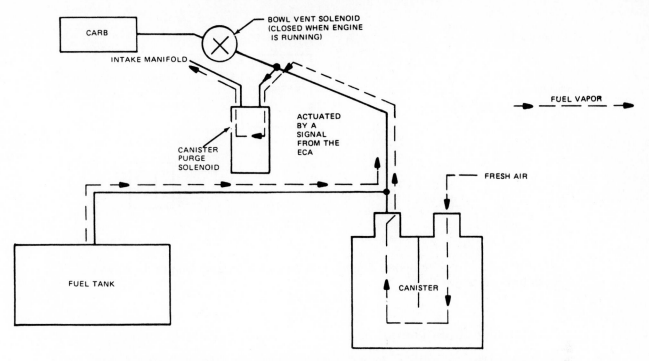

FIGURE 8-38. Canister Purge Solenoid Valve and System. (Courtesy of Ford Motor Co.)

FIGURE 8-39. Throttle Kicker Actuator. (Courtesy of Ford Motor Co.)

FIGURE 8-40. Throttle Kicker Solenoid. (Courtesy of Sun Electric Corp.)

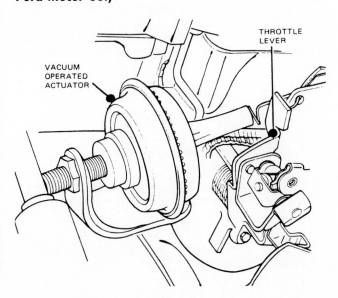

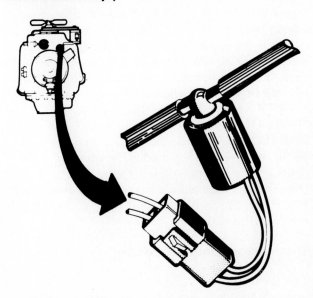

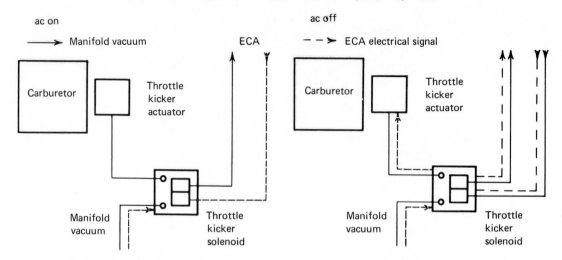

FIGURE 8-41. Throttle Kicker Solenoid (TKS) Valve Vacuum Diagrams. (Courtesy of Ford Motor Co.)

Exhaust Gas Recirculation (EGR) Control

The EGR valve in the EEC II system is operated by manifold vacuum rather than by thermactor pump pressure, as in the EEC I system. The EGR valve has an EGR valve position (EVP) sensor similar to the EVP sensor in the EEC I system. The vacuum-operated EGR valve used in the EEC II system is shown in Figure 8-42.

The ECA operates the EGR vent solenoid and vacuum solenoid to control the EGR valve position. In the increase-EGR-flow mode, the ECA energizes both solenoids, with the result that manifold vacuum is applied through the EGR vacuum solenoid to the EGR valve. In the energized position, the EGR vent solenoid is closed and the vacuum is trapped in the EGR vacuum hoses, as illustrated in Figure 8-43.

In the maintain-EGR-flow mode, the ECA energizes the EGR vent solenoid and deenergizes the EGR vacuum solenoid. Thus, the manifold vacuum becomes trapped in the EGR vacuum system and the EGR valve position is maintained.

When the input sensors signal the ECA to reduce EGR flow, the ECA deenergizes both solenoids in the decrease-EGR-flow mode. The vacuum solenoid shuts off the vacuum applied to the EGR valve, and the vent solenoid vents any vacuum in the system. Under this condition, the EGR valve moves toward the closed position.

Thermactor Pump Air Control

The electronic control assembly (ECA) operates the dual air control solenoids to control the thermac-

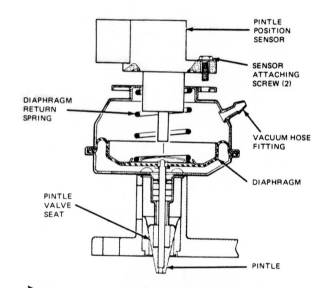

FIGURE 8-42. Vacuum-Operated Exhaust Gas Recirculation (EGR) Valve. (Courtesy of Ford Motor Co.)

tor pump airflow. Vacuum applied to the thermactor air bypass valve and diverter valve is turned on and off by the dual air control solenoids. The dual air control solenoids are illustrated in Figure 8-44 and the air bypass and diverter valve in Figure 8-45.

The thermactor pump airflow has three possible routings:

1. Downstream air, which is injected into the catalytic converter.

2. Upstream air, which is injected into the exhaust manifolds.

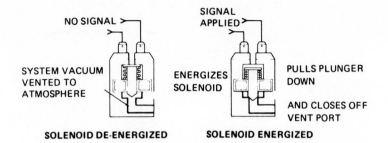

"NORMALLY OPEN" EGR SOLENOID 'VENT VALVE' OPERATION

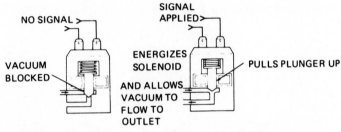

"NORMALLY CLOSED" EGR SOLENOID CONTROL VALVE OPERATION

FIGURE 8-43. EGR Solenoid Operation. (Courtesy of Ford Motor Co.)

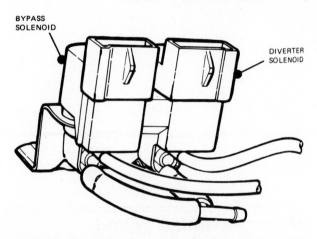

FIGURE 8-44. Dual Air Control Solenoids. (Courtesy of Ford Motor Co.)

3. Bypass air, which is bypassed into the atmosphere.

Figure 8-46 illustrates the complete thermactor pump system.

In the bypass-air mode the ECA deactivates both solenoids. Under this condition, vacuum is not applied to the air bypass valve or the air diverter valve and thermactor pump air is bypassed to the atmosphere, as shown in Figure 8-47.

The bypass-air mode will occur when idling is prolonged, the throttle is wide open, or the exhaust gas oxygen (EGO) sensor signal is extremely low.

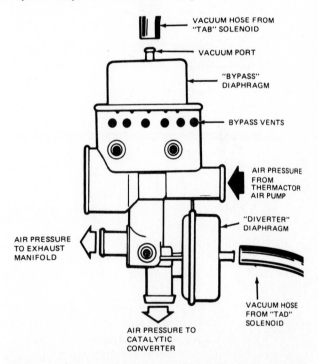

FIGURE 8-45. Air Bypass and Diverter Valve. (Courtesy of Ford Motor Co.)

When the engine coolant is cold, the ECA activates the bypass solenoid and the diverter solenoid. If both solenoids are activated, manifold vacuum is applied through the solenoids to the air bypass and air diverter valves. Under this condition,

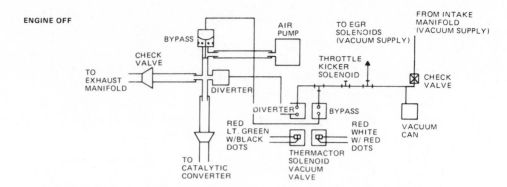

FIGURE 8-46. Thermactor Pump System. (Courtesy of Ford Motor Co.)

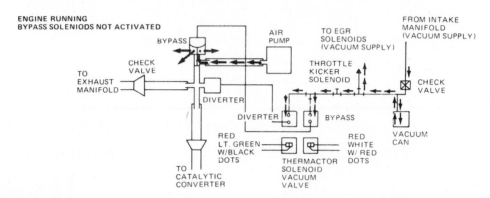

FIGURE 8-47. Thermactor System Bypass-Air Mode. (Courtesy of Ford Motor Co.)

thermactor air is diverted to the exhaust manifold, as shown in Figure 8-48.

Air from the thermactor pump is diverted to the exhaust manifolds to lower hydrocarbon (HC) and carbon monoxide (CO) emissions during the engine warm-up period.

When the input sensors call for downstream thermactor pump air injection, the ECA activates the bypass solenoid and deactivates the diverter solenoid. Manifold vacuum is maintained through the activated bypass solenoid to the bypass valve, but the manifold vacuum to the diverter valve is shut off by the deactivated diverter solenoid. Under this condition, thermactor pump air flows through the bypass valve and diverter valve to the catalytic converter, as illustrated in Figure 8-49.

FORD ELECTRONIC ENGINE CONTROL III (EEC III) SYSTEMS

Design

The electronic engine control III (EEC III) system is very similar to the EEC II system. One of the main differences in the EEC III system is that the feedback carburetor actuator (FBCA) controls the amount of air that enters the main system air bleed rather than the pressure in the float bowl, as in the other systems. The air bleed passage from the FBCA into the main system is pictured in Figure 8-50.

The electronic control assembly (ECA) operates

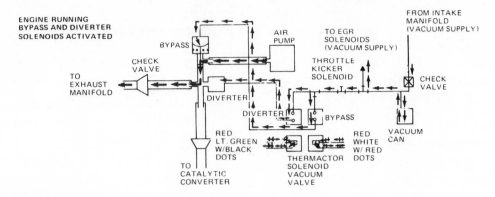

ENGINE RUNNING
BYPASS AND DIVERTER
SOLENOIDS ACTIVATED

AIR FLOW
ENGINE VACUUM
ECA ELECTRICAL FLOW

FIGURE 8-48. Thermactor Air Diverted to Exhaust Manifolds. (Courtesy of Ford Motor Co.)

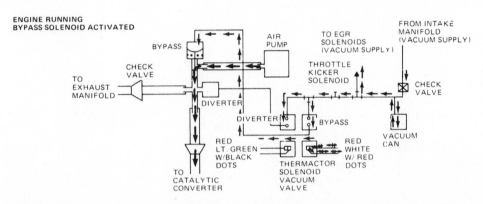

ENGINE RUNNING
BYPASS SOLENOID ACTIVATED

FIGURE 8-49. Thermactor Air Directed to Catalytic Converter. (Courtesy of Ford Motor Co.)

FIGURE 8-50. Feedback Carburetor Actuator (FBCA) System. (Courtesy of Ford Motor Co.)

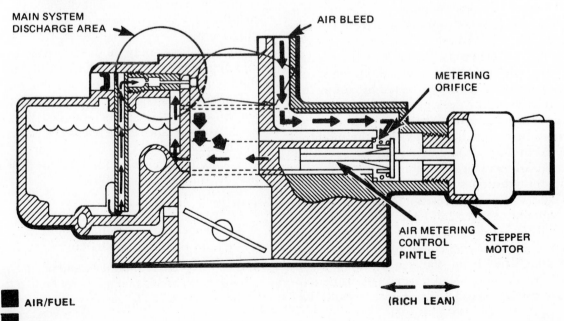

AIR/FUEL

FUEL

AIR

the FBCA to provide the air-fuel ratio of 14.7:1. The input sensors and output control functions are basically the same in the EEC II and EEC III systems. An EEC III system is also available with electronic fuel injection (EFI). The wiring of the EEC II and EEC III systems is shown in Figure 8-51.

DIAGNOSIS OF EEC II AND EEC III SYSTEMS

Complete Circuit

The initial timing and the advance timing tests for the EEC I, EEC II, and EEC III systems are all the same. (See Chapter 7 for a description of these tests on the EEC I system.) Always refer to the underhood specifications. If the initial timing and advance timing are fixed at 10° before top dead center (BTDC), the system is operating in the limited-operation-strategy (LOS) mode. A defect exists in the electronic control assembly (ECA) or in the input sensor circuit if the system is in the LOS mode.

When performance or economy complaints are being diagnosed, the throttle position (TP) sensor, barometric (Baro) sensor, EGR valve position (EVP) sensor, and the manifold absolute pressure (MAP) sensor may be tested by checking the reference voltage and the modified output voltage at each sensor. This same procedure for testing the sensors was used on the EEC I system (see Chapter 7). The reference voltage at each sensor may be tested by connecting a voltmeter from the orange-white wire at each sensor to ground. When the ignition switch is on and the engine is not running, the reference voltage should be 8.5–9.5 V. If the reference voltage is not within specifications, the ECA or the orange-white wire is defective. The modified reference voltage from each sensor may be tested by connecting a voltmeter from the dark green–light green wire at each sensor to ground. When the ignition switch is on and the engine is not running, the modified reference voltages should be within the specifications that are provided in Table 8-1.

With the ignition switch on, a voltmeter connected across the engine coolant temperature (ECT) sensor should register 2.0–4.4 V. The exhaust gas oxygen (EGO) sensor test may be performed with the sensor wire disconnected and a digital voltmeter connected from the sensor terminal to ground. With the engine at normal operating temperature and idle speed, the voltmeter should read

TABLE 8-1.

	Elevation	Modified Output Voltage
Barometric pressure and manifold absolute pressure sensors	0–1000 ft	6.7–7.8
	1000–2000 ft	6.5–7.4
	2000–3000 ft	6.2–7.3
	3000–4000 ft	6.0–7.0
	4000–5000 ft	5.8–6.8
	5000–6000 ft	5.6–6.6
	6000–7000 ft	5.3–6.4
Throttle position sensor		1.8–2.4
EGR valve position sensor		1.4–1.9

0.2–0.5 V. When the carburetor cold enrichment rod is moved upward, the voltmeter reading should increase to 0.6–1.1 V. If the voltage readings are not within specifications, the sensor should be replaced. The EGO sensor will be ruined if an ordinary voltmeter is used for test purposes. The carburetor cold enrichment rod is illustrated in Figure 8-52.

The throttle position (TP) sensor may be adjusted by loosening the attaching screws and rotating the sensor until the modified output voltage is within specifications. The engine should be at normal operating temperature and the throttle in the idle position.

The crankshaft position (CP) sensor is not adjustable, but it must be bottomed properly in the holding fixture to obtain a clearance of 0.011–0.080 in (0.279–2.03 mm) between the sensor and the pulse ring teeth.

An ohmmeter may be used to test the solenoid windings. The resistance values for the various solenoid windings are provided in Table 8-2.

Ohmmeter tests may be performed on the CP sensor, ECT sensor, TP sensor, and the EVP sensor. The specified resistance value of each sensor is provided in Table 8-3.

The feedback carburetor actuator (FBCA) motor may be diagnosed as follows:

1. Remove the FBCA from the carburetor.

2. Reconnect the wiring harness to the FBCA, and turn on the ignition switch.

3. The FBCA shaft should extend when the ignition switch is turned on and retract when the switch is turned off. The shaft movement should be

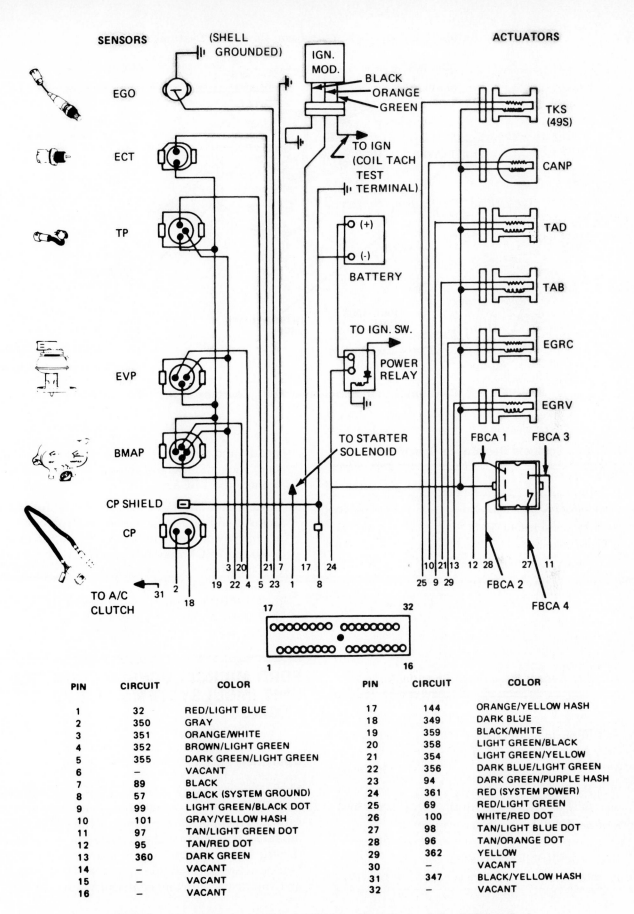

PIN	CIRCUIT	COLOR	PIN	CIRCUIT	COLOR
1	32	RED/LIGHT BLUE	17	144	ORANGE/YELLOW HASH
2	350	GRAY	18	349	DARK BLUE
3	351	ORANGE/WHITE	19	359	BLACK/WHITE
4	352	BROWN/LIGHT GREEN	20	358	LIGHT GREEN/BLACK
5	355	DARK GREEN/LIGHT GREEN	21	354	LIGHT GREEN/YELLOW
6	—	VACANT	22	356	DARK BLUE/LIGHT GREEN
7	89	BLACK	23	94	DARK GREEN/PURPLE HASH
8	57	BLACK (SYSTEM GROUND)	24	361	RED (SYSTEM POWER)
9	99	LIGHT GREEN/BLACK DOT	25	69	RED/LIGHT GREEN
10	101	GRAY/YELLOW HASH	26	100	WHITE/RED DOT
11	97	TAN/LIGHT GREEN DOT	27	98	TAN/LIGHT BLUE DOT
12	95	TAN/RED DOT	28	96	TAN/ORANGE DOT
13	360	DARK GREEN	29	362	YELLOW
14	—	VACANT	30	—	VACANT
15	—	VACANT	31	347	BLACK/YELLOW HASH
16	—	VACANT	32	—	VACANT

FIGURE 8-51. EEC II and EEC III Wiring. (Courtesy of Ford Motor Co.)

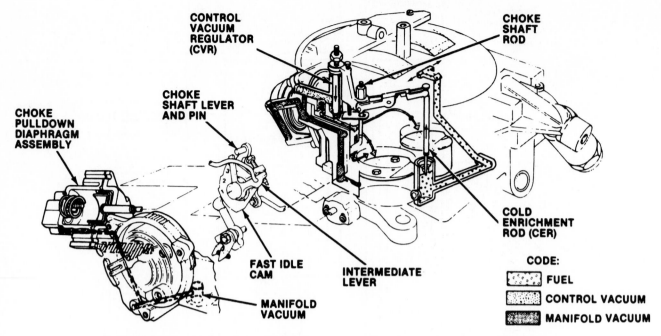

FIGURE 8-52. Carburetor Cold Enrichment Rod. (Courtesy of Ford Motor Co.)

TABLE 8-2. Solenoid Resistance Specifications

	Ohms
Canister purge (CANP) solenoid	50–100
Throttle kicker solenoid (TKS)	50–100
Dual EGR solenoids	35– 85
Thermactor bypass (TAB) solenoid	50–100
Thermactor air diverter (TAD) solenoid	50–100

TABLE 8-3. Sensor Resistance Specifications

	Ohms
Crankshaft position (CP) sensor	100–640
Engine coolant temperature (ECT) sensor	1500–6000
Throttle position (TP) sensor	
Orange-white to black-white wires	3000–5000
Dark green–light green to black-white wires	560–130
EGR valve position (EVP) sensor	
Black-white to brown–light green wires	500–900

steady and the shaft should not rotate. If the shaft movement is uneven or the shaft rotates, replace the FBCA.

4. If the motor shaft does not move, the following items could be the cause of the problem:
   a. The power feed wire from the ignition switch to the FBCA may be defective.
   b. The wires from the FBCA to the ECA may have an open circuit.
   c. The FBCA motor may be defective.
   d. The ECA may be defective.

FORD MICROPROCESSOR CONTROL UNIT (MCU) SYSTEMS IN 2.3L ENGINES

Operation

The microprocessor control unit (MCU) systems use an MCU module to control the carburetor air-fuel ratio and the thermactor pump airflow. Some MCU modules also control the canister purge system, throttle kicker system, and the spark retard system. The Dura-Spark II ignition system and conventional distributor vacuum and centrifugal advance mech-

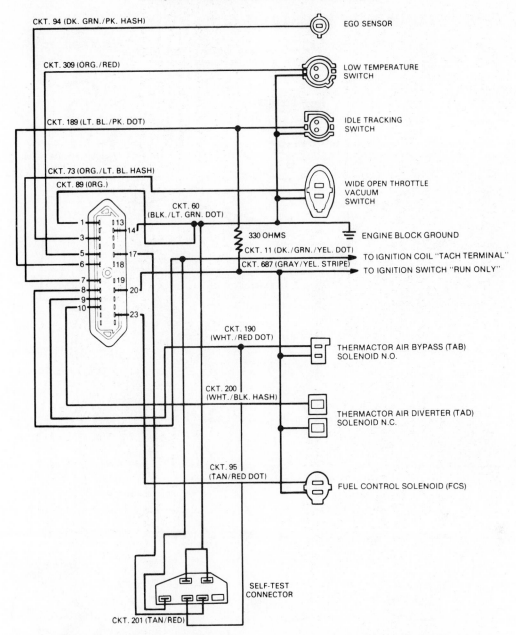

CKT. 94 (DK. GRN./PK. HASH) — EGO SENSOR

CKT. 309 (ORG./RED) — LOW TEMPERATURE SWITCH

CKT. 189 (LT. BL./PK. DOT) — IDLE TRACKING SWITCH

CKT. 73 (ORG./LT. BL. HASH)
CKT. 89 (ORG.) — WIDE OPEN THROTTLE VACUUM SWITCH

CKT. 60 (BLK./LT. GRN. DOT)

330 OHMS — ENGINE BLOCK GROUND
CKT. 11 (DK./GRN./YEL. DOT) — TO IGNITION COIL "TACH TERMINAL"
CKT. 687 (GRAY/YEL. STRIPE) — TO IGNITION SWITCH "RUN ONLY"

CKT. 190 (WHT./RED DOT) — THERMACTOR AIR BYPASS (TAB) SOLENOID N.O.

CKT. 200 (WHT./BLK. HASH) — THERMACTOR AIR DIVERTER (TAD) SOLENOID N.C.

CKT. 95 (TAN/RED DOT) — FUEL CONTROL SOLENOID (FCS)

SELF-TEST CONNECTOR

CKT. 201 (TAN/RED)

FIGURE 8-53. Wiring of MCU System, 2.3L Engine. (Courtesy of Ford Motor Co.)

anisms are used with the MCU system. The wiring for the MCU system on the 2.3L engine is illustrated in Figure 8-53.

The MCU components and their respective locations in the 2.3L engine are illustrated in Figure 8-54.

The exhaust gas oxygen (EGO) sensor and the coolant low-temperature switch in the MCU system operate in the same way as they do in the EEC II system. Thermactor pump airflow is controlled by the thermactor air bypass (TAB) solenoid, thermac-

tor air diverter (TAD) solenoid, and the thermactor air valve. The MCU module operates the TAB and TAD solenoids in the same way as the electronic control assembly (ECA) in the EEC II system described earlier in the chapter.

The idle tracking switch (ITS) sends an electrical signal to the MCU module to indicate a closed-throttle condition. When the engine is decelerating or operating at idle speed, the throttle linkage opens the ITS switch contacts.

When the throttle is in the wide-open position,

TYPICAL COMPONENT LOCATIONS (2.3L)

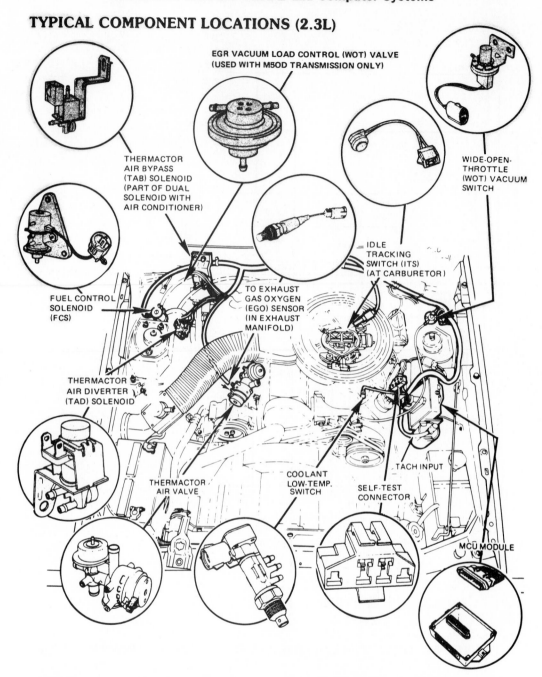

FIGURE 8-54. Location of MCU Components in 2.3L Engine. (Courtesy of Ford Motor Co.)

an electrical signal is sent from the wide-open throttle (WOT) vacuum switch to the MCU module. The WOT vacuum switch contains a set of contacts that are normally closed but that will open when the manifold vacuum drops below 3 in Hg (10 kPa) on vehicles with an automatic transmission, or below 5 in Hg (16.5 kPa) on vehicles with a standard transmission.

The exhaust gas recirculation (EGR) WOT valve is not part of the MCU system. (This valve was de-scribed in Chapter 2.) An input signal from the "tach" terminal of the coil provides the MCU module with an RPM signal on all MCU systems.

Air-Fuel Ratio Control

The fuel control solenoid (FCS) contains an electrically operated vacuum solenoid and three vacuum ports, as shown in Figure 8-55. Manifold

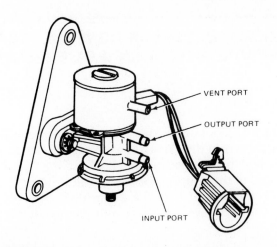

FIGURE 8-55. Fuel Control Solenoid (FCS). (Courtesy of Ford Motor Co.)

vacuum is connected to the input port and the output port is connected to the piston and diaphragm assembly in the carburetor. A vacuum hose connects the vent port to a clean air source at the top of the carburetor. The vacuum solenoid winding in the FCS is energized ten times per second by the MCU module. Output vacuum from the FCS is determined by the "on time" of the solenoid wind-

ing. The MCU module varies the "on time" of the FCS winding to provide an output vacuum of 2–5 in Hg (6.8–17 kPa), which controls the piston and diaphragm assembly in the carburetor, as shown in Figure 8-56.

A tapered metering rod is operated by the piston and diaphragm assembly. Fuel flows past the tapered metering rod as well as through the main metering jet. Once the exhaust gas oxygen (EGO) sensor signal to the MCU module indicates that the air-fuel ratio is excessively lean, the MCU module changes the "on time" of the FCS winding to produce a lower output vacuum from the FCS. The piston and diaphragm assembly and the tapered metering rod in the carburetor then move downward to provide a richer air-fuel ratio.

A rich mixture signal from the EGO sensor to the MCU module increases FCS vacuum output so that the piston and diaphragm assembly and the tapered metering rod move upward, and thus a leaner air-fuel ratio is provided. The MCU module and the FCS solenoid control the piston and diaphragm assembly in the carburetor to provide an air-fuel ratio of 14.7:1. With the engine operating at normal temperature and idle speed, the vacuum output

FIGURE 8-56. Carburetor Vacuum-Controlled Piston and Diaphragm Assembly. (Courtesy of Ford Motor Co.)

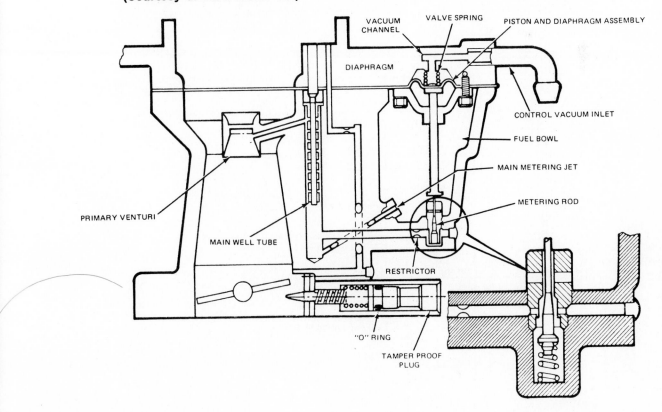

from the FCS should be pulsating between 1.9–3.15 in Hg (6.4–10.6 kPa), if the system is working normally.

MCU systems operate in the open-loop mode during engine warm-up, on acceleration, deceleration, and prolonged idle conditions. In this mode, the MCU module controls the FCS to provide an air-fuel ratio that is slightly richer. The MCU system will operate in the closed-loop mode when the engine is operating at normal temperature and light-load, part-throttle conditions. In the closed-loop mode, the MCU module controls the FCS from the EGO sensor signal to provide an air-fuel ratio of 14.7:1.

MICROPROCESSOR CONTROL UNIT (MCU) SYSTEMS ON 4.9L ENGINES

Operation

The MCU module on the 4.9L engine receives input signals from the exhaust gas oxygen (EGO) sensor, coolant low-temperature switch, and the "tach" terminal of the ignition coil. These same input signals were used on the MCU system in the 2.3L engine. In the MCU system of the 4.9L engine, a vacuum

FIGURE 8–57. MCU System Wiring, 4.9L Engine. (Courtesy of Ford Motor Co.)

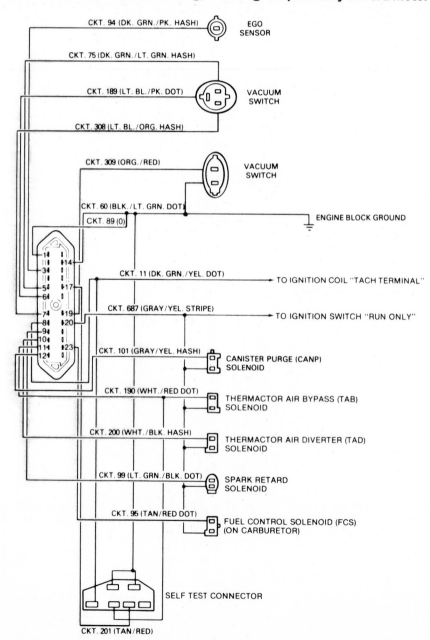

switch assembly sends an electrical signal to the MCU module in relation to manifold vacuum. On receiving this signal, the MCU module adjusts air-fuel ratio and thermactor pump airflow. The wiring for the MCU system on the 4.9L engine is outlined in Figure 8-57. The locations of the MCU components in 4.9L engines are shown in Figure 8-58.

The distributor vacuum advance hoses are routed through the spark retard solenoid. When engine operating conditions are such that engine detonation could occur, the MCU module sends an electrical signal to the spark retard solenoid that causes the solenoid to bleed off the vacuum applied to the vacuum advance. The MCU module operates the canister purge (CANP) solenoid valve to control the vapor canister system. Activation of the CANP solenoid valve will occur when the following conditions are present:

1. The engine coolant temperature is above a specific cold temperature and below a set overheat value.

2. The engine speed is above a calibrated RPM.

FIGURE 8-58. Locations of MCU Components in 4.9L Engines. (Courtesy of Ford Motor Co.)

TYPICAL COMPONENT LOCATIONS (4.9L)

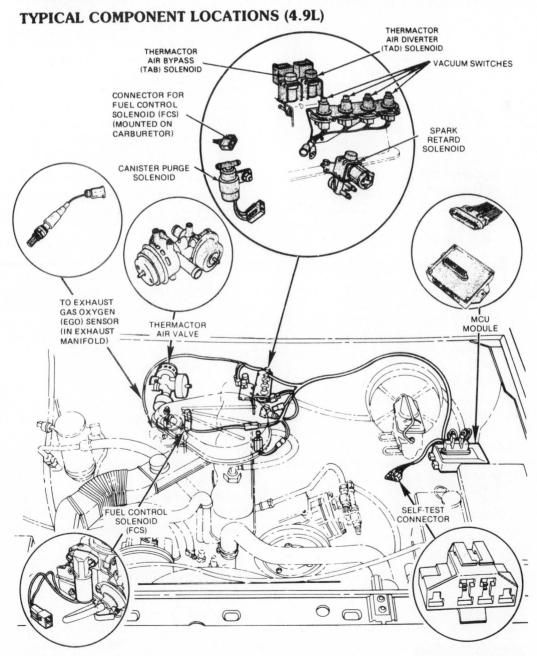

3. A set delay period has been completed following an engine start.

The MCU module controls thermactor pump airflow by operating the thermactor air bypass and thermactor air diverter solenoids in the normal manner. The fuel control solenoid (FCS) is activated ten times per second by the MCU module. An exact air fuel ratio of 14.7:1 is provided as the MCU module controls the "on time" of the FCS.

MICROPROCESSOR CONTROL UNIT (MCU) SYSTEMS ON 4.2L, 5.0L, AND 5.8L EIGHT-CYLINDER ENGINES

Operation

The exhaust gas oxygen (EGO) sensor and the "tach" connection from the ignition coil are the same as those used in other MCU systems. A coolant mid-temperature switch and dual-temperature switch send electrical signals to the MCU module in relation to coolant temperature. These signals are used by the MCU module to control the air-fuel ratio, spark retard, thermactor air, and canister purge system. Three zoned vacuum switches send electrical signals to the MCU module in relation to manifold vacuum. When engine detonation occurs, the knock sensor signals the MCU module to retard the ignition timing. The MCU module can retard the ignition spark advance through the yellow–light green wire connected to the ignition module, as shown in Figure 8-59. Figure 8-60 shows the locations of the MCU components of eight-cylinder systems.

The thermactor pump and canister purge of the MCU system in eight-cylinder engines operate in the same way as in other MCU systems or in EEC II systems. The MCU module activates the throttle kicker solenoid, which applies vacuum to the throttle kicker actuator and increases the idle speed when any of the following conditions are present:

1. Engine coolant is below a specific temperature.
2. Engine coolant is above an overheat temperature.
3. Air-conditioning compressor clutch is engaged.

Air-Fuel Ratio Control

The MCU module operates the feedback carburetor actuator (FBCA) motor to control the air-fuel ratio. A metering orifice in the carburetor is opened and closed by the tapered valve on the FBCA motor shaft. The metering orifice controls the amount of air that enters the main system air bleed in the carburetor, as outlined in Figure 8-61.

The MCU module can energize four different windings in the FBCA motor to move the motor shaft and tapered metering valve through 120 steps, which provide a total shaft travel of 0.400 in (10.16 mm). When the MCU module receives a signal from the exhaust gas oxygen (EGO) sensor that indicates a rich air-fuel ratio, the module activates the FBCA motor and opens the tapered metering valve. This movement allows additional air into the main system and provides a leaner air-fuel ratio. A lean mixture signal from the EGO sensor to the MCU module causes the module to activate the FBCA motor and close the tapered metering valve to provide a richer air-fuel ratio. In the closed-loop mode, the MCU module controls the FBCA motor to provide an air-fuel ratio of 14.7:1.

DIAGNOSIS OF MICROPROCESSOR CONTROL UNIT (MCU) SYSTEMS

If the engine fails to start, the Dura-Spark II Ignition system used with the MCU system may be diagnosed as outlined in Chapter 6. The MCU module has the capability to diagnose the MCU system. When a performance or an economy complaint is being diagnosed, make sure all other possible causes are diagnosed before the MCU system. Other causes of performance and economy complaints could include: low engine compression, defective ignition components, and improper operation of the exhaust gas recirculation (EGR) system. A self-test connector is provided in the wiring harness of each MCU system. When the system is being diagnosed, a voltmeter and a jumper wire should be connected to the self-test connector, as shown in Figure 8-62.

Position the voltmeter switch on the 20-V scale. When the ignition switch is turned off, the voltmeter will indicate battery voltage. A STAR tester provided by the car manufacturer may be connected to the self-test connector in place of the voltmeter, as illustrated in Figure 8-63.

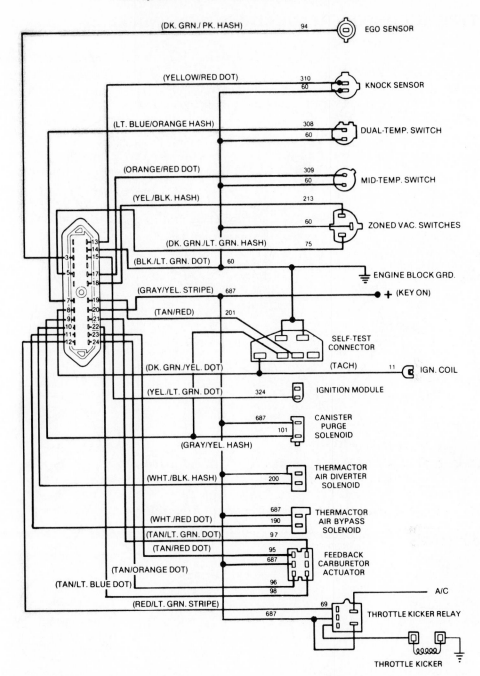

FIGURE 8-59. MCU System Wiring, Eight-Cylinder Engines. (Courtesy of Ford Motor Co.)

Ignition On Engine-Not-Running Functional Test

When the ignition switch is turned on, the voltmeter reading will drop to zero and within 30 seconds any service codes will be indicated by pulsations of the voltmeter needle. One pulsation followed quickly by a second pulsation on the voltmeter indicates a code 11. The service codes will be repeated twice. A digital reading for the service codes is provided on the STAR tester. The service codes for the different MCU systems are listed in Table 8-4.

When the ignition switch is turned on, if the voltmeter shows a steady reading or the STAR tester indicates 00, the self-test is not functional. Code 11 should always be displayed even if there are no de-

TYPICAL COMPONENT LOCATIONS (8-CYLINDER)

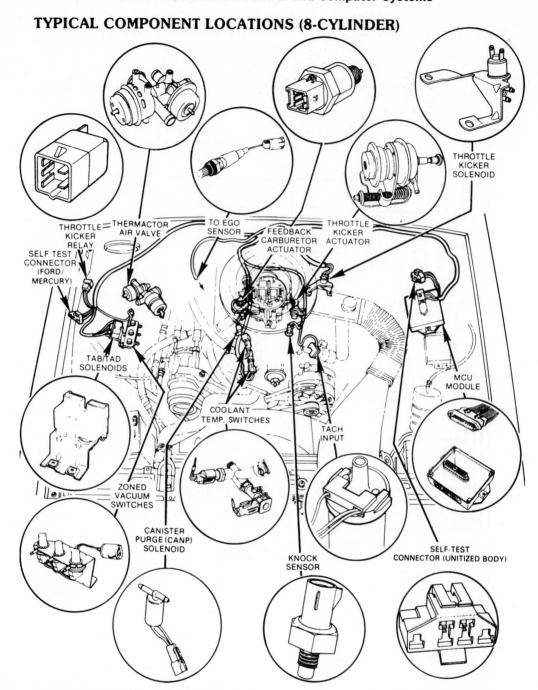

FIGURE 8-60. Locations of MCU Components in Eight-Cylinder Engines. (Courtesy of Ford Motor Co.)

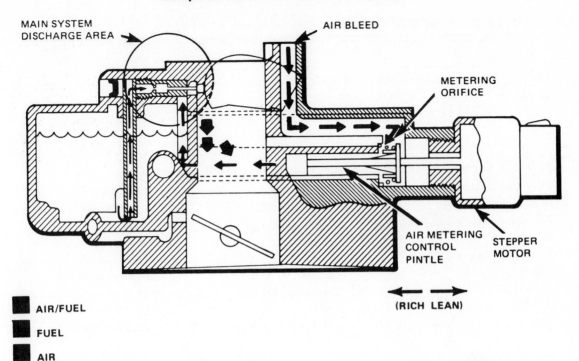

MAIN SYSTEM
DISCHARGE AREA

AIR BLEED

METERING
ORIFICE

AIR METERING
CONTROL
PINTLE

STEPPER
MOTOR

(RICH LEAN)

AIR/FUEL

FUEL

AIR

FIGURE 8-61. Feedback Carburetor Actuator (FBCA), Eight-Cylinder Engines. (Courtesy of Ford Motor Co.)

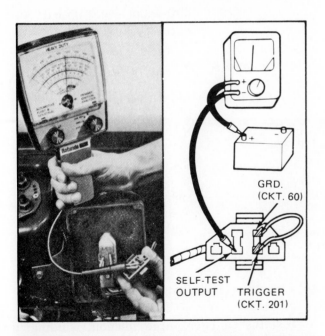

GRD.
(CKT. 60)

SELF-TEST
OUTPUT

TRIGGER
(CKT. 201)

FIGURE 8-62. Voltmeter Connections to Self-Test Connector. (Courtesy of Ford Motor Co.)

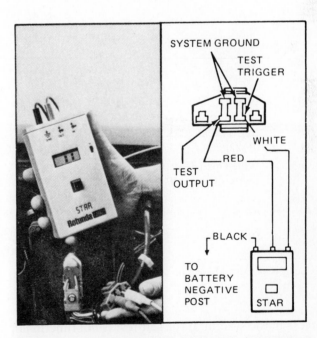

SYSTEM GROUND

TEST
TRIGGER

WHITE

RED

TEST
OUTPUT

BLACK

TO
BATTERY
NEGATIVE
POST

STAR

FIGURE 8-63. STAR Tester Connections to Self-Test Connector. (Courtesy of Ford Motor Co.)

TABLE 8-4. Service Codes for MCU Systems

Code	2.3L	4.9L	8-Cyl.
—	Self-Test Not Functional (P. 32)	Self-Test Not Functional (P. 44)	Self-Test Not Functional (P. 60)
11	System OK	System OK	System OK
12	N.A.	N.A.	Idle Speed Incorrect (P. 56)
25	N.A.	N.A.	Knock Detection Inoperative (P. 59)
33	RUN Test Not Initiated (P. 31)	RUN Test Not Initiated (P. 31)	N.A.
41	EGO Always Lean (P. 24)	EGO Always Lean (P. 38)	EGO Always Lean (P. 52)
42	EGO Always Rich (P. 26)	EGO Always Rich (P. 39)	EGO Always Rich (P. 54)
44	Thermactor Air System Problem (P. 28)	Thermactor Air System Problem (P. 40)	Thermactor Air System Problem (P. 57)
45	Thermactor Air Always Upstream (P. 30)	Thermactor Air Always Upstream (P. 42)	Thermactor Air Always Upstream (P. 58)
46	Thermactor Air Not Bypassing (P. 31)	Thermactor Air Not Bypassing (P. 43)	Thermactor Air Not Bypassing (P. 58)
51	LOW-Temp. Switch Open (P. 21)	LOW-Temp. Vacuum Switch Open (P. 36)	HI/LO Vacuum Switch(es) Open (P. 50)
52	Idle Tracking Switch Open (P. 22)	Wide-Open-Throttle Vacuum Switch Open (P. 37)	N.A.
53	Wide-Open-Throttle Vacuum Switch Open (P. 23)	Crowd Vacuum Switch Open (P. 37)	DUAL-Temp. Switch Open (P. 50)
54	N.A.	N.A.	MID-Temp. Switch Open (P. 50)
55	N.A.	N.A.	MID-Vacuum Switch Open (P. 50)
56	N.A.	Closed-Throttle Vacuum Switch Open (P. 37)	N.A.
61	N.A.	N.A.	HI/LO Vacuum Switch(es) Closed (P. 51)
62	Idle Tracking Switch Closed (P. 22)	Wide-Open-Throttle Vacuum Switch Closed (P. 37)	N.A.
63	Wide-Open-Throttle Vacuum Switch Closed (P. 23)	Crowd Vacuum Switch Closed (P. 37)	N.A.
65	N.A.	N.A.	Mid-Vacuum Switch Closed (P. 51)
66	N.A.	Closed-Throttle Vacuum Switch Closed (P. 37)	N.A.

Courtesy of Ford Motor Co.

fects in the system. If the self-test is not functional, the following items should be tested using the MCU wiring diagrams provided earlier in this chapter:

1. Check the ground wire from the MCU module to the self-test connector and the engine block.
2. Check the wire connected from the MCU module to the self-test connector.
3. Check the wire from the ignition coil "tach" terminal to the MCU module.
4. Test the canister purge solenoid winding; it should have a resistance of 50-100 Ω.
5. Check the wires from the canister purge solenoid to the MCU module.
6. Test for 12 V at the power supply wire from the battery positive to the MCU module.
7. If no defects have been found in steps 1–6, replace the MCU module.

Engine-Running Functional Test, 2.3L and 4.9L Engines

If the vehicle is equipped with a purge control valve on the charcoal canister, the purge hose to the canister should be disconnected and plugged during this test. The following diagnostic steps may be followed to complete the "engine running" tests:

1. Operate the engine until the coolant reaches normal temperature.

2. Stop and restart the engine. Increase the speed to 2,500–3,000 RPM within 20 seconds.

3. Initialization pulses will appear on the voltmeter or STAR tester. Two initialization pulses will be indicated on the voltmeter for 2.3L engines and three pulses for 4.9L engines. The STAR tester will read 20 on 2.3L engines and 30 on 4.9L engines. When the initialization pulses appear, hold the engine speed steady for 40 seconds. The initialization pulses will be followed by any service codes available, and then the initialization code will be repeated.

4. Return the throttle to idle speed and observe the voltmeter or STAR tester for additional service codes.

5. Reconnect the purge hose to the canister purge control valve.

The engine must be running before some service codes will be provided. The service codes will only be provided when the defect is present. A service code indicates a malfunction in a specific circuit. When a service code indicates a defective exhaust gas oxygen (EGO) circuit, the EGO sensor, connecting wire from the sensor to the MCU module, or the MCU module may be defective.

Engine-Running Functional Test, Eight-Cylinder Engines

The PCV valve should be removed from the rocker arm cover before proceeding with this test. The necessary steps in the test procedure are:

1. Operate the engine until the coolant reaches normal operating temperature. Operate the engine at 2,000 RPM for at least 2 minutes.

2. Stop and restart the engine. Operate the engine at idle speed and watch for initialization pulses on the voltmeter. The voltmeter should indicate four initialization pulses, or a STAR tester should display code 40.

3. If the system is equipped with a knock sensor, tap the engine casting near the sensor with a bar and hammer. Observe the voltmeter for additional service codes, which will be provided four seconds after the initialization code. When the initialization code is repeated, all the codes have been reported.

FORD ELECTRONIC ENGINE CONTROL IV (EEC IV) SYSTEMS IN 2.3L HIGH-SWIRL COMBUSTION (HSC) ENGINES

Some EEC IV systems use fuel injectors in place of a carburetor, while other versions of the system have a feedback control solenoid (FBCS) and a carburetor. Many similarities exist between EEC IV systems and other EEC systems. The differences between the EEC IV systems and the other EEC systems are outlined in this section.

Input Sensors

Most of the input sensors in the EEC IV system (all are shown in Figure 8-64) are similar to the input sensors in the EEC III systems. The only one that is different in the EEC IV system is the profile ignition pickup (PIP) sensor. A Hall Effect device is used in place of the distributor pickup coil, as illustrated in Figure 8-65.

A plate containing a number of metal tabs is attached to the distributor shaft. The number of tabs is matched to the number of engine cylinders. Each time a tab rotates through the Hall Effect device, a signal is sent to the electronic control assembly (ECA). The spark advance is determined by the ECA, and the conventional vacuum and centrifugal advances are not used in the EEC IV distributor. Ignition operation and diagnosis are discussed in Chapter 6.

Output Controls

Some of the output devices controlled by the ECA are the same as those used in other EEC or MCU systems. The output devices are also shown in Figure 8-64. The feedback control solenoid (FBCS) is basically the same as the one in the MCU system in 2.3L engines. (See Chapter 8 for a complete description of the FBCS.) A thick film integrated (TFI) ignition module is used in this system. The sig-

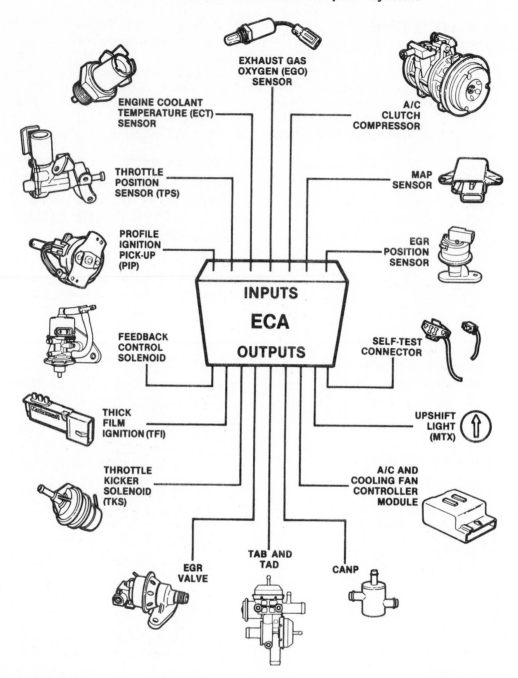

FIGURE 8-64. EEC IV Input Sensors and Output Controls. (Courtesy of Ford Motor Co.)

nal from the profile ignition pickup (PIP) is sent through the ECA to the TFI module. In this way, the ECA can control the spark advance.

The throttle kicker is the same as the one used in EEC II or EEC III systems. The ECA operates a throttle kicker solenoid (TKS), which in turn controls the vacuum that is applied to the throttle kicker actuator.

An exhaust gas recirculation EGR valve with an EGR valve position (EVP) sensor is used in the EEC IV system. This EVP is basically the same as that in the EEC II or EEC III systems. The EVP sensor and the EGR valve are serviced as two separate parts. The ECA controls the EGR valve by operating an EGR vent (EGRV) solenoid and EGR control (EGRC) solenoid.

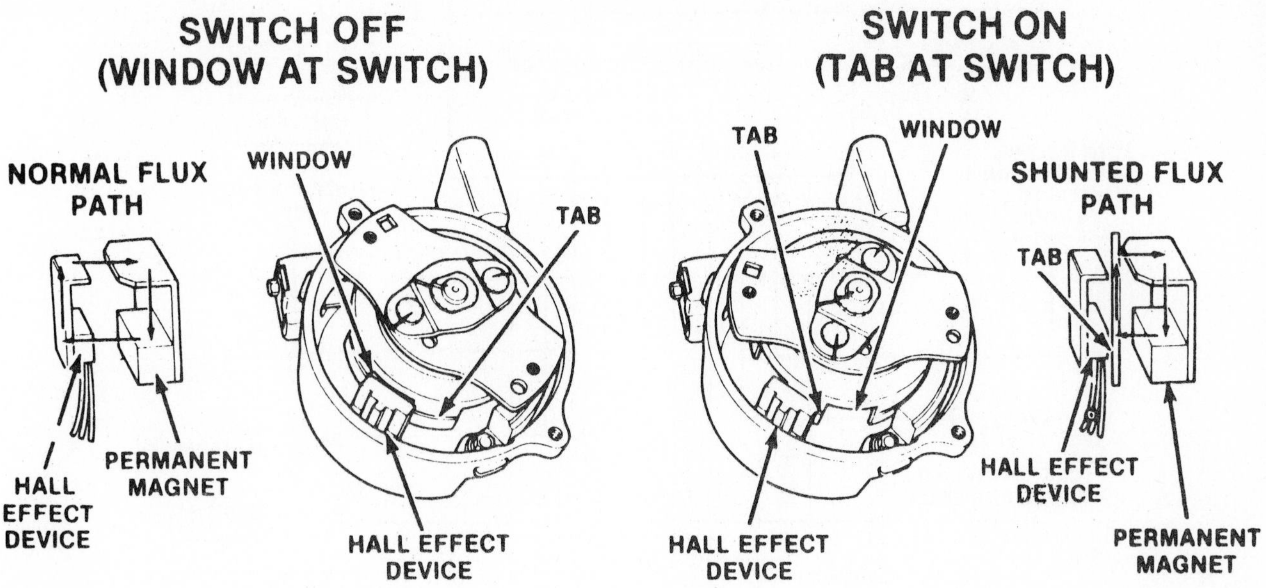

SWITCH OFF (WINDOW AT SWITCH)

SWITCH ON (TAB AT SWITCH)

FIGURE 8-65. Hall Effect Device. (Courtesy of Ford Motor Co.)

The ECA controls the airflow from the thermactor pump by operating thermactor air bypass (TAB) and thermactor air divert (TAD) solenoids. These solenoids control the vacuum that is applied to the TAB and TAD valves. Each time the engine is started, the TAB valve allows thermactor pump air to be directed to the atmosphere. Once the engine has been operating for a few seconds, the TAB valve directs the thermactor air to the TAD valve, which then directs thermactor airflow to the exhaust ports during open-loop operation and to the catalytic converter during closed-loop operation. (See Chapter 8 for a complete description of the thermactor air system in the EEC II system.)

The canister purge system is the same as the one used in the EEC II or EEC III systems.

Air Conditioning (A/C) and Cooling Fan Controller Module

A cooling fan and A/C controller module are used in the EEC IV system of the 2.3L engine to control the operation of the cooling fan and the A/C compressor clutch. The module is mounted under the right side of the instrument panel. Input signals are sent to the module from the coolant temperature switch, the ECA, and the stop lamp switch. The stop lamp switch signal is used only on vehicles that are equipped with an automatic transaxle (ATX). Output signals from the module control the engine cooling fan and the A/C compressor clutch, as illustrated in Figure 8-66.

On receiving a signal from the clutch cycling pressure switch, the module will activate the A/C compressor clutch. Whenever the throttle is wide open, the ECA sends a signal to the module that disables the signal to the compressor clutch. This WOT signal from the ECA also deenergizes the cooling fan motor if the engine coolant temperature is below 210°F (97°C). The coolant temperature switch applies a ground signal to the module if the coolant temperature exceeds 210°F (97°C). The ground signal overrides the WOT signal from the ECA and prevents the cooling fan from shutting off.

The ECA also contains a time-out feature for the WOT signal. After approximately 30 seconds, the WOT signal is stopped even if the WOT condition is still present. However, slow recycling from WOT to part throttle and back to WOT will again shut off the compressor clutch for 30 seconds.

When the brake pedal is applied on vehicles that are equipped with an automatic transaxle (ATX), an input signal is sent from the brake lamp switch to the module. In response, a disable signal is sent from the module to the cooling fan and the A/C compressor clutch for 3–5 seconds to prevent engine stalling on deceleration. When the compressor clutch and the cooling fan at WOT are disabled, the engine is provided with more power.

Control Modes

The *crank mode* is used by the ECA while the engine is being cranked or if the engine stalls and the ignition switch is left in the start position. When the

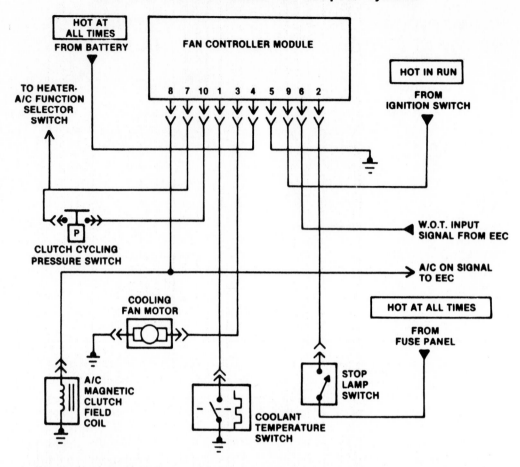

FIGURE 8-66. Cooling Fan and A/C Compressor Controller Module. (Courtesy of Ford Motor Co.)

engine is being cranked, fuel control is in open loop, and the ECA sets ignition timing at 10–15° before top dead center (BTDC). The EGR solenoids are not energized, and therefore the EGR valve is closed. Thermactor air is upstream to the exhaust ports and the canister purge system is off. The crank mode is a special program that is used to aid engine starting.

The *underspeed mode* is used to enrich the fuel and increase airflow to the engine to help it recover from a stumble. If the engine speed drops below 600 RPM, the ECA enters the underspeed mode. Operation in this mode is similar to the crank mode.

In the *closed-throttle mode*, the ECA evaluates the various input sensor signals and then determines the correct air-fuel ratio and ignition timing. As long as the exhaust gas oxygen (EGO) sensor signal is maintained, the system remains in closed loop. If the EGO sensor cools off and its signal is no longer available, the system goes into open-loop operation. The ECA activates the throttle kicker system if the engine coolant is above a predetermined temperature, or if the air conditioning is on. The

throttle kicker is also activated for a few seconds after the engine is started. During the closed-throttle mode, the EGR valve and the canister purge system are turned off. The closed-throttle mode is used during idling or deceleration.

The *part-throttle mode* is entered during moderate cruising speed conditions. In this mode, the system operates in closed loop as long as the EGO sensor signal is available. The throttle kicker is activated to provide a dashpot function when the engine is decelerated. The ECA activates the EGR solenoids and the EGR valve opens.

Power Relay Circuit

The power relay circuit is used to supply voltage to the ECA when the ignition switch is in the run or the start position. When the ignition switch is turned to the run position, the power relay winding is energized and the relay points close to supply power to terminal 57 on the ECA (see Figure 8-67).

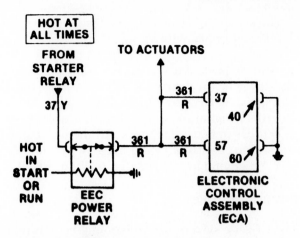

FIGURE 8-67. EEC IV Power Relay Circuit. (Courtesy of Ford Motor Co.)

The complete wiring of the EEC IV system used in the 2.3L engine is shown in Figure 8-68.

FORD ELECTRONIC ENGINE CONTROL IV (EEC IV) SYSTEMS ON 2.8L V6 ENGINES

Input Sensors

Many of the same sensors are used in the EEC IV system of the 2.8L V6 engine and the 2.3L engine. The knock sensor idle tracking switch (ITS), air charge temperature (ACT) sensor, and the neu-

FIGURE 8-68. EEC IV 2.3L Engine Wiring. (Courtesy of Ford Motor Co.)

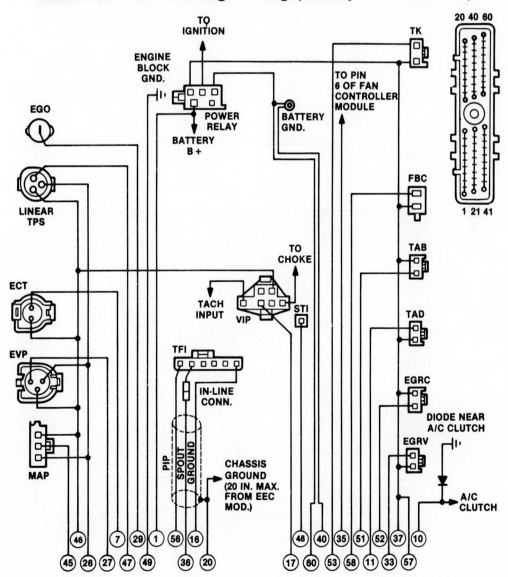

tral/start switch provide additional input signals on the 2.8L V6 engine. The knock sensor sends a signal to the ECA if the engine begins to detonate. When the ECA receives this signal, it retards the ignition spark advance. The amount of retard depends on the severity and duration of the detonation; the maximum amount of retard is 8°.

The idle tracking switch (ITS) is an integral part of the idle speed control motor. This switch signals the ECA when the throttle is closed. At this point, the ITS switch opens and the ECA operates the idle speed control motor to control the idle speed. When the throttle is opened, the ITS switch closes and the ECA no longer operates the idle speed control motor.

The air charge temperature (ACT) sensor is mounted in the air cleaner. The signal it sends to the ECA is based on air intake temperature. The ACT sensor signal is used by the ECA to control the choke opening when the air intake temperature is cold. Once the engine is warmed up, the ACT signal is used by the ECA to control idle speed. When the air intake temperature drops below 55°F (13°C), the ACT signal causes the ECA to direct thermactor air to the atmosphere.

The neutral/start switch signal is used in vehicles that are equipped with an automatic transmission. This switch is closed when the transmission is in neutral or park, and it is open in all other transmission selector positions. The ECA increases the idle speed 50 RPM to improve idle quality when it receives a signal indicating that the transmission selector is in neutral or park. All the input signals and output controls in the EEC IV system of the 2.8L V6 engine are illustrated in Figure 8-69.

Output Controls

The EEC IV system of the 2.8L engine and that of the 2.3L engine have similar output controls. Additional output controls that might be found on the 2.8L engine include the temperature-compensated accelerator pump (TCP) solenoid, variable voltage choke (VVC), and the idle speed control (ISC) motor. A different type of feedback control solenoid (FBS) is used in the 2150A two-barrel carburetor of the 2.8L V6 engine. The ECA operates the feedback control plunger, which controls the amount of airflow into the idle and main circuit air bleed passages, as pictured in Figure 8-70.

FIGURE 8-69. EEC IV Input Signals and Output Controls, 2.8L V6 Engine. (Courtesy of Ford Motor Co.)

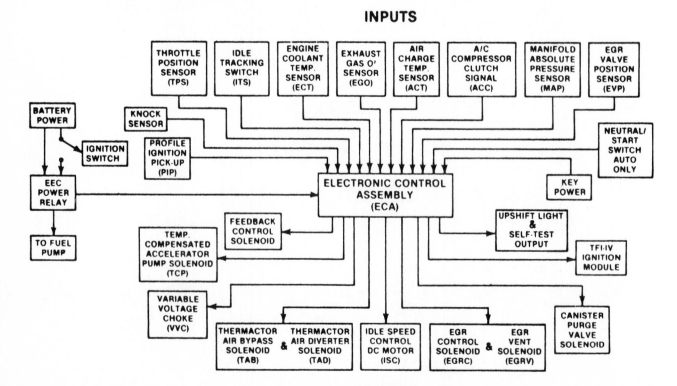

INPUTS

OUTPUTS

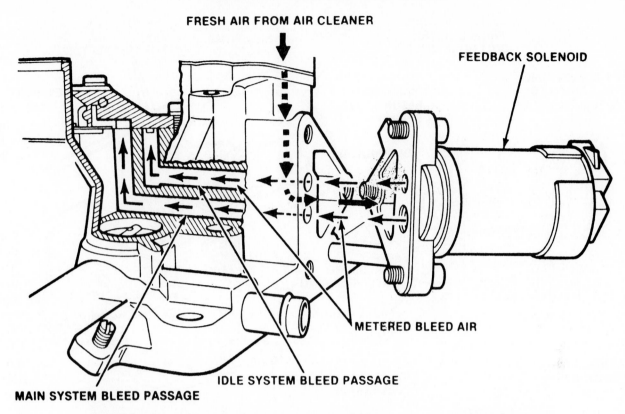

FRESH AIR FROM AIR CLEANER

FEEDBACK SOLENOID

METERED BLEED AIR

IDLE SYSTEM BLEED PASSAGE

MAIN SYSTEM BLEED PASSAGE

FIGURE 8-70. Feedback Control Solenoid (FBCS), 2.8L V6 Engine. (Courtesy of Ford Motor Co.)

When the exhaust gas oxygen (EGO) sensor signals the ECA that the air–fuel ratio is too rich, the ECA operates the FBCS plunger to allow more air into the idle and main circuits to provide a leaner air-fuel ratio. The ECA operates the FBCS to maintain an air-fuel ratio of 14.7:1 when the engine is at normal operating temperature and the system is operating in closed loop.

The temperature-compensated accelerator pump (TCP) allows delivery of a large pump capacity to improve cold-engine performance, and a smaller pump capacity during warm-engine operation. The TCP contains a bypass bleed valve that is operated by a vacuum diaphragm. A vacuum solenoid controlled by the ECA applies vacuum to the vacuum diaphragm in the TCP. When the engine coolant is cold, the vacuum solenoid shuts off the vacuum applied to the TCP and thus allows the bypass valve to remain closed. Under this condition, full accelerator pump capacity is delivered through the accelerator pump system. The TCP system is illustrated in Figure 8-71.

At normal engine coolant temperature, the ECA energizes the TCP solenoid and applies vacuum to the TCP diaphragm. The bypass valve then opens and some fuel is returned past the bypass valve to the float bowl each time the engine is accelerated.

The variable voltage choke (VVC) is an electric choke that is controlled by the ECA and a solid state power relay. A variable duty cycle is used by the ECA and power relay to control the opening of the choke. The VVC and the power relay are pictured in Figure 8-72.

As the air intake temperature decreases, the ECA and the power relay reduce the "on time" of the voltage supplied to the choke. As a result, the choke is able to remain on longer. When atmospheric temperature is above 80°F (27°C), the ECA and the power relay supply voltage to the choke continuously to assure that the choke remains open.

The idle speed control (ISC) motor is a direct current (DC) electric motor that is used to control idle speed after the choke is open. An ISC motor is shown in Figure 8-73.

A conventional fast idle cam that is operated by the choke spring provides the fast idle speed when the engine coolant is cold. The ISC motor will not move past preset limits, and it will not move if the idle tracking switch (ITS) signal to the ECA indicates that the throttle is not in the idle position.

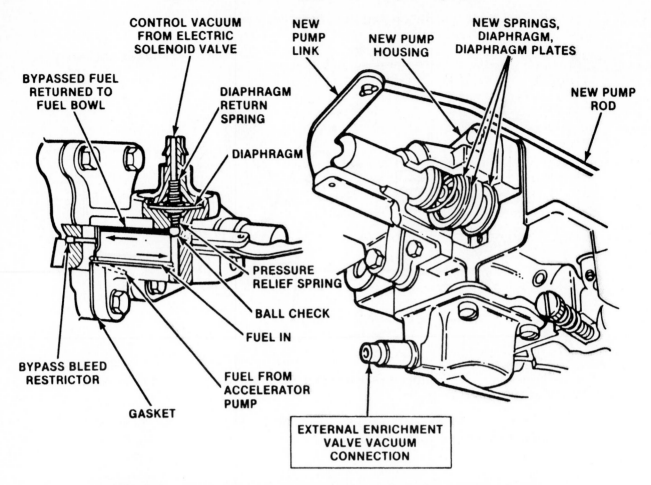

FIGURE 8-71. Temperature-Compensated Accelerator Pump (TCP). (Courtesy of Ford Motor Co.)

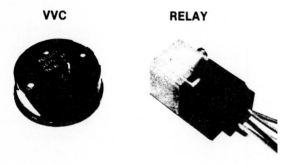

VVC   RELAY

FIGURE 8-72. Variable Voltage Choke (VVC) and Power Relay. (Courtesy of Ford Motor Co.)

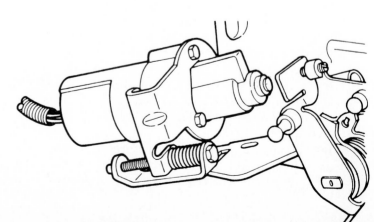

FIGURE 8-73. Idle Speed Control (ISC) Motor. (Courtesy of Ford Motor Co.)

When the engine is shut off, the ISC motor retracts and closes the throttle to prevent the engine from dieseling. After the engine stops running, the ISC motor extends to its maximum travel for restart purposes. A time-delay circuit in the power relay supplies voltage to the ECA for 10 seconds after the ignition switch is turned off. This allows the ECA to extend the ISC motor after the ignition switch is turned off. When the engine is started, the ISC motor extends for a brief period of time to provide a faster idle speed.

The amount of time the ISC motor is extended depends on coolant temperature. Under cruising speed conditions, the ISC motor extends to provide a dashpot action when the engine is decelerated. If the engine overheats, or if the battery voltage is low,

the ECA operates the ISC motor to increase the idle speed. The idle speed is also increased by the ISC motor if the air conditioner is turned on, or if the automatic transmission is placed in neutral or park.

Most of the solenoids and relays in the EEC IV system are located behind a plastic cover on the firewall or on the fender shield. The ECA is located in the passenger compartment under the right kick pad. The complete wiring of the EEC IV system used in the 2.8L V6 engine is illustrated in Figure 8-74.

Whenever a transmission shift to the next highest gear will provide optimum fuel economy, the upshift light in the instrument panel is illuminated by the ECA. This signal tells the driver to make the necessary shift. The upshift light is used on vehicles

FIGURE 8-74. EEC IV Wiring, 2.8L V6 Engine. (Courtesy of Ford Motor Co.)

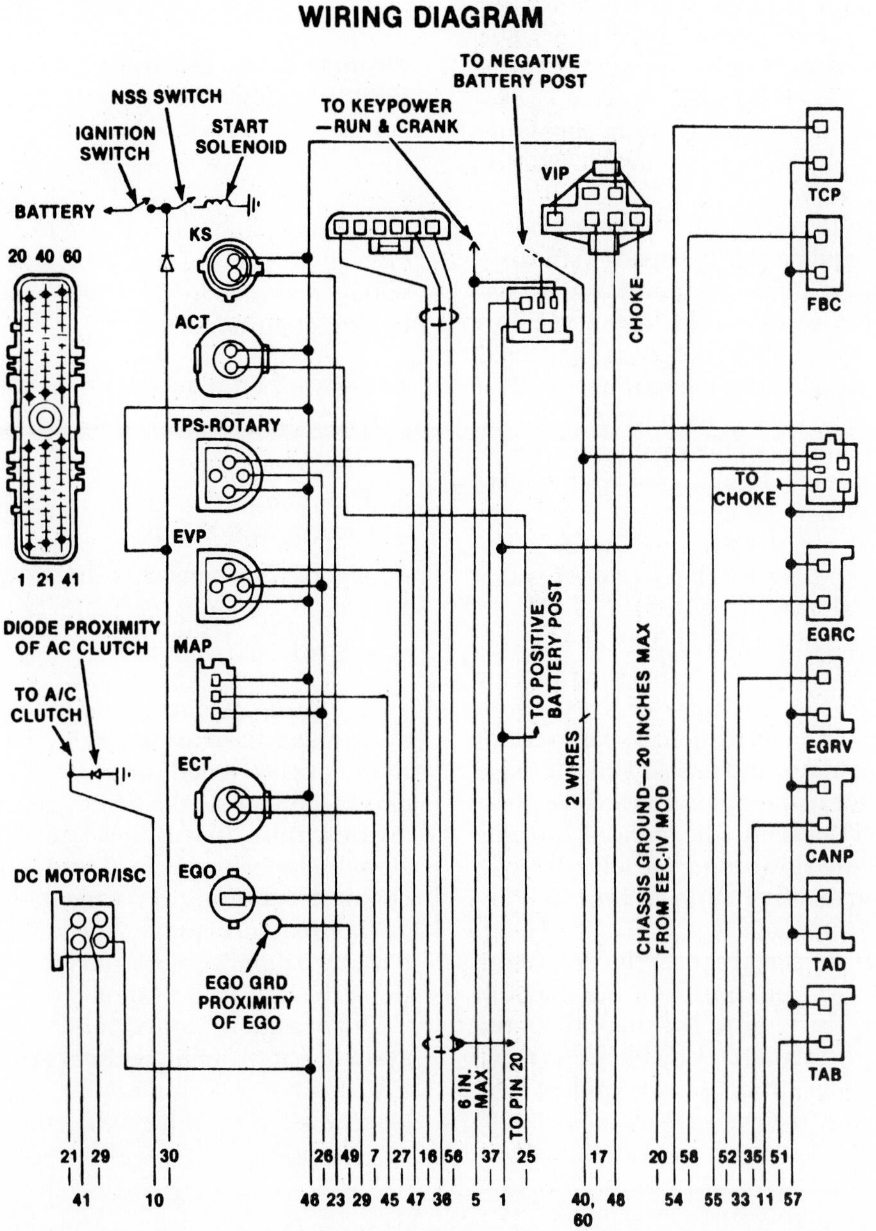

that are equipped with manual transmissions or manual transaxles. The ECA will not illuminate the upshift light when the transmission is in the top gear, or when the engine coolant is cold.

Since the diagnosis of EEC IV carbureted engines and fuel-injected engines is basically the same, this diagnosis is explained in Chapter 9.

CHRYSLER ELECTRONIC FEEDBACK CARBURETOR (EFBC) SYSTEMS

Operation

In the electronic feedback carburetor system (EFBC), the feedback carburetor controller is combined with the electronic spark control computer that is mounted on the side of the air cleaner. (See Chapter 7 for a description of the electronic spark advance system.) The electronic feedback carburetor controller activates the vacuum regulator winding ten times per second. Figure 8-75 illustrates the EFBC system.

The EFBC controller varies the "on time" of the vacuum regulator winding as it responds to the oxygen sensor signal. An input vacuum hose is connected from the intake manifold to the vacuum regulator, and the output vacuum hose is connected from the vacuum regulator to the carburetor. The vacuum regulator vent hose is connected to the air cleaner. Output vacuum from the vacuum regulator is used to control two diaphragms in the carburetor. One diaphragm is connected to a feedback-controlled idle air bleed and the second diaphragm operates the feedback-controlled main system fuel jet, as shown in Figure 8-76.

When the oxygen sensor signal indicates an excessively rich air-fuel ratio, the feedback carburetor controller increases the "on time" of the vacuum regulator winding, and consequently the vacuum output from the regulator increases. This increase moves the carburetor diaphragms upward, with the result that less fuel flows into the main carburetor system and more air flows into the idle system. Therefore, a leaner air-fuel ratio is provided regardless of engine speed.

A lean oxygen sensor signal causes the EFBC controller to reduce the "on time" of the vacuum regulator winding. Thus, output vacuum from the regulator is lower, the carburetor diaphragms move down, and the air-fuel ratio becomes richer. The output vacuum from the vacuum regulator will

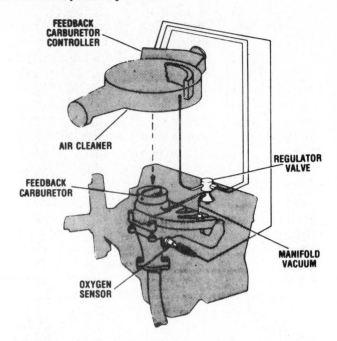

FIGURE 8-75. Electronic Feedback Carburetor (EFBC) System. (Courtesy of Chrysler Canada Ltd.)

vary between 1 and 4 in Hg (3.3–13.5 kPa) to provide an air-fuel ratio of 14.7:1. An open-loop mode will be provided by the EFBC system under the following conditions:

1. The coolant temperature is below 150°F (66°C).
2. The oxygen sensor temperature is below 660°F (350°C).
3. The manifold vacuum is low under wide-open throttle conditions.
4. The oxygen sensor is defective.
5. A few seconds have elapsed after a hot-engine restart.

In the open-loop mode, the EFBC controller maintains a fixed air-fuel ratio that is slightly richer than normal. When the system is operating in the closed-loop mode, the EFBC controller controls the air-fuel ratio in response to the oxygen sensor signal. The system is in closed-loop mode when the engine is operating at normal temperature and idle or lightload cruising conditions. Figure 8-77 illustrates the input sensors and the output control functions in the EFBC system.

Dual catalytic converters are used in the EFBC system. The air pump system is not operated by the ESC computer. A similar air pump system that uses an air-switching valve was discussed in Chapter 3.

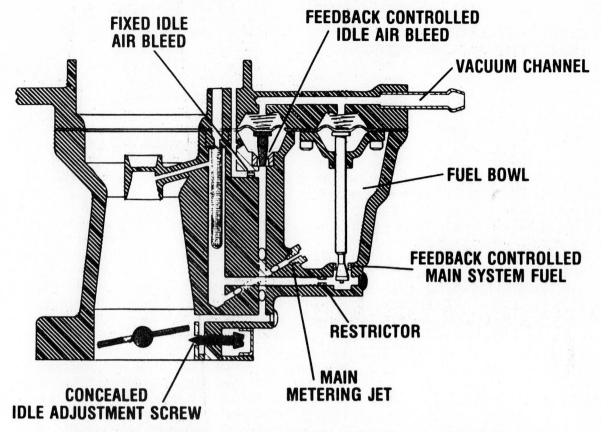

FIGURE 8-76. Electronic Feedback Carburetor Diaphragms. (Courtesy of Chrysler Canada Ltd.)

FIGURE 8-77. Electronic Feedback (EFBC) Carburetor System. (Courtesy of Sun Electric Corp.)

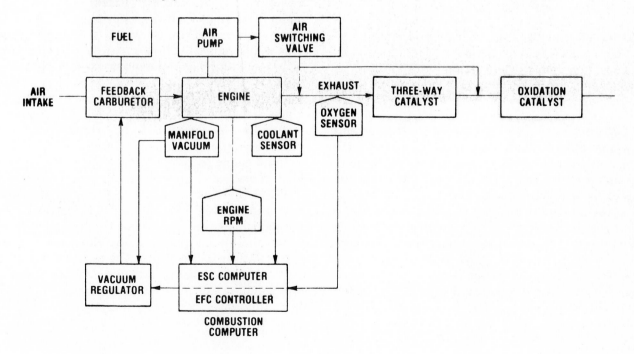

DIAGNOSIS OF ELECTRONIC FEEDBACK CARBURETOR (EFBC) SYSTEMS

When Control Vacuum Is above 4 in Hg (13.5 kPa)

A "T" fitting should be used to connect a vacuum gauge to the vacuum hose between the vacuum regulator and the carburetor. When the engine is operating at normal temperatures, the vacuum gauge reading should be oscillating between 1 and 4 in Hg (3.3–13.5 kPa). Each time the engine is started, a fixed vacuum gauge reading of 2.5 in Hg (8.4 kPa) is provided for approximately 100 seconds while the system is operating in open loop. If the vacuum gauge reading exceeds 4 in Hg (13.5 kPa), the EFBC system is trying to provide a leaner air-fuel ratio, and therefore an excessively rich oxygen sensor signal may be emitted. To locate the defective component, follow these tests:

Carburetor Test. Remove the positive crankcase ventilation (PCV) hose from the PCV valve and cover the end of the hose with your thumb. Gradually uncover the hose with the engine idling. If the control vacuum reading decreases, the carburetor is providing an excessively rich mixture. Therefore the carburetor must be replaced or repaired. If the vacuum gauge reading does not change when the PCV hose is uncovered, some other component in the system is defective.

Vacuum Regulator Test. When the electrical connector is disconnected from the vacuum regulator, the vacuum gauge reading should drop to zero. If the vacuum gauge reading does not drop to zero, the regulator is defective.

Electronic Feedback Carburetor Controller and Oxygen Sensor Test. Disconnect the oxygen sensor wire and hold the oxygen sensor wiring harness terminal with the fingers of one hand while the other hand touches the battery ground terminal. The vacuum gauge reading should slowly drop to zero within 15 seconds. If the vacuum gauge reading does not decrease to zero, replace the EFBC controller. If the vacuum gauge registers zero, the EFBC controller is operating normally and the oxygen sensor must be defective. The EFBC controller and electronic spark advance computer are replaced as a unit. When the oxygen sensor is being tested, the sensor wire must never be connected to ground or to the battery positive terminal. The test connections must always be completed at the wiring harness side of the oxygen sensor connector.

Diagnosis When Control Vacuum Is below 1 in Hg (3.3 kPa)

The vacuum gauge should be connected in the same way as it was in the previous tests. If the vacuum gauge reading is below 1 in Hg (3.3 kPa), follow these tests:

Carburetor Test. With the engine at normal operating temperature, position the fast idle speed screw on the lowest step of the fast idle cam. Gradually move the choke until it is 75 percent closed. If the vacuum gauge reading increases to 5 in Hg (16.9 kPa), the carburetor is providing an excessively lean mixture. Check all the possible sources of intake manifold vacuum leaks. If there are no vacuum leaks, the carburetor may be defective. Another cause of the control vacuum being less than 1 in Hg (3.3 kPa) could be a defective air-switching system that is pumping air into the exhaust ports at all times, which will provide a lean oxygen sensor signal. The air-switching system should direct air from the air pump to the exhaust ports when the engine coolant is cold, and the air should be pumped into the catalytic converter when the engine is at normal temperature.

Air-Switching System Test. If the air-switching system pumps air into the exhaust ports with the engine at normal operating temperature, an excessively lean oxygen sensor signal may be emitted. This would cause the EFBC controller to reduce the "on time" of the vacuum regulator winding, and the regulator control vacuum would be reduced to less than 1 in Hg (3.3 kPa) as the system tried to provide a rich air-fuel mixture. Disconnect the air injection hose at the rear of the cylinder head and plug the outlet from the cylinder head. If the vacuum gauge reading remains below 1 in Hg (3.3 kPa), replace or repair the carburetor.

When the vacuum gauge reading increases to over 5 in Hg (16.9 kPa), the air-switching valve or the coolant-controlled vacuum switch connected between the intake manifold and the air-switching valve is defective. Disconnect the vacuum hose from the coolant-controlled vacuum switch to the air-switching valve. If vacuum is present at the air-switching valve hose with the engine at normal temperature, the coolant-controlled vacuum switch is defective. The air-switching valve is defective if airflow from the air pump is directed through the valve to the cylinder head air outlet with the vacuum hose disconnected from the air-switching valve.

Vacuum Regulator Test. Disconnect the electrical connector from the regulator. Connect a jumper wire from the battery positive terminal to one regulator terminal, and connect the other regulator terminal to ground. If the control vacuum does not increase to 5 in Hg (16.9 kPa) or more, replace the vacuum regulator.

Electronic Feedback Carburetor Controller and Oxygen Sensor Test. Disconnect the oxygen sensor wire and hold the oxygen sensor wiring harness terminal with the fingers of one hand while the other hand touches the battery positive terminal. If the control vacuum increases gradually to 5 in Hg (16.9 kPa) or more within 15 seconds, replace the oxygen sensor. When the control vacuum does not increase, the EFC controller is defective. Never connect the oxygen sensor wire to the battery positive terminal or to ground. The test connections must be made at the wiring harness side of the oxygen sensor connector.

CHRYSLER OXYGEN FEEDBACK CARBURETOR SYSTEMS

Operation

The oxygen feedback carburetor system is similar to the electronic feedback carburetor (EFBC) system. In the oxygen feedback carburetor system, the vacuum regulator is discontinued and the electronic fuel control computer controls the "on time" of the carburetor feedback solenoid winding. When the oxygen sensor sends a signal to the electronic fuel control computer that indicates an excessively rich air-fuel ratio, the computer increases the "on time" of the carburetor feedback solenoid and provides a leaner mixture in the main carburetor system, as illustrated in Figure 8-78.

The return spring on the carburetor feedback solenoid holds the solenoid plunger in the upward position. When the electronic fuel control computer energizes the carburetor feedback solenoid winding, the solenoid plunger is held in the downward position, as indicated in Figure 8-79. If the oxygen

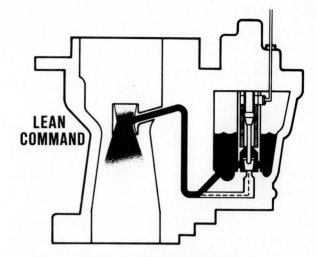

FIGURE 8-78. Carburetor Feedback Solenoid—Lean Command. (Courtesy of Chrysler Canada Ltd.)

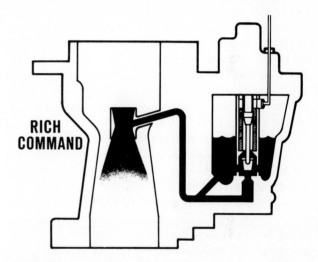

FIGURE 8-79. Carburetor Feedback Solenoid—Rich Command. (Courtesy of Chrysler Canada Ltd.)

sensor sends a lean air-fuel ratio signal to the electronic fuel control computer, the computer decreases the "on time" of the solenoid winding. This allows the solenoid plunger to remain in the upward position for a longer period of time and thus provides a richer air-fuel ratio, as shown in Figure 8-79.

The carburetor feedback solenoid also controls the airflow into the idle system. Downward movement of the solenoid plunger allows more air to flow into the idle system, and a leaner air-fuel ratio is thus provided at idle speed. The airflow into the idle system is reduced when the solenoid plunger is in the upward position, and therefore a richer air-fuel ratio occurs at idle speed. In the closed-loop mode, the electronic fuel control computer operates

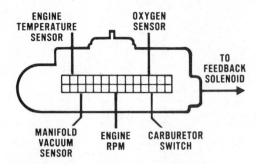

FIGURE 8-80. Computer Input Sensor Signals. (Courtesy of Chrysler Canada Ltd.)

the carburetor feedback solenoid in response to the oxygen sensor signal to provide an air-fuel ratio of 14.7:1. The electronic fuel control computer and the electronic spark advance computer are combined in one unit. The input sensor signals used by the combined computers are pictured in Figure 8-80.

The air pump system uses an air-switching valve to direct the airflow from the pump to the exhaust ports when the engine coolant is cold. The air-switching valve directs airflow from the air pump to the catalytic converters when the engine is at normal operating temperature. A coolant-controlled vacuum switch applies manifold vacuum to the air-switching valve when the engine coolant is cold. This type of air-switching system is also used on the electronic feedback (EFBC) carburetor system. (See Chapter 3.)

In some engines that are equipped with an oxygen feedback system, the electronic fuel control computer operates a vacuum solenoid winding to control the manifold vacuum that is applied to the air-switching valve. When the engine coolant is cold, the electronic fuel control computer energizes the vacuum solenoid winding, and thus allows vacuum through the solenoid to the air-switching valve. When manifold vacuum is applied to the air-switching valve, the airflow from the air pump is directed to the exhaust manifold, as shown in Figure 8-81.

FIGURE 8-81. Air-Switching Valve. (Courtesy of Chrysler Canada Ltd.)

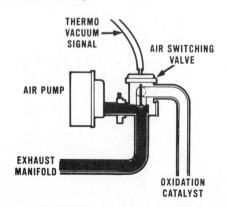

The electronic fuel control computer deenergizes the vacuum solenoid when the engine coolant is at normal operating temperature. Under this condition, the vacuum solenoid shuts off the vacuum applied to the air-switching valve, and the airflow from the pump is directed through the air-switching valve to the catalytic converter.

DIAGNOSIS OF OXYGEN FEEDBACK CARBURETOR SYSTEMS

When diagnosing an oxygen feedback carburetor system, make sure that all other sources of performance and economy complaints—such as the electronic spark advance system and the compression of the engine—are checked first.

Temperature Sensors

Oxygen feedback carburetor systems may use a dual-terminal thermistor temperature sensor, or a single-terminal sensor to send a coolant temperature signal to the electronic fuel control computer. A charge temperature switch may be mounted in the intake manifold; its signal to the computer will be based on the temperature of the air-fuel mixture. When an ohmmeter is connected across the terminals on the dual-terminal temperature sensor, the ohmmeter should read 500–1,000 Ω if the engine coolant is cold, or 1,300 Ω or more if the coolant is at the normal operating temperature, as shown in Figure 8-82.

If a single-terminal temperature switch is used, connect an ohmmeter from the switch terminal to

FIGURE 8-82. Dual-Terminal Temperature Sensor Test. (Courtesy of Chrysler Canada Ltd.)

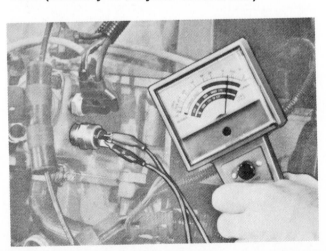

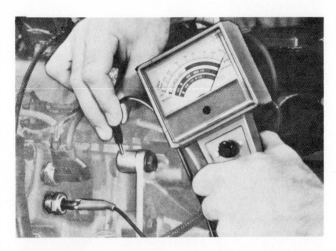

FIGURE 8-83. Single-Terminal Temperature Sensor Test. (Courtesy of Chrysler Canada Ltd.)

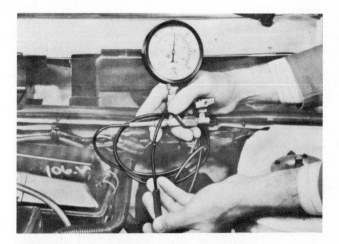

FIGURE 8-84. Charge Temperature Sensor Test. (Courtesy of Chrysler Canada Ltd.)

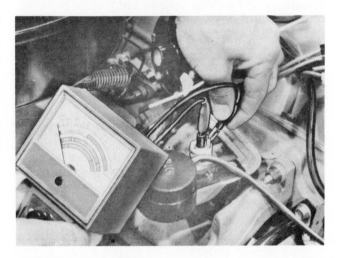

FIGURE 8-85. Vacuum Test at Air-Switching Valve. (Courtesy of Chrysler Canada Ltd.)

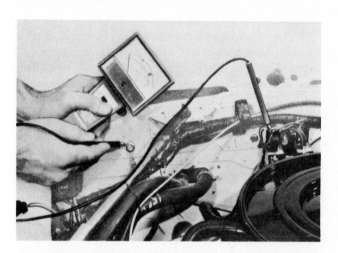

FIGURE 8-86. Air-Switching System Vacuum Solenoid Test. (Courtesy of Chrysler Canada Ltd.)

ground. The ohmmeter should read zero if the engine coolant is cold, as illustrated in Figure 8-83.

When the engine coolant is at the normal operating temperature, the ohmmeter reading should be infinite. A charge temperature sensor may be tested in the same way as the single-terminal coolant temperature sensor, as shown in Figure 8-84.

If the engine is cold, the ohmmeter should indicate zero ohms, and if the engine is at normal temperature, it should register an infinite reading.

Air Pump System

A computer-operated vacuum solenoid or a coolant-controlled vacuum switch may be used to switch the vacuum on and off between the intake manifold and the air-switching valve. In either case, the vacuum gauge should be connected to the air-switching valve vacuum hose with a "T" fitting. Operate the engine at idle speed and observe the vacuum gauge. If the engine is cold, the gauge should indicate manifold vacuum, as pictured in Figure 8-85.

When the engine coolant is at normal temperature, the vacuum gauge should read zero. If a coolant-controlled vacuum switch is used, the switch is defective if the vacuum does not drop to zero with the engine at normal operating temperature.

If the system has a computer-controlled vacuum switch, the computer or the vacuum switch is defective when the vacuum gauge reading does not decrease to zero with a warm engine. Without disconnecting the solenoid electrical terminal, connect a voltmeter from the green wire terminal at the solenoid to ground, as shown in Figure 8-86.

When the engine coolant is cold, the voltmeter reading should be less than 1 V when the engine is operating at idle speed. The computer should be replaced if the voltmeter reading exceeds 1 V. When the engine is at normal operating temperature, the voltmeter should register 12 V or more after the engine has been running for 100 seconds. If the voltmeter reading is not correct, disconnect the vacuum solenoid electrical connector and connect an ohmmeter across the solenoid terminals. If the ohmmeter registers below 20 Ω or above 30 Ω, replace the solenoid.

The air-switching valve may be tested by disconnecting the vacuum hose and the air outlet hoses from the switching valve. When a hand-operated vacuum pump is used to apply 16 in Hg (54 kPa) to the air-switching valve, as shown in Figure 8-87, airflow should be directed from the outlet port that is connected to the exhaust manifold.

When the vacuum pump is disconnected, airflow should be directed from the outlet port that is connected to the catalytic converter. If the air-switching valve does not operate as specified, the valve should be replaced.

Carburetor Feedback Solenoid

The following test procedure should be used to diagnose the carburetor feedback solenoid:

1. Connect a voltmeter from the blue wire that is connected to the carburetor feedback solenoid and ground, as illustrated in Figure 8-88. The voltmeter should indicate 12 V with the engine running. The blue wire is connected from the ig-

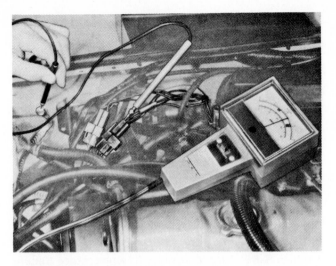

FIGURE 8-88. Carburetor Idle Stop Solenoid Test. (Courtesy of Chrysler Canada Ltd.)

nition switch to the feedback solenoid. In some systems a set of oil pressure switch contacts is connected in series with the feedback solenoid. If the voltmeter reading is less than specified, the oil pressure switch, or the blue wire, is defective.

2. Apply 16 in Hg (54 kPa) to the vacuum transducer on the computer. Operate the engine until it reaches normal temperature and install an insulator between the carburetor switch and the throttle linkage. The insulator must be thick enough to provide 2,000 RPM on four-cylinder engines, or 1,500 RPM on six- and eight-cylinder engines, as illustrated in Figure 8-89.

3. Disconnect the connector from the carburetor feedback solenoid. The engine speed should in-

FIGURE 8-87. Air-Switching Valve Test. (Courtesy of Chrysler Canada Ltd.)

FIGURE 8-89. Preparation for Carburetor Feedback Solenoid Test. (Courtesy of Chrysler Canada Ltd.)

FIGURE 8-90. Carburetor Feedback Solenoid Test Connector When Disconnected. (Courtesy of Chrysler Canada Ltd.)

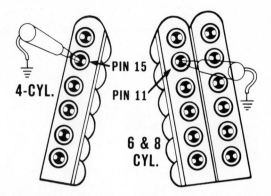

FIGURE 8-91. Computer Wiring Test. (Courtesy of Chrysler Canada Ltd.)

crease 50 RPM, as pictured in Figure 8-90, because the feedback solenoid will move upward and provide a richer air-fuel ratio.

4. With the blue wire connected to the feedback solenoid, connect a jumper wire from the green wire terminal on the solenoid to ground. The engine speed should decrease 100 RPM, compared with the engine speed with the connector disconnected, because the feedback solenoid will move to the downward position and provide a leaner air-fuel ratio. If the speed changes in steps 3 and 4 are not within specifications, the carburetor should be repaired or replaced.

Computer Wiring

In the case of four-cylinder engines, disconnect the six-terminal single connector and connect terminal 15 to ground with the engine idling. If the vehicle is equipped with a six- or eight-cylinder engine, disconnect the twelve-terminal double connector and connect terminal 11 to ground, as shown in Figure 8-91.

Terminal 15 on four-cylinder engines or terminal 11 on six- and eight-cylinder engines is connected through the green wire to the carburetor feedback solenoid. When either of these terminals is connected to ground, the feedback solenoid should move downward and the engine speed should decrease 50 RPM. If there is no change in engine speed and tests at the feedback solenoid were satisfactory, repair the open circuit or the grounded condition

in the green wire that is connected from terminal 11 or 15 to the feedback solenoid.

Fuel Control Computer

When testing the fuel control computer, operate the engine at normal temperature and 1,500 RPM on six- and eight-cylinder engines, or 2,000 RPM if the vehicle has a four-cylinder engine. Use a hand vacuum pump to apply 16 in Hg (54 kPa) to the vacuum transducer in the computer. Use a probe to connect the positive voltmeter lead to the green wire at the feedback solenoid, and connect the negative voltmeter lead to ground. Disconnect the oxygen sensor connector while placing one hand on the sensor wire from the computer and touching the negative battery terminal with the other hand, as illustrated in Figure 8-92.

When the oxygen sensor wire is grounded, the voltmeter should register over 9 V and the engine speed should increase 50 RPM. If there is no change in engine speed and the voltmeter reads less than 9 V, touch the oxygen sensor wire connected to the computer with one hand, and the positive battery terminal with the other hand. The engine speed should decrease at least 50 RPM and the voltmeter should read less than 3 V. If there is no change in voltage and engine speed, replace the computer.

The Oxygen Sensor

The engine speed and test conditions must be the same when testing the oxygen sensor or the fuel control computer. Reconnect the oxygen sensor wire and move the choke to the 75 percent closed position with the air cleaner removed. The computer should react within 10 seconds to provide a

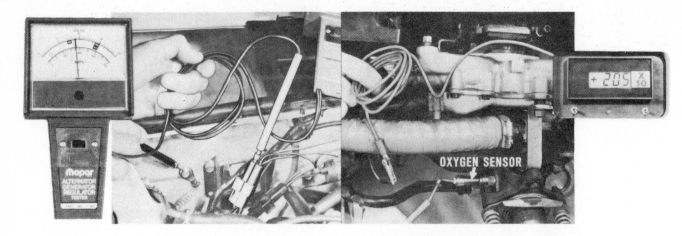

FIGURE 8-92. Fuel Control Computer Test. (Courtesy of Chrysler Canada Ltd.)

voltmeter reading of 3 V or less, as shown in Figure 8-93.

Reinstall the air cleaner and remove the positive crankcase ventilation (PCV) hose from the PCV valve. Cover the PCV hose with your thumb and slowly uncover the hose to create a large manifold vacuum leak. Within 10 seconds the computer should react to provide a voltmeter reading of 9 V or more. Each oxygen sensor test should be limited to 90 seconds. If the voltmeter readings are not correct, the oxygen sensor should be replaced.

FIGURE 8-93. Oxygen Sensor Test. (Courtesy of Chrysler Canada Ltd.)

CHRYSLER O_2 FEEDBACK WITH SELF-DIAGNOSTIC CAPABILITIES

Diagnosis

On 1985 and later model O_2 feedback systems, the computer has self-diagnostic capabilities. If certain defects occur in the system, fault codes are stored in the computer memory. When the system is being diagnosed, a digital readout tester must be connected to a diagnostic connector near the right front strut tower. This tester is also used to diagnose electronic fuel injection (EFI) systems. A complete wiring diagram of the O_2 feedback system with self-diagnostic capabilities is illustrated in Figure 8-94, and the readout tester connection is shown in Figure 8-95.

Fault Code Diagnosis

When a driveability complaint is being diagnosed, the engine compression and the ignition system should be checked prior to a diagnosis of the O_2 feedback system. The wiring harness connections and vacuum hoses should always be checked before the fault code diagnosis. Proceed with the fault code diagnosis as follows:

1. With the engine at normal operating temperature and the ignition switch in the off position, connect the digital readout tester to the diagnostic connection.

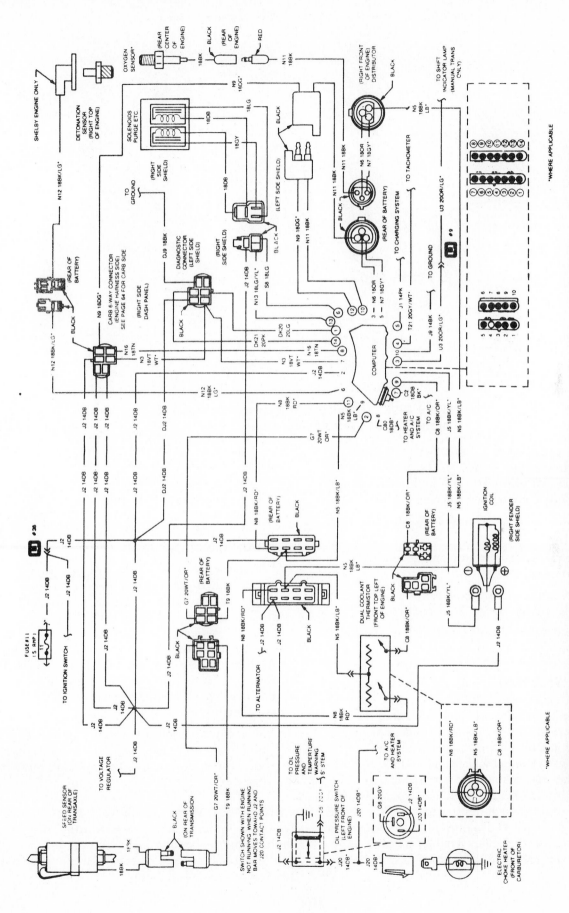

FIGURE 8-94. Chrysler O₂ Feedback System with Self-Diagnostic Capabilities. (Courtesy of Chrysler Canada Ltd.)

191

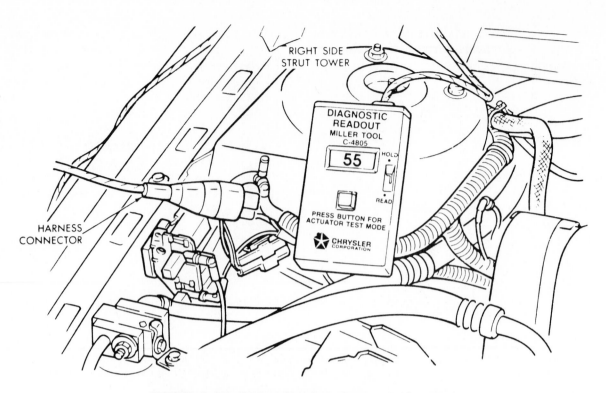

**FIGURE 8-95. Digital Readout Tester Connection.
(Courtesy of Chrysler Canada Ltd.)**

2. Place the fast idle screw on the highest step of the fast idle cam to open the carburetor switch.

3. Move the read/hold tester switch to the read position.

4. Turn on the ignition switch and do not cycle the switch on and off.

5. When 00 appears on the tester display, set the read/hold switch to the hold position.

6. Record all the fault codes indicated on the tester display.

Actuation Test Mode. A fault code indicates a defect in a certain circuit. For example, code 17 proves that a problem exists in electronic throttle control solenoid circuit. When this code is provided, it indicates a defect in this solenoid, or in the connecting wires between the solenoid and the computer. This code may also indicate that the computer is not capable of operating the electronic throttle control solenoid. The actuation test mode may be used to diagnose a certain component in the system when a fault code indicates a defect in this component. Proceed with the actuation test mode as follows:

1. When the fault code diagnosis is completed, the tester will display 55. At this time, press and hold the actuation test mode button on the tester, which causes different actuation test mode (ATM) codes to be displayed on the tester.

2. When the desired ATM code appears on the tester, release the ATM button. This action will cause the computer to operate the component indicated by the ATM code. For example, if a fault code 17 indicates a problem in the electronic throttle control solenoid, ATM code 97 may be used to activate this solenoid. In the ATM mode the technician can listen to the component or perform specific voltage tests to locate the exact defects.

3. The computer operates each component indicated by an ATM code for 5 minutes. Each ATM code test is stopped if the ignition switch is turned off.

Switch Test Mode. Certain switches in the O_2 feedback system can be tested in the switch test mode. This test mode must be entered after the fault code test mode. The procedure for the switch test mode is the following:

192

1. When 55 is displayed on the tester at the end of the fault code diagnosis, turn off a switch input to the computer such as the air conditioning switch.

2. Press the ATM button and immediately set the read/hold switch to the read position.

3. When 00 appears on the tester display, turn on the air conditioning switch.

4. If this switch input is received by the computer, 88 will appear on the display. The display will return to 00 when the air conditioning switch is turned off. Each computer input switch may be tested in the same way.

Fault codes may be erased from the computer memory by disconnecting the 14 terminal connector for 10 seconds with the ignition switch off.

A sensor test mode is available on 1986 and later model Chrysler O_2 feedback systems. The sensor test mode is the same on O_2 feedback and Chrysler EFI systems. To access the sensor test mode, press the ATM button at the end of the fault code diagnosis with the read/hold switch in the read position. When the desired sensor test code appears, release the ATM button and move the read/hold switch to the hold position. This action changes the ATM code to a sensor test code, and the tester will now display the output of the desired sensor. The same numbers are used for ATM codes and sensor test codes, as shown in the list of codes in Figure 8-96.

FIGURE 8-96. Diagnostic Codes for O_2 Feedback System. (Courtesy of Chrysler Canada Ltd.)

Code	Type	Engine	Circuit	When Monitored By The Computer	When Put Into Memory	ATM Test Code	Sensor Test Code
11	Fault	3.7 - 5.2 with O_2 Feedback	Carburetor O_2 Solenoid	All the time when the engine is running.	If the O_2 solenoid circuit does not respond to computer commands.	91	None
12	Fault	3.7 - 5.2 Federal	Transmission unlock relay circuit	All the time when the engine is running.	If the Transmission unlock relay does not respond to computer commands.	92	None
13	Fault	3.7 - 5.2	Air Switching Solenoid	All the time when the engine is running.	If the air switching solenoid circuit does not turn on and off at the correct time.	93	None
14	Indication	3.7 - 5.2	Battery Feed for Computer Memory	All the time when the ignition switch is on.	If the battery was disconnected within the last 20-40 engine starts.	None	None
17	Fault	3.7 - 5.2	Electronic Throttle Control Solenoid	All the time when the engine is running.	If the throttle circuit control solenoid does not turn on and off at the correct time.	97	None
18	Fault	3.7 - 5.2	EGR or Canister Purge Control Solenoid	All the time when the engine is running.	If the EGR control or canister purge solenoid circuit does not turn on and off at the correct time.	98	None
21	Fault	3.7 - 5.2	Distributor Pickup Coil	During engine cranking.	If there is no distributor signal input at the computer.	None	None
22	Fault	3.7 - 5.2 with O_2 Feedback	Oxygen Feedback System	All the time when the engine is running in closed loop operation.	If the oxygen feedback system stays rich or lean longer than 5 minutes during closed loop operation.	None	None
24	Fault	3.7 - 5.2	Computer	All the time when the engine is running.	If the vacuum transducer fails.	None	91
25	Fault	3.7	Charge Temperature Switch	All the time when the engine is running.	If the charge temperature switch is open with the temperature below 35°F as determined by the engine coolant sensor.	None	93

Code	Type	Engine	Circuit	When Monitored By The Computer	When Put Into Memory	ATM Test Code	Sensor Test Code
26	Fault	3.7	Engine Coolant Sensor	All the time when the engine is running.	If the engine coolant or charge temperature sensor circuit does not read 100°F after 30 minutes from when the engine started. Also, if the circuit is shorted or changes too fast to be real.	None	92
		5.2	Charge Temperature Sensor.				
28	Fault Manual Trans Only	3.7 without O$_2$ Feedback	Speed Sensor	During vehicle deceleration from 2000 to 1800 rpm for 3 seconds, engine temperature above 150°F and vacuum above 21.5 inches.	If speed sensor circuit does not indicate between 2 and 150 mph.	None	96
31	Fault	3.7 - 5.2	Battery Feed for Computer Memory	All the time when the ignition switch is on.	If the engine has not been cranked since the battery was disconnected.	None	None
32 33	Fault	3.7 - 5.2	Computer	Upon entry into the diagnostic mode.	If computer fails.	None	None
55	Indication	3.7 - 5.2			Indicates end of diagnostic mode.	None	None
88	Indication	3.7 - 5.2			Indicates start of diagnostic mode. **NOTE:** This code must appear first in the diagnostic mode or fault codes will be inaccurate. Indicates switch is on in switch test mode.	None	None
00	Indication	3.7 - 5.2			Indicates that the diagnostic readout box is powered up. Indicates switch is off in switch test mode.	None	None

FIGURE 8-96 (continued). Diagnostic Codes for O$_2$ Feedback System. (Courtesy of Chrysler Canada Ltd.)

QUESTIONS

Catalytic Converters

1. A rodium catalytic converter reduces _____ _____ levels.

2. A three-way catalytic converter requires an air-fuel ratio of _____ in order to control hydrocarbon (HC), carbon monoxide (CO), and nitrous oxide (NOx) emission levels.

Computer Command Control Systems

1. When the air-fuel ratio becomes richer, the oxygen sensor voltage _____ .

2. Leaded gasoline may be used in a vehicle that has an oxygen sensor in the exhaust system.

     T   F

3. If the electronic control module (ECM) increases the "on time" of the mixture control solenoid winding, the air-fuel ratio will become _____ .

4. The ECM controls the spark advance while the engine is being started.   T   F

5. Airflow from the air pump should be directed to the catalytic converter when the engine is operating at normal temperature and under light-load cruising speed conditions.   T   F

6. If the check-engine light comes on when the engine is running, it indicates a defect in the 3C system.   T   F

7. When basic ignition timing is being checked, the _____ must be disconnected.

8. If the 3C system is working normally, the mixture control solenoid dwell reading will vary continuously between 45° and 55°.   T   F

9. A mixture control solenoid dwell reading that is fixed at 55° indicates that the 3C system is trying to provide a _____ air-fuel ratio.

10. The distance that the mixture control solenoid plunger can travel is adjustable.   T   F

Electronic Engine Control II (EEC II) Systems

1. The EEC II system controls the carburetor air-fuel ratio.   T   F

2. The crankshaft position (CP) sensor is located at the _____ of the crankshaft.

3. When the atmospheric pressure in the float bowl is reduced, the air-fuel ratio will become _____.

4. The exhaust gas recirculation (EGR) valve is opened by _____.

5. When the engine coolant is cold, thermactor pump airflow will be directed downstream to the catalytic converters.   T   F

6. A defective electronic control (ECA) assembly could result in a "no start" complaint. T   F

Microprocessor Control Unit (MCU) Systems

1. The MCU module controls distributor spark advance.   T   F

2. In the MCU system used in the 2.3L engine, the MCU module controls a fuel control solenoid in the carburetor.   T   F

3. Airflow into the main system air bleed is controlled by the movement of the tapered valve on the feedback carburetor actuator (FBCA) motor.   T   F

4. When the oxygen sensor signal indicates that the air-fuel ratio is too rich, the MCU module will operate the FBCA motor to allow _____ air into the main system.

5. When the self-diagnostic system is being used, a STAR tester or a _____ may be connected to the self-test connector.

6. If a code 11 is received in the self-diagnostic procedure, the MCU system is satisfactory.   T   F

Electronic Feedback Carburetor (EFBC) Systems

1. The feedback carburetor controller is combined with the _____ in the EFBC system.

2. When the feedback carburetor controller increases the "on time" of the vacuum regulator winding, the vacuum output from the regulator will _____.

3. A decrease in the vacuum output from the vacuum regulator will cause the air-fuel ratio to become _____.

4. The system operates in the closed-loop mode when the engine coolant is 100°F (38°C). T   F

5. If the float level in the float bowl is excessively high, continuous vacuum regulator output could drop below 1 in Hg (3.3 kPa).   T   F

6. The wire attached to the oxygen sensor may be connected to the positive battery terminal when the system is tested.   T   F

Oxygen Feedback Carburetor Systems

1. The electronic fuel control computer controls the "on time" of the _____ _____.

2. When manifold vacuum is applied to the air-switching valve, the airflow from the air pump is diverted to the _____.

3. The manifold vacuum that is applied to the air-switching valve is controlled by a coolant-controlled vacuum switch or a _____.

9

COMPUTER-CONTROLLED FUEL INJECTION SYSTEMS

GENERAL MOTORS THROTTLE BODY INJECTION SYSTEMS

Throttle Body Injection Assemblies

In a throttle body injection system, the carburetor is replaced with a throttle body injector assembly. This assembly may contain a single injector, as illustrated in Figure 9-1. When a single injector is used, the injection system is referred to as electronic fuel injection (EFI) or throttle body injection (TBI).

Some larger displacement engines have dual injectors in the throttle body assembly, as shown in Figure 9-2.

Such systems are referred to as digital fuel injection (DFI) or digital electronic fuel injection (DEFI). Other throttle body injection systems use a pair of single injectors that are spaced in an offset position on a special intake manifold, as pictured in Figure 9-3.

FIGURE 9-1. Throttle Body Injection (TBI) Assembly. (Courtesy of GM Product Service Training, General Motors Corporation)

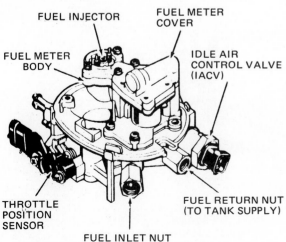

FUEL INJECTOR

FUEL METER COVER

FUEL METER BODY

IDLE AIR CONTROL VALVE (IACV)

THROTTLE POSITION SENSOR

FUEL RETURN NUT (TO TANK SUPPLY)

FUEL INLET NUT (FROM FUEL PUMP AND TANK SUPPLY)

FIGURE 9-2. Digital Fuel Injection (DFI), Throttle Body Injector Assembly. (Reference taken from SAE Paper No. 800164. Reprinted with Permission © 1980 Society of Automotive Engineers)

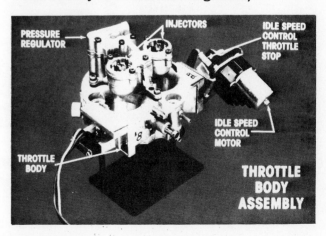

PRESSURE REGULATOR

INJECTORS

IDLE SPEED CONTROL THROTTLE STOP

IDLE SPEED CONTROL MOTOR

THROTTLE BODY

THROTTLE BODY ASSEMBLY

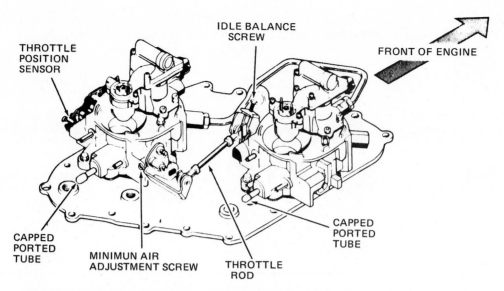

FIGURE 9-3. Crossfire Injection (CFI) Throttle Body Injector Assembly. (Courtesy of Chevrolet Motor Division, General Motors Corporation)

When offset single injectors are used, the injection system is called crossfire injection (CFI).

Regardless of the type of throttle body injector assembly used, the internal structure is similar: Gasoline is forced into the fuel inlet from the fuel pump, and then flows past the injector(s) to the pressure regulator. When fuel pressure reaches 10 PSI (69 kPa), the pressure regulator spring compresses, the pressure regulator valve opens, and any excess fuel is returned to the fuel tank. Pressurized fuel surrounds the injectors at all times, as illustrated in Figure 9-4.

A return spring on the injector plunger holds the injector closed. The injector opens when the solenoid coil surrounding the plunger is energized by the electronic control module (ECM). When the injector opens, fuel sprays into the airstream above the throttles, as shown in Figure 9-5.

FIGURE 9-4. Internal Design of Throttle Body Injector Assembly. (Courtesy of GM Product Service Training, General Motors Corporation)

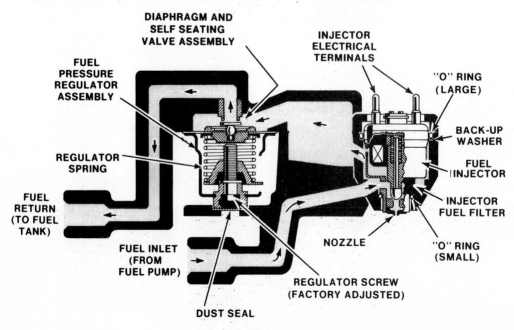

198

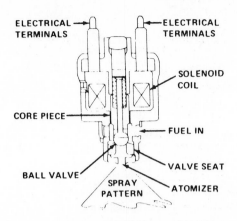

FIGURE 9-5. Injector Assembly. (Courtesy of GM Product Service Training, General Motors Corporation)

Complete Fuel System

A twin turbine fuel pump is mounted on the fuel tank with the gauge-sending unit. The fuel pump forces fuel through the filter under the vehicle to the throttle body injector assembly. Excess fuel from the injector assembly flows through the return fuel line to the fuel tank, as illustrated in Figure 9-6.

The ECM energizes the fuel pump relay winding when the ignition switch is turned on, and the relay points close. Thus, battery voltage is allowed to go through the relay points to the fuel pump, which then supplies fuel to the injectors. If no attempt is made to start the vehicle within 2 seconds from the time the ignition switch is turned on, the ECM de-energizes the fuel pump relay winding and the relay points open the circuit to the fuel pump. The ECM energizes the fuel pump relay winding as long as the engine is running. An alternate circuit is provided through a special set of oil pressure switch contacts to the fuel pump, as illustrated in Figure 9-7.

If the fuel pump relay becomes defective, power is still supplied through the oil pressure switch contacts to the fuel pump so that it is still possible to drive the vehicle.

Throttle Body Injection (TBI) System Components

The electronic control module (ECM) is the "brain" of the TBI system. The ECM receives information from the sensors and performs the necessary con-

FIGURE 9-6. Complete Throttle Body Fuel System. (Courtesy of GM Product Service Training, General Motors Corporation)

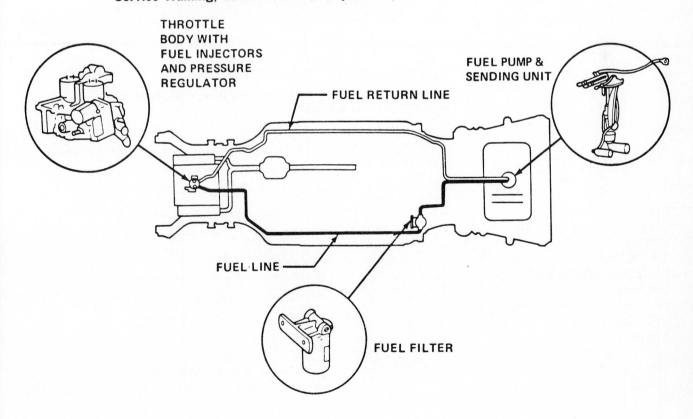

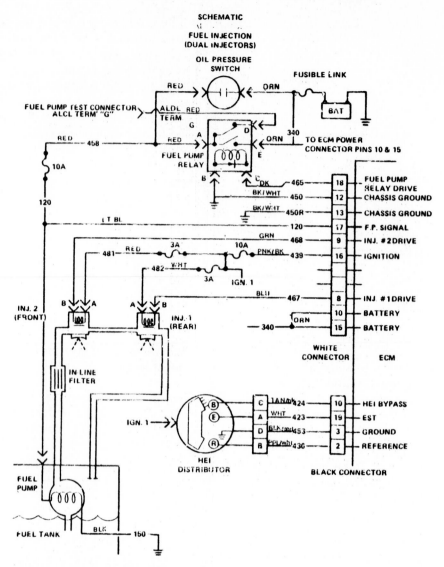

FIGURE 9-7. Wiring of Electronic Fuel Pump System. (Courtesy of GM Product Service Training, General Motors Corporation)

trol functions. A removable programmable read-only memory (PROM) is installed in the ECM, as pictured in Figure 9-8.

The PROM contains calibration data for each engine, transmission, and rear axle ratio. This information is always retained in the permanent memory of the PROM. Special tools are available for PROM removal and replacement. The technician must be careful to install the PROM in the correct position, and to be sure the pins on the PROM are not bent during installation. An alignment notch is used to position the PROM correctly, as shown in Figure 9-8.

The most important electronic components of the TBI systems are:

1. Manifold absolute pressure (MAP) sensor
2. Barometric pressure (Baro) sensor
3. Throttle position sensor (TPS)
4. Coolant temperature sensor (CTS)
5. Manifold air temperature (MAT) sensor
6. Oxygen (O_2) sensor
7. Distributor reference pulses
8. Vehicle speed sensor (VSS)
9. Park/neutral (P/N) switch
10. Air conditioning clutch switch

The input sensors are basically the same as the sensors used in the computer command control (3C)

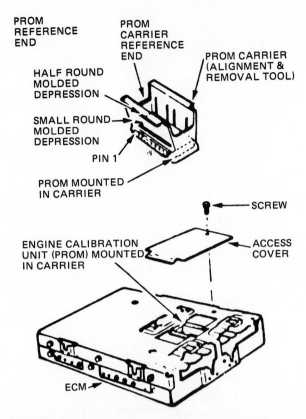

PROM
REFERENCE
END

PROM
CARRIER
REFERENCE
END

PROM CARRIER
(ALIGNMENT &
REMOVAL TOOL)

HALF ROUND
MOLDED
DEPRESSION

SMALL ROUND
MOLDED
DEPRESSION

PIN 1

PROM MOUNTED
IN CARRIER

SCREW

ENGINE CALIBRATION
UNIT (PROM) MOUNTED
IN CARRIER

ACCESS
COVER

ECM

FIGURE 9-8. Electronic Control Module (ECM) with Programmable Read-Only Memory (PROM). (Courtesy of GM Product Service Training, General Motors Corporation)

system. (See Chapter 8 for a complete description of the 3C system.) All the components listed above may not be used in every throttle body injection system.

Electronic Control Module (ECM) Control Functions

Fuel. The ECM maintains the correct air-fuel ratio of 14.7:1 by controlling the time that the injectors are open. The ECM varies the open time or "on time" of the injectors from 1 to 2 milliseconds at idle or 6 to 7 milliseconds at wide-open throttle. As the "on time" of the injector increases, the amount of fuel injected increases proportionately.

The injectors are operated by the ECM in two different modes: synchronized, and nonsynchronized. In the synchronized mode, the injector is energized by the ECM each time a distributor reference pulse is received by the ECM. On systems with dual throttle body injectors, the injectors are pulsed alternately. In the nonsynchronized mode of opera-

tion, the injectors are energized every 12.5 or 6.5 milliseconds, depending on the application. This pulse time is controlled by the ECM and it is independent of distributor reference pulses. The nonsynchronized pulses take place under the following conditions:

1. When the injector "on time" becomes less than 1.5 milliseconds.
2. During the delivery of prime pulses that are used to charge the intake manifold with fuel when the engine is being started or just before starting.
3. During acceleration, when a richer air-fuel mixture is required.
4. During deceleration, when a leaner air-fuel mixture is required.

If the engine coolant temperature is cold, the ECM will deliver "prime pulses" to the injector while the engine is being cranked and the throttle is less than 80 percent open. This system eliminates the need for a conventional choke. As the coolant temperature decreases, the ECM increases the "on time" or pulse width of the injectors. At $-33°F$ $(-36°C)$ coolant temperature, the ECM increases the injector pulse width to provide an air-fuel ratio of 1.5:1 for cold-start purposes. If dual throttle body injectors are used, the ECM may energize both injectors simultaneously during the prime pulses.

Should a cold engine become flooded, the ECM has the capability to clear this condition. To clear a flooded engine, the driver must depress the accelerator pedal to the wide-open position. The ECM then reduces the injector pulse width to deliver an air-fuel ratio of 20:1. This ratio is maintained as long as the throttle is held wide open and the engine speed is below 600 RPM.

When the engine is running at a speed above 600 RPM, the system enters the open-loop mode. In this mode the ECM ignores the signal from the oxygen (O_2) sensor and calculates the injector pulse width on the basis of the input signals from the coolant temperature sensor (CTS) and the manifold absolute pressure (MAP) sensor. During the open-loop mode, the ECM analyzes the following information before it enters the closed-loop mode:

1. Oxygen sensor voltage output must be present.
2. Temperature of the CTS sensor (it must be at or above normal).
3. The time that has elapsed from engine start-up (a specific period is required.)

When the ECM is satisfied with the input signals, the system enters the closed-loop mode. In this mode the ECM modifies the injector "on time" on the basis of the signal from the O_2 sensor to provide the ideal air-fuel ratio of 14.7:1.

If the engine is accelerated suddenly, the manifold vacuum decreases rapidly. The ECM senses the increase in throttle opening from the TPS sensor, and it senses the decrease in manifold vacuum from the MAP sensor. When the ECM receives input signals indicating sudden acceleration of the engine, it increases the injector "on time" to provide a slightly richer air-fuel ratio. This acceleration enrichment prevents hesitation when the engine is accelerated.

If the engine is decelerated, a leaner air-fuel ratio is required to reduce CO and HC emission levels. A sudden increase in manifold vacuum occurs when the engine is decelerated. The ECM senses engine deceleration from the MAP sensor signal and the TPS sensor input. When the ECM receives signals indicating engine deceleration, it reduces the injector "on time" to provide a leaner air-fuel ratio and prevent high emissions of HC and CO.

If a defect occurs in the system that makes it possible for the ECM to operate in the normal modes, a throttle body backup (TBB) circuit in the ECM takes over operation of the injectors. When the TBB circuit is in operation, the "on time" of the injectors is increased so that the air-fuel ratio becomes richer than normal, but the vehicle can be driven until the necessary repairs are made.

Electronic Spark Timing (EST). The spark advance is controlled by the ECM in response to the input signals that it receives. (Refer to Chapter 6 for a complete description of the ignition systems used with General Motors fuel injection systems.)

Exhaust Gas Recirculation (EGR). The ECM controls a vacuum solenoid that is used to open and close the vacuum circuit to the EGR valve. The ECM energizes the vacuum solenoid and shuts off the vacuum to the EGR valve if the coolant temperature is below 150°F (64°C), or if the engine is operating at idle speed or under heavy-load, wide-open throttle conditions. Under part-throttle conditions with the coolant temperature above 150°F (64°C), the ECM de-energizes the solenoid and allows vacuum to open the EGR valve. The EGR solenoid and related circuit are shown in Figure 9-9.

Canister Purge. Vapors from the fuel tank are collected in the charcoal canister. The ECM operates

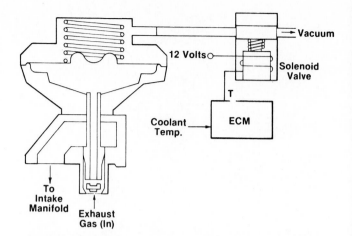

FIGURE 9-9. EGR Solenoid Circuit. (Courtesy of GM Product Service Training, General Motors Corporation)

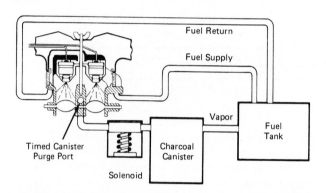

FIGURE 9-10. Canister Purge Solenoid Circuit. (Courtesy of GM Product Service Training, General Motors Corporation)

a solenoid in the purge hose between the canister and the intake manifold, as illustrated in Figure 9-10.

The ECM operates the purge solenoid and allows canister purging if the engine coolant temperature is warm, the throttle is above idle speed, and the vehicle speed is above 10 MPH (16 KPH).

Idle Speed. On some throttle body injection systems, the ECM operates an idle speed control (ISC) motor to control idle speed under slow or fast idle conditions. The same ISC motor is used in some computer command control (3C) systems. An ISC motor and the motor test procedure are illustrated in Figure 9-11.

Some throttle body injection systems use an idle air control (IAC) motor to control idle speed. The IAC motor is operated by the ECM and it controls idle speed by opening or closing an air passage in-

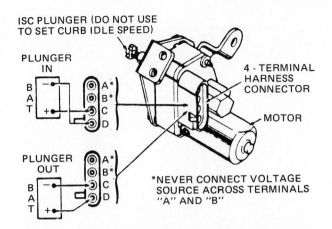

FIGURE 9-11. Idle Speed Control (ISC) Motor. (Courtesy of GM Product Service Training, General Motors Corporation)

to the intake manifold. Airflow through the IAC passage bypasses the throttle, as shown in Figure 9-12.

The IAC motor is used to control fast idle and slow idle speed. If the coolant temperature sensor (CTS) sends a signal to the ECM that indicates cold engine coolant, the ECM operates the IAC motor and opens the IAC passage to increase the idle speed. When the engine is at normal operating temperature, the IAC motor provides a faster idle speed for a few seconds each time the engine is started. When dual throttle body assemblies are used, as in the crossfire injection system, an IAC motor is located in each throttle body assembly.

Air Injection. The air injection reactor (AIR) system is basically the same for throttle body injection systems or computer command control (3C) systems. (See Chapter 8 for a description of 3C AIR systems.) A complete AIR system is illustrated in Figure 9-13.

The ECM operates two electric solenoids in the air control valve and air-switching valve to control airflow from the air pump to the air cleaner, exhaust ports, or catalytic converter. On many throttle body injection systems, the ECM operates the air control valve to bypass air to the cleaner for a few seconds each time the engine is started. Once the engine has been running for a brief period of

FIGURE 9-12. Idle Air Control (IAC) Motor. (Courtesy of GM Product Service Training, General Motors Corporation)

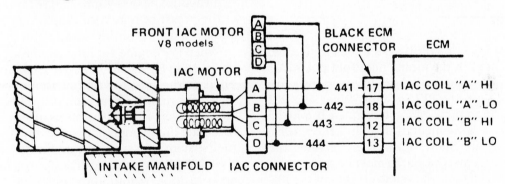

FIGURE 9-13. Air Injection Reactor (AIR) System. (Courtesy of GM Product Service Training, General Motors Corporation)

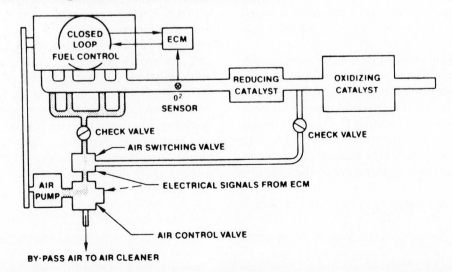

time, the air control valve directs the airflow from the air pump to the air-switching valve. During the engine warm-up time, the air-switching valve directs airflow from the air pump to the exhaust ports. When the coolant reaches normal operating temperature and the system enters the closed-loop mode, the air-switching valve directs the airflow from the pump to the catalytic converter.

Torque Converter Clutch. The vehicle speed sensor (VSS) signal is used by the ECM to control the torque converter lockup time. The VSS and the torque clutch system are the same on the throttle body injection systems and computer command control (3C) systems. (See Chapter 8 for a description of the VSS and torque converter clutch in the 3C system.) When the VSS on throttle body injection systems sends a speed signal above 35 MPH (56 KPH) to the ECM, the ECM maintains the idle air control motors in the extended position. A defective VSS can cause erratic idling.

GENERAL MOTORS PORT FUEL INJECTION

Fuel System

The port fuel injection (PFI) system has individual injectors located in the intake manifold ports. Air injection pumps and heated air inlet systems are not required with the PFI systems. When the PFI system is used on turbocharged engines, it may be referred to as a sequential fire injection (SFI) system. There are many similarities between the PFI and SFI systems. One of the most important differences between the two systems is that in the PFI system all the injectors are energized once for each crankshaft revolution by the electronic control module (ECM), whereas in the SFI system the injectors are energized individually. A naturally aspired 186 CID (3.0L) V6 engine with PFI is shown in Figure 9-14.

The electric fuel pump circuit is basically the same as the circuit used on the General Motors throttle body injection systems. (An explanation of the fuel pump circuit was provided earlier in this chapter.) Fuel is forced through the fuel filter and accumulator to the injector fuel rail by the electric fuel pump. An "O" ring seal is located on each end of the injectors to seal the injectors to the fuel rail and the intake manifold.

The pressure regulator limits the fuel pressure

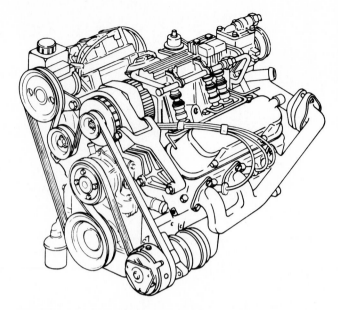

FIGURE 9-14. Non-Turbocharged V6 with PFI. (Courtesy of GM Product Service Training, General Motors Corporation)

to 26–46 PSI (179–317 kPa), and excess fuel is returned from the regulator through the fuel return line to the tank. The fuel system components are shown in Figure 9-15.

A vacuum hose is connected from the lower side of the pressure regulator diaphragm to the intake manifold as shown in Figure 9-16.

When the engine is idling, manifold vacuum will be 16–18 in Hg (54–61 kPa). This low pressure is applied to the injector tips in the intake manifold. When the manifold vacuum is applied to the pressure regulator, less fuel pressure is required to force the diaphragm and valve downward and the fuel system pressure will be limited to 26–28 PSI (179–193 kPa). As the throttle approaches the wide-open position, manifold vacuum decreases to 2–3 in Hg (6.8–10 kPa), and this increase in pressure is sensed at the injector tips. If the manifold vacuum applied to the regulator diaphragm is only 2–3 in Hg (6.8–10 kPa), fuel pressure must increase to force the diaphragm and valve downward, and fuel system pressure increases to 33–38 PSI (227–262 kPa). When the intake manifold pressure increases at wider throttle openings, the pressure regulator increases the fuel pressure at the injectors to maintain a constant pressure drop across the injectors and provide more precise air-fuel ratio control.

Later model port fuel injection systems have multec injectors. A ball valve on these injector tips resists gum deposit buildup, and thus improves engine performance. A multec injector is shown in Figure 9-17.

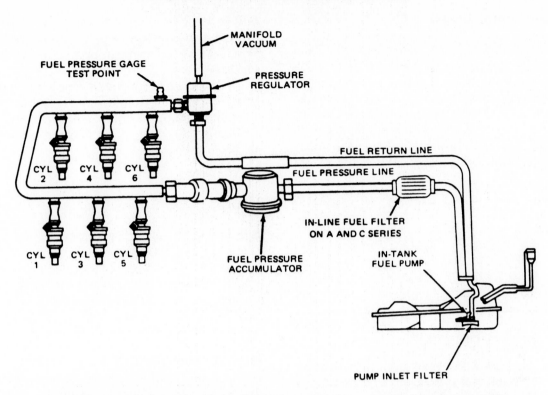

FIGURE 9-15. Fuel System Components. (Courtesy of GM Product Service Training, General Motors Corporation)

FIGURE 9-16. Pressure Regulator. (Courtesy of Buick Motor Division, General Motors Corporation)

FIGURE 9-17. Multec Injector. (Courtesy of GM Product Service Training, General Motors Corporation)

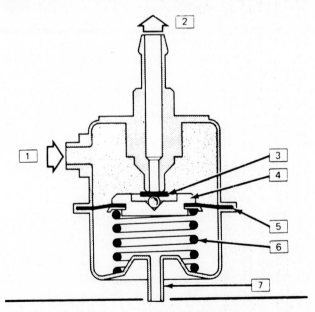

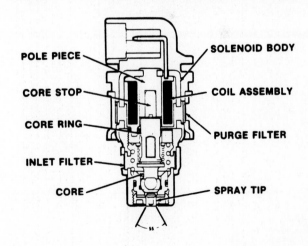

1 FUEL INLET
2 FUEL RETURN OUTLET
3 VALVE
4 VALVE HOLDER
5 DIAPHRAGM
6 COMPRESSION SPRING
7 VACUUM CONNECTION

Electronic Control Module (ECM)

The ECM is located under the instrument panel and it is similar in appearance to those used with throttle body injection systems. Removable programmable read-only memory (PROM) and calibration package (CALPAK) units are located under an access cover in the ECM, as indicated in Figure 9-18.

The CALPAK unit allows fuel delivery if other parts of the ECM are damaged. When the PROM

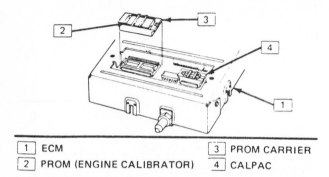

1	ECM	3	PROM CARRIER
2	PROM (ENGINE CALIBRATOR)	4	CALPAC

FIGURE 9-18. ECM with PROM and CALPAK Units. (Courtesy of Buick Motor Division, General Motors Corporation)

FIGURE 9-19. PROM and CALPAK Removal and Replacement. (Courtesy of Buick Motor Division, General Motors Corporation)

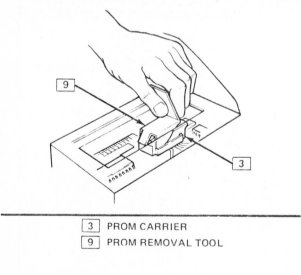

3	PROM CARRIER
9	PROM REMOVAL TOOL

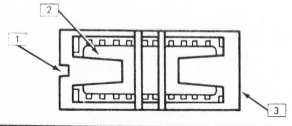

1	REFERENCE END
2	PROM
3	PROM CARRIER

unit is replaced, it must be installed in the original direction and the connectors must not be bent. The removal and replacement procedure is pictured in Figure 9-19.

When the ECM connectors are being removed or installed, the ignition switch should be in the off position. The operating conditions sensed by the ECM and the systems that it controls are outlined in Table 9-1.

Inputs

Similarities to Other Systems. The throttle position sensor (TPS) oxygen (O_2) sensor, coolant temperature sensor, manifold absolute pressure (MAP) sensor, and the park/neutral switch are similar to the sensors used on General Motors 3C and throttle body injection systems. (Refer to Chapter 8 for an explanation of these sensors in the 3C systems.)

Mass Air Flow (MAF) Sensor. The MAF sensor is used in place of the MAP sensor on many engines. Turbocharged engines always use the MAF sensor. The MAF sensor is located in the air intake between the air cleaner and the throttle body assembly, as indicated in Figure 9-20.

The MAF sensor contains a resistor that measures the temperature of the incoming air. A heated film in the sensor is a nickel grid coated with Kapton, a high-temperature material. An electronic module in the top of the sensor maintains the temperature of the heated film at 167°F (75°C). If more energy is required to maintain the heated film at 167°F (75°C), the incoming mass of air has increased. This information is then sent to the ECM, and the ECM uses this input to provide a very precise air-fuel ratio control. The MAF sensor is shown in Figure 9-21.

Vehicle Speed Sensor (VSS). The VSS is a signal generator located in the transaxle. It is rotated mechanically, and generates an electrical signal in relation to vehicle speed. This signal is sent through the VSS signal buffer to the ECM, as indicated in Figure 9-22.

Knock Sensor. A knock sensor is used on some of the PFI systems. The knock sensor signal is sent through the electronic spark control (ESC) module to the ECM. When the engine detonates, the knock sensor signal to the ESC module and ECM causes the ECM to retard the spark advance. The knock sensor circuit is shown in Figure 9-23.

TABLE 9-1. ECM Operating Conditions Sensed and Systems Controlled.

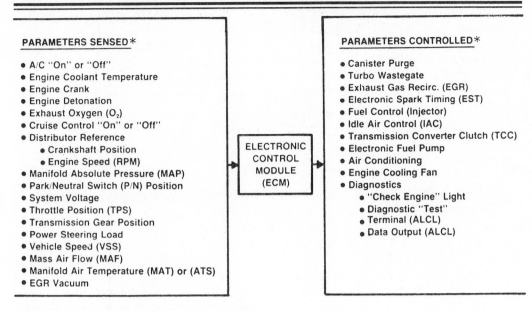

PARAMETERS SENSED*		PARAMETERS CONTROLLED*
• A/C "On" or "Off"		• Canister Purge
• Engine Coolant Temperature		• Turbo Wastegate
• Engine Crank		• Exhaust Gas Recirc. (EGR)
• Engine Detonation		• Electronic Spark Timing (EST)
• Exhaust Oxygen (O₂)	ELECTRONIC	• Fuel Control (Injector)
• Cruise Control "On" or "Off"	CONTROL	• Idle Air Control (IAC)
• Distributor Reference	MODULE	• Transmission Converter Clutch (TCC)
• Crankshaft Position	(ECM)	• Electronic Fuel Pump
• Engine Speed (RPM)		• Air Conditioning
• Manifold Absolute Pressure (MAP)		• Engine Cooling Fan
• Park/Neutral Switch (P/N) Position		• Diagnostics
• System Voltage		• "Check Engine" Light
• Throttle Position (TPS)		• Diagnostic "Test"
• Transmission Gear Position		• Terminal (ALCL)
• Power Steering Load		• Data Output (ALCL)
• Vehicle Speed (VSS)		
• Mass Air Flow (MAF)		
• Manifold Air Temperature (MAT) or (ATS)		
• EGR Vacuum		

*NOT ALL SYSTEMS USED ON ALL ENGINES.

Courtesy of GM Product Service Training, General Motors Corporation

FIGURE 9-20. MAF Sensor Location. (Courtesy of Buick Motor Division, General Motors Corporation)

FIGURE 9-21. MAF Sensor. (Courtesy of GM Product Service Training, General Motors Corporation)

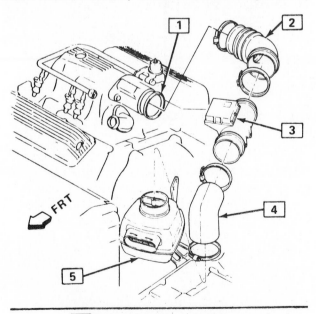

1	THROTTLE BODY ASM.
2	REAR AIR INTAKE DUCT
3	MASS AIR FLOW (MAF) SENSOR
4	INT. AIR INTAKE DUCT
5	AIR CLEANER ASM.

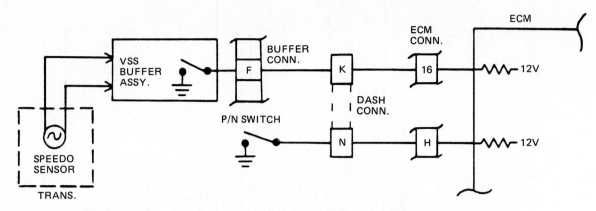

FIGURE 9-22. VSS Sensor. (Courtesy of Buick Motor Division, General Motors Corporation)

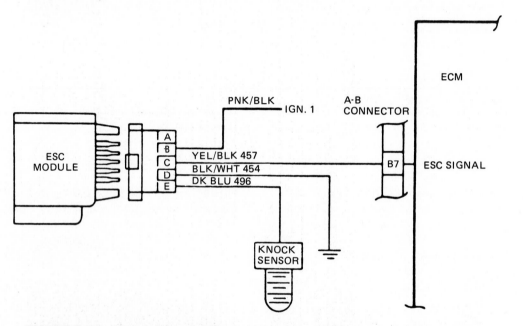

FIGURE 9-23. Knock Sensor Circuit. (Courtesy of GM Product Service Training, General Motors Corporation)

Exhaust Gas Recirculation (EGR) Vacuum Diagnostic Switch. The ECM operates a vacuum solenoid which applies vacuum to the EGR valve. When vacuum is applied to the EGR valve, the EGR vacuum diagnostic switch signals the ECM that the EGR valve is operating. If the EGR valve is operating when the necessary input signals are not present, the EGR vacuum diagnostic switch signals the ECM and code 32 is set in the ECM memory. The EGR vacuum diagnostic switch and the EGR solenoid circuits are illustrated in Figure 9-24.

Manifold Air Temperature (MAT) Sensor. Some PFI systems use a MAT sensor which sends a signal to the ECM in relation to the temperature of the air in the intake manifold or in the air cleaner.

Distributor Reference Signal. Many PFI systems use the same ignition system as General Motors 3C systems. (These ignition systems are described in Chapter 6.) The distributor reference signal from the high energy ignition (HEI) module to the ECM is used as a crankshaft position signal and engine speed signal.

ECM Outputs

Spark Advance. When the engine is running, the distributor pickup coil signal is sent through the HEI module and the distributor reference circuit to the ECM. This pickup signal travels through the electronic spark timing (EST) circuit in the ECM

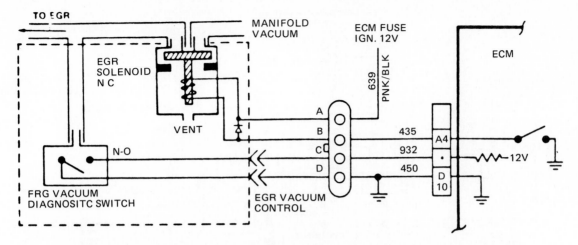

FIGURE 9-24. EGR Vacuum Diagnostic Switch and Solenoid. (Courtesy of GM Product Service Training, General Motors Corporation)

and then out to the HEI module through the EST wire. When the HEI module receives this pickup signal via the EST wire, it switches off the primary ignition circuit, which results in magnetic field collapse in the ignition coil and spark plug firing. The EST circuit in the ECM varies the pickup coil signal to provide the exact spark advance required by the engine. Figure 9-25 illustrates the spark advance control circuit.

(Refer to Chapter 6 for General Motors high energy ignition (HEI) systems and computer-controlled coil ignition (C³I) systems.)

Coolant Fan. Several different coolant fan circuits are used on PFI-equipped vehicles. In these circuits, the ECM operates the coolant fan under certain conditions. One coolant fan circuit is outlined in Figure 9-26.

Some heavy-duty cooling sytems have an optional cooling fan located beside the standard fan. The low-speed fan relay winding is grounded by the ECM if the coolant temperature is above 208°F (98°C) and the air conditioning (A/C) system pressure is below 260 PSI (1,793 kPa), and the vehicle speed is below 45 MPH (72 KPH). The "LO" contacts

FIGURE 9-25. HEI Module and EST Circuit in ECM. (Courtesy of GM Product Service Training, General Motors Corporation)

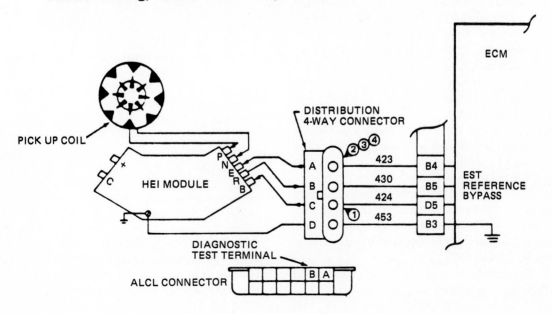

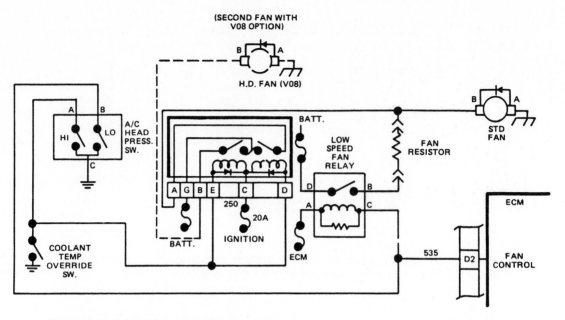

FIGURE 9-26. Coolant Fan Circuit without Timer. (Courtesy of GM Product Service Training, General Motors Corporation)

in the A/C head pressure switch ground the low-speed fan relay contacts if the A/C is turned on and the pressure is above 260 PSI (1,793 kPa). When the low-speed fan relay contacts are closed, voltage is supplied through the resistor to the standard fan motor, which causes the fan to run at low speed.

If coolant temperature exceeds 223°F (106°C), the coolant temperature override switch closes, which grounds both windings in the dual contact relay. One set of points supplies voltage to the standard fan and the other contacts complete the circuit to the heavy-duty fan so that both fans operate at high speed. When the A/C is on and the pressure in the system exceeds 300 PSI (2,068 kPa), the "HI" contacts in the A/C head pressure switch close, which grounds both windings in the dual relay and causes both fans to operate at high speed. The standard fan operates at low speed when terminals A and B are connected together in the assembly line diagnostic link (ALDL).

The low-speed fan relay is mounted behind a plastic panel on the firewall with most of the other relays in the PFI system, and the dual relay is located near the brake booster.

Some coolant fan circuits have a timer which is used to operate the coolant fan after the engine is shut off. This type of circuit has one coolant fan that operates at two speeds. The coolant fan circuit is illustrated in Figure 9-27.

The ECM operates the coolant fan at low speed if the following conditions are present.

1. Coolant temperature above 208°F (98°C).
2. Vehicle speed under 45 MPH (72 KPH).
3. A/C system pressure under 260 PSI (1,793 kPa)

The A/C head pressure switch "LO" contacts close and operate the coolant fan at low speed if the A/C system pressure exceeds 260 PSI (1,793 kPa). When the coolant temperature exceeds 223°F (106°C), the coolant temperature switch closes, which energizes the high-speed relay winding and supplies full voltage through the high-speed relay contacts to coolant fan motor. This operates the coolant fan at high speed. If the A/C system pressure exceeds 300 PSI (2,068 kPa), the "HI" contacts in A/C head pressure switch close, which also causes the coolant fan to operate at high speed.

When the ignition is turned off and the coolant temperature is above 223°F (106°C), the coolant temperature override switch remains closed, which grounds the timer relay winding. This supplies full voltage to the coolant fan motor through the timer relay contacts, so that the coolant fan operates at high speed. The coolant fan is shut off after 10 minutes by the timer, or when the coolant temperature drops below 223°F (106°C) and the coolant temperature override switch contacts open.

A/C Compressor Clutch. When the A/C control switch is turned on and the cycling switch is closed, voltage is supplied to terminal B8 on the ECM and to the A/C clutch control relay contacts. Under this

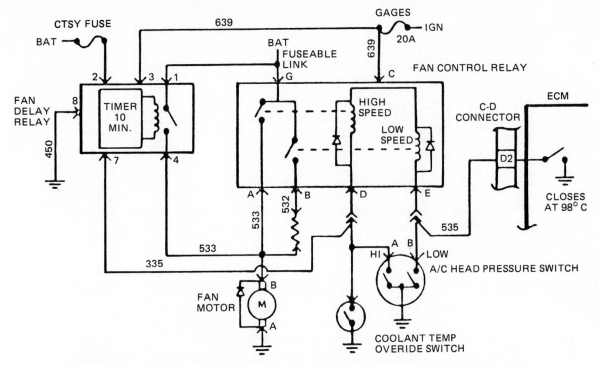

FIGURE 9-27. Coolant Fan Circuit with Timer. (Courtesy of GM Product Service Training, General Motors Corporation)

condition, the ECM grounds the A/C clutch control relay winding and the relay contacts close, which supplies voltage to the A/C compressor clutch winding. If the power steering (P/S) pressure is high, such as when a full turn of the steering is completed, the P/S switch opens the A/C clutch control relay winding circuit. When this occurs, the relay contacts open and the compressor clutch is disengaged. The ECM also disengages the compressor clutch when the engine is operating at wide-open throttle (WOT). Figure 9-28 shows the compressor clutch circuit.

Idle Air Control (IAC) Motor. The ECM operates the IAC motor on the throttle body assembly to control idle speed under all operating conditions. (This motor was explained earlier in this chapter in the section on General Motors throttle body injection systems.) The IAC motor connections to the ECM are shown in Figure 9-29.

Transmission Converter Clutch (TCC). The purpose of the TCC system is to eliminate torque converter power loss when the vehicle is operating at

FIGURE 9-28. Compressor Clutch Circuit. (Courtesy of GM Product Service Training, General Motors Corporation)

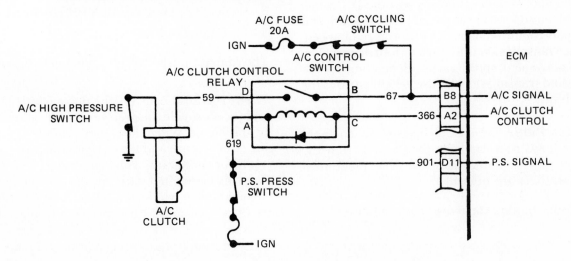

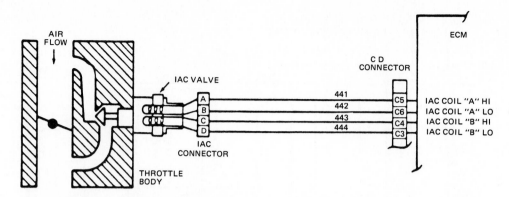

FIGURE 9-29. IAC Motor Circuit. (Courtesy of GM Product Service Training, General Motors Corporation)

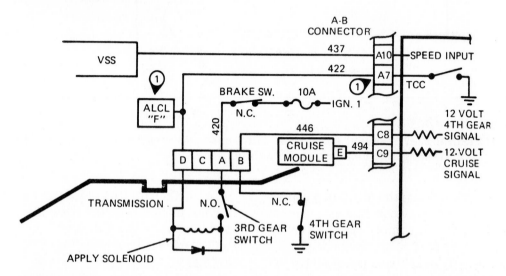

FIGURE 9-30. TCC Circuit. (Courtesy of Buick Motor Division, General Motors Corporation)

moderate cruising speed. The ECM grounds the apply solenoid winding and engages the TCC when the engine is operating at normal temperature and light-load conditions above a specific road speed, and the transmission is in third or fourth gear. The TCC engagement speed varies, depending on the engine and transmission. Figure 9-30 illustrates the TCC circuit.

Canister Purge. The ECM operates a solenoid which allows fuel vapors to be purged from the canister into the intake manifold when the following conditions are present:

1. Engine running for more than 1 minute.

2. Coolant temperature above 165°F (80°C).

3. Vehicle speed above 10 MPH (16 KPH).

4. Throttle above idle speed.

The canister purge circuit is shown in Figure 9-31.

Fuel. The ECM controls the pulse width of the injectors to provide the precise air-fuel ratio required by the engine. When the engine is at normal operating temperature and the engine is operating at idle speed or moderate cruising speed, the ECM will control the air-fuel ratio at, or very close to, the stoichiometric value of 14.7:1. The ECM provides mixture enrichment by increasing the injector pulse width when the engine coolant is cold, at wide-open throttle, and on sudden acceleration. The ECM stops energizing the injectors when the engine is decelerated in order to lower emission levels and improve fuel economy.

Turbo Wastegate. The ECM operates a solenoid to limit the turbo boost pressure on turbocharged engines. When the ECM energizes the wastegate solenoid, some of the pressure supplied to the wastegate diaphragm is vented through the solenoid. This reduces the wastegate opening and increases the boost pressure, so that the ECM closes the wastegate solenoid. The ECM uses the MAF sensor and engine speed signals to limit the boost pressure to approximately 7.8 PSI (55.5 kPa). The wastegate solenoid circuit is illustrated in Figure 9-32.

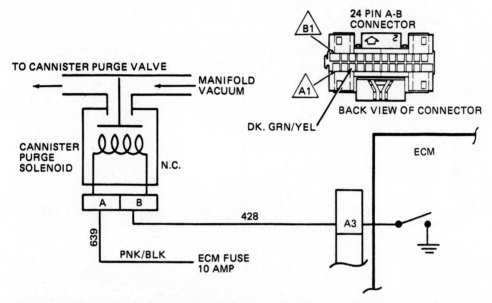

FIGURE 9-31. Canister Purge Circuit. (Courtesy of Buick Motor Division, General Motors Corporation)

FIGURE 9-32. Wastegate Solenoid Circuit. (Courtesy of GM Product Service Training, General Motors Corporation)

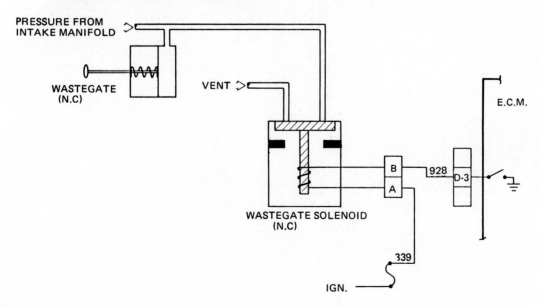

GENERAL MOTORS TUNED PORT INJECTION

Design

The tuned port injection (TPI) system is used on some 5.7L (350 CID) and 5.0L (305 CID) V8 engines. These engines have an air plenum and tubular runners in the intake manifold which provide smooth unobstructed airflow into the cylinders. The TPI system is similar to the PFI system, with a fuel injector located in each intake port. A manifold air temperature (MAT) sensor is used in the TPI system, and a separate cold-start injector is operated by the ECM to inject additional fuel when a cold engine is being started. An engine with a TPI system is illustrated in Figure 9-33.

GENERAL MOTORS COMPUTER SYSTEMS WITH BODY COMPUTER MODULE AND SERIAL DATALINE

System Design

Many 1986 and later model General Motors cars are equipped with a body computer module (BCM) and a serial data line which interconnects the ECM, BCM, instrument panel cluster (IPC), electronic climate control (ECC), voice/chime module, and heater ventilation and air conditioning (HVAC) programmer. The serial data line between these components is referred to as an 800 circuit, as indicated in Figure 9-34.

A permanent connector in the assembly line diagnostic link (ALDL) is part of the serial data line.

FIGURE 9-33. Tuned Port Injection System. (Courtesy of GM Product Service Training, General Motors Corporation)

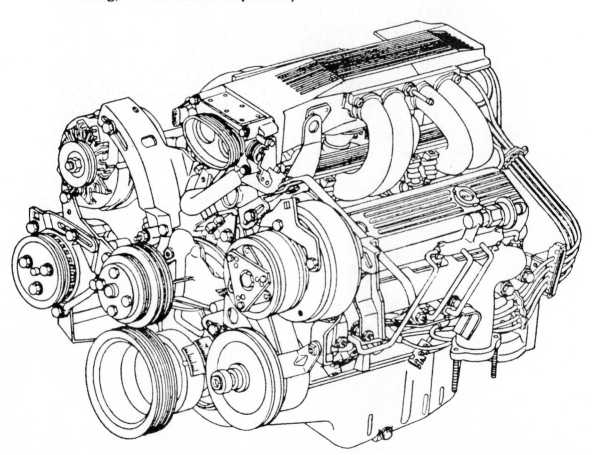

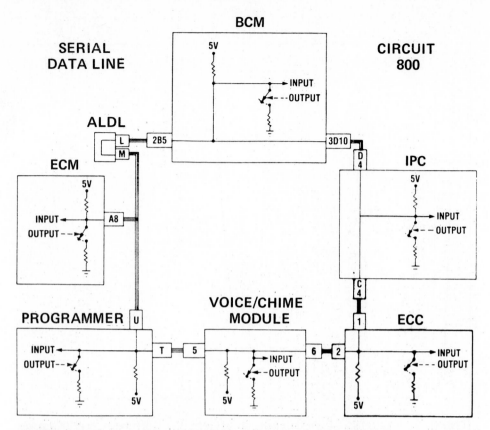

FIGURE 9-34. Serial Data Line and Interconnected Components. (Courtesy of GM Product Service Training, General Motors Corporation)

The ALDL is similar to the ALDL in other systems. Since the serial data line is bidirectional, the system will still function if the permanent connector is removed from the ALDL. The serial data line may be compared to a party or conference telephone connection—when data is being transmitted between two components, the other components can listen to the data. For example, as the ECM sends data to the BCM, such as A/C clutch status, coolant temperature, and engine speed, the other components have access to this information. The IPC uses the engine speed information to display engine revolutions per minute (RPM) for the driver.

Electronic Control Module Functions

The electronic control module (ECM) is similar to the ECM in other port fuel injection (PFI) systems. The input and output control functions are listed in Table 9-2, and the location of the ECM and the

TABLE 9-2. ECM Input and Output Control Functions

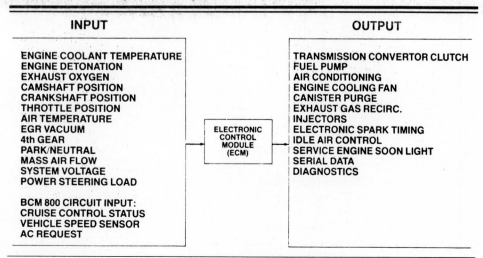

INPUT	OUTPUT
ENGINE COOLANT TEMPERATURE	TRANSMISSION CONVERTOR CLUTCH
ENGINE DETONATION	FUEL PUMP
EXHAUST OXYGEN	AIR CONDITIONING
CAMSHAFT POSITION	ENGINE COOLING FAN
CRANKSHAFT POSITION	CANISTER PURGE
THROTTLE POSITION	EXHAUST GAS RECIRC.
AIR TEMPERATURE	INJECTORS
EGR VACUUM	ELECTRONIC SPARK TIMING
4th GEAR	IDLE AIR CONTROL
PARK/NEUTRAL	SERVICE ENGINE SOON LIGHT
MASS AIR FLOW	SERIAL DATA
SYSTEM VOLTAGE	DIAGNOSTICS
POWER STEERING LOAD	
BCM 800 CIRCUIT INPUT:	
CRUISE CONTROL STATUS	
VEHICLE SPEED SENSOR	
AC REQUEST	

Courtesy of GM Product Service Training, General Motors Corporation

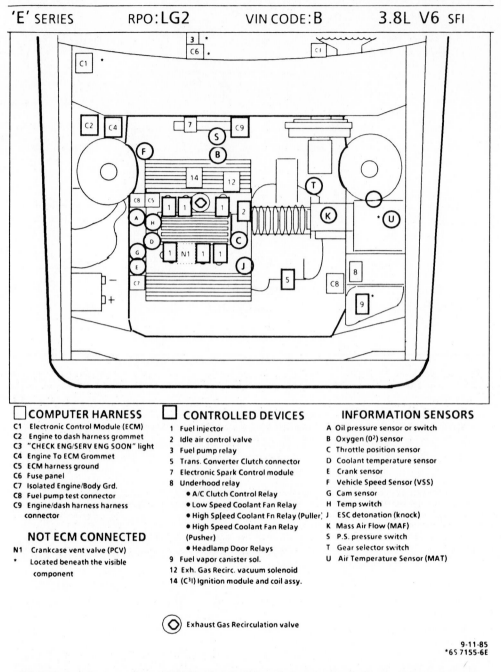

☐ COMPUTER HARNESS

C1 Electronic Control Module (ECM)
C2 Engine to dash harness grommet
C3 "CHECK ENG/SERV ENG SOON" light
C4 Engine To ECM Grommet
C5 ECM harness ground
C6 Fuse panel
C7 Isolated Engine/Body Grd.
C8 Fuel pump test connector
C9 Engine/dash harness harness
     connector

NOT ECM CONNECTED

N1 Crankcase vent valve (PCV)
*   Located beneath the visible
     component

☐ CONTROLLED DEVICES

1 Fuel injector
2 Idle air control valve
3 Fuel pump relay
5 Trans. Converter Clutch connector
7 Electronic Spark Control module
8 Underhood relay
    • A/C Clutch Control Relay
    • Low Speed Coolant Fan Relay
    • High Sp[eed Coolant Fn Relay (Puller]
    • High Speed Coolant Fan Relay
      (Pusher)
    • Headlamp Door Relays
9 Fuel vapor canister sol.
12 Exh. Gas Recirc. vacuum solenoid
14 (C³I) Ignition module and coil assy.

INFORMATION SENSORS

A Oil pressure sensor or switch
B Oxygen (O²) sensor
C Throttle position sensor
D Coolant temperature sensor
E Crank sensor
F Vehicle Speed Sensor (VSS)
G Cam sensor
H Temp switch
J ESC detonation (knock)
K Mass Air Flow (MAF)
S P.S. pressure switch
T Gear selector switch
U Air Temperature Sensor (MAT)

◇ Exhaust Gas Recirculation valve

9-11-85
*6S 7155-6E

FIGURE 9-35. Location of ECM and Related Inputs and Outputs. (Courtesy of Oldsmobile Division, General Motors Corporation)

related inputs and outputs are shown in Figure 9-35.

The input and output control functions are similar to the ones described previously in this chapter, and the fuel system and fuel pump relay circuit are basically the same as those used in earlier PFI systems. Sequential energizing of the injectors is provided by the ECM. Since the electric cooling fan circuit is somewhat different from those described previously, this circuit is illustrated in Figure 9-36.

When a heavy-duty cooling system is ordered with the vehicle, an optional high-speed pusher fan is located in front of the radiator in addition to the conventional two-speed fan behind the radiator. The operation of the cooling fan circuit may be summarized as follows:

1. At 208°F (98°C) coolant temperature, the ECM grounds the low-speed fan relay winding. This action closes the relay contacts and supplies voltage through the relay contacts and the re-

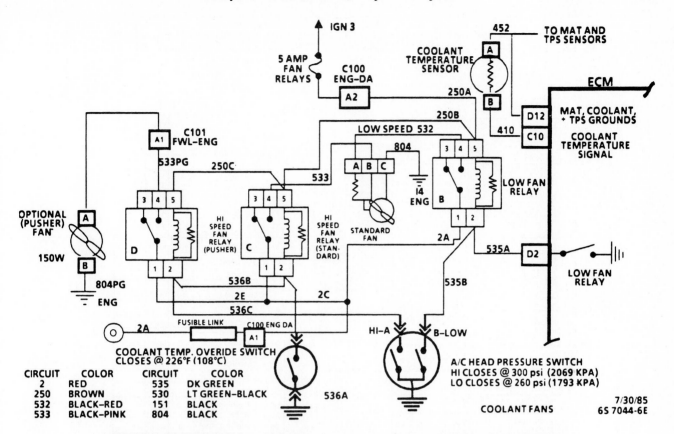

FIGURE 9-36. Electric Cooling Fan Circuit. (Courtesy of GM Product Service Training, General Motors Corporation)

sistor to the low-speed fan motor. The resistor lowers the voltage at the motor, which results in low-speed fan operation.

2. Low-speed fan operation also occurs if the low pressure contacts close in the A/C head pressure switch, before the ECM grounds the low-speed fan relay winding. These low pressure contacts close when the A/C refrigerant pressure reaches 260 PSI (1,793 kPa).

3. The coolant temperature override switch closes and grounds both high-speed fan relays if the coolant temperature reaches 226°F (108°C). When this occurs, both relay contacts close so that the standard fan relay supplies voltage directly to the low-speed fan and the pusher fan relay supplies voltage to the pusher fan. This action results in high-speed operation of both fans.

4. When the A/C head pressure switch high pressure contacts close at 300 PSI (2,069 kPa) refrigerant pressure, before the coolant temperature override switch closes, both high-speed fan relay windings are grounded through the high pressure contacts and high-speed operation of both fans occurs.

A distributorless direct ignition system (DIS) is used in these systems which has a different camshaft position sensor and coil assembly compared to the computer-controlled coil ignition C³I systems. (These ignition systems are described in Chapter 6.) The ECM terminals and the voltage readings at each terminal are provided in Figure 9-37.

Central Power Supply

The central supply (CPS) supplies 12 V and 7 V to the BCM, IPC, voice/chime module, and other electronic components regardless of input voltage variations. This 12-V supply is available continuously from the battery through the CPS to the BCM, IPC, and voice/chime module to maintain the computer memories and operate the IPC. When the ignition switch is off, the 12-V supply operates the voice/chime module. The 7-V power supply is reduced to 5 V by the BCM and then sent out as a reference voltage to the BCM input sensors. These 7-V circuits only function when the CPS turns them on in response to a "wake-up" signal from the BCM. The ground circuits from the BCM, IPC, HVAC pro-

ECM CONNECTOR IDENTIFICATION

This ECM voltage chart is for use with a digital voltmeter to further aid in diagnosis. The voltages you get may vary due to low battery charge or other reasons, but they should be very close.

THE FOLLOWING CONDITIONS MUST BE MET BEFORE TESTING:
- Engine at operating temperature   • Engine idling in closed loop (for "Engine Run" column)
- Test terminal not grounded   • ALCL tool not installed • A/C Off

VOLTAGE				
KEY "ON"	ENG. RUN	OPEN CKT.	CIRCUIT	PIN
.01	B+	0	FUEL PUMP RELAY	A1
.01	0	0	A/C CLUTCH CONTROL	A2
B+	B+	0	CANISTER PURGE CONTROL	A3
B+	B+	0	EGR SOLENOID	A4
.79	B+	0	SERVICE ENGINE SOON LIGHT	A5
B+	B+	B+	IGN #1 (ISO)	A6
B+	B+	B+	TCC CONTROL	A7
2.3 3.6	2.3 3.6	3.7	SERIAL DATA	A8
5.0	4.9	5.0	DIAG. TERM	A9
B+	B+	B+	VSS SIGNAL	A10
0	6.8	0	CAM HIGH	A11
0	.02	0	GROUND	A12

(④ Fuel Pump Relay, ⑤ A/C Clutch Control, ① VSS Signal)

KEY "ON"	ENG. RUN	OPEN CKT.	CIRCUIT	PIN
			NOT USED	C1
			NOT USED	C2
.86 11.8	12.3	3.8 10.5	IAC-B-LO	C3
.86 11.8	.86	B+ .5	IAC-B-HI	C4
.86 11.8	.86	B+ .5	IAC-A-HI	C5
.86 11.8	12.3	.5 B+	IAC-A-LOW	C6
			NOT USED	C7
0	0	B+	4TH GEAR SIGNAL	C8
			NOT USED	C9
2.24	1.75	5.0	COOLANT TEMP SIGNAL	C10
2.2	1.75	5.0	AIR TEMP	C11
B+	B+	0	INJECTOR 6	C12
.45	.45	0	TPS SIGNAL	C13
5.0	4.9	5.0	TPS 5V REF	C14
B+	B+	0	INJECTOR 2	C15
B+	B+	.5	B+	C16

(② Coolant Temp Signal)

BACK VIEW OF CONNECTOR — A1, B1 — 24 PIN A-B CONNECTOR

WHEN TWO VALUES ARE GIVEN, THE VOLTAGE SIGNAL WILL CYCLE BETWEEN THE TWO VALUES

BACK VIEW OF CONNECTOR — C1, D1 — 32 PIN C-D CONNECTOR

PIN	CIRCUIT	VOLTAGE		
		KEY "ON"	ENG RUN	OPEN CKT.
B1	NOT USED			
B2	NOT USED			
B3	CRANK REF LOW	0	.0	3.0
B4	EST CONTROL	.04	1.2	0
B5	CRANK REF HI	0	4.0	0
B6	MASS AIR FLOW SENSOR SIGNAL	2.0 3.5	2.6	5.0
B7	ESC SIGNAL	9.2	9.2	0
B8	NOT USED			
B9	NOT USED			
B10	PARK/NEUTRAL SW. SIGNAL	.0	.0	B+
B11	NOT USED			
B12	INJECTOR 5	B+	B+	0
D1	GROUND	.0	.0	.0
D2	COOLING FAN CONTROL	B+	1.57	.45
D3	NOT USED			
D4	NOT USED			
D5	EST BYPASS SIGNAL	.0	3.75 4.85	.0
D6	GRND (O_2) LOW	0	0	1,8
D7	O_2 SENSOR SIGNAL	.42	.25 1.0	.42
D8	NOT USED			
D9	EVRV FBK SIGNAL	B+	B+	B+
D10	GROUND	.0	.0	.0
D11	POWER STEERING SW SIGNAL	B+	B+	0
D12	MAT (AIR TEMP) & COOLANT TPS GROUND	.0	0	.0
D13	NOT USED			
D14	INJECTOR 1	B+	B+	0
D15	INJECTOR 3	B+	B+	
D16	INJECTOR 4	B+	B+	0

(③ Mass Air Flow Sensor Signal, ③ O_2 Sensor Signal)

① Varies from .60 to battery voltage depending on position of drive wheels.
② Normal operating temperature.
③ Varies
④ 12 V only for first 2 seconds unless engine is cranking or running.
⑤ 6.62 with A/C On (System will not reenergize if fuel rail pressure is high)

8-8-85
6S7008-6E

FIGURE 9-37. Electronic Control Module (ECM) Terminal Identification. (Courtesy of Oldsmobile Division, General Motors Corporation)

grammer, voice/chime module, and the EEC panel are completed through the CPS. This ground circuit is connected directly to the negative battery terminal. Therefore, these ground circuits are not shared with any other components, so that electromagnetic interference (EMI) is reduced. The CPS output voltages and ground circuits are illustrated in Figure 9-38.

Normally the BCM triggers the CPS "wake-up" circuit when it receives an input signal from a door handle switch when a door is opened. On cars without illuminated entry, this input signal is supplied from the doorjamb switch. Alternate input signals could also trigger the wake-up circuit, as indicated in Figure 9-39.

When the BCM receives a wake-up signal from

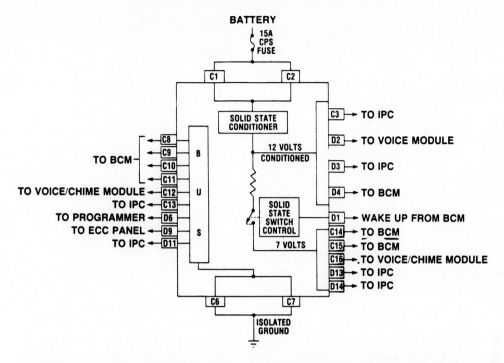

FIGURE 9-38. Central Power Supply Output Voltages and Ground Circuits. (Courtesy of GM Product Service Training, General Motors Corporation)

FIGURE 9-39. Wake-Up Circuit to Body Computer Module and Central Power Supply. (Courtesy of GM Product Service Training, General Motors Corporation)

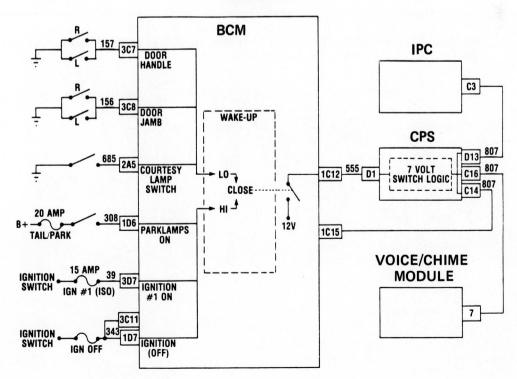

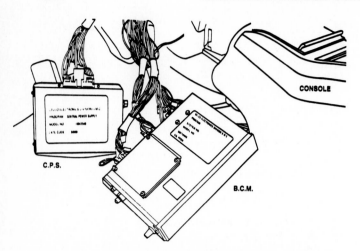

FIGURE 9-40. Body Computer Module (BCM) and Central Power Supply (CPS) Location. (Courtesy of GM Product Service Training, General Motors Corporation)

a door handle switch, the BCM supplies 12 V to the CPS. This causes the CPS to activate the various 7-V circuits to the BCM, IPC, and voice/chime module. The BCM and the CPS are usually located under the dash. The BCM and the CPS are pictured in Figure 9-40, and the location of the BCM, ECM, HVAC module, and voice/chime module are shown in Figure 9-41.

Body Computer Module

General Functions. The body computer module (BCM) is the center of communications for the multiple computer system. It performs the following functions:

1. Operates the electronic climate control system.
2. Operates the cruise control system.
3. Controls the fuel level display and performs fuel data calculations.
4. Controls vehicle status messages such as "door ajar" warning.
5. Provides vehicle speed data.
6. Remembers and updates odometer information.
7. Provides transaxle shifter position data for display. (This feature was not available on 1986 early production models.)
8. Calculates English/metric conversions for displays.
9. Controls the courtesy lights, including the optional illuminated entry.
10. Controls the optional twilight sentinel.
11. Controls instrument panel dimming.

FIGURE 9-41. Module and Computer Location. (Courtesy of GM Product Service Training, General Motors Corporation)

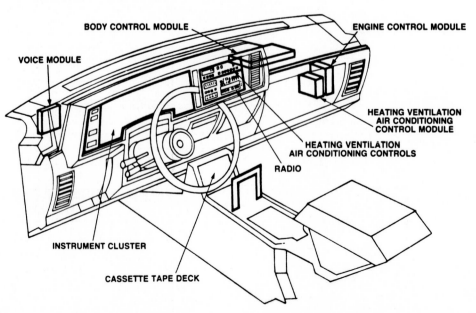

12. Recognizes and compensates for BCM system failures and stores diagnostic codes which can be displayed for diagnostic purposes.

The BCM functions above are from a 1986 Toronado. The BCM functions vary, depending on the application and year of vehicle.

Cruise Control Operation. The BCM has taken over the function of the previous external cruise control electronic controller. A cruise set speed display appears on the IPC when the cruise is engaged or resumed, or when the speed is changed. The BCM receives input signals from the cruise on/off switch, set/coast switch, and resume/accelerate switch on the turn signal lever. These normally open switches provide a voltage input signal to the BCM when they are closed by the driver. If the brake pedal is depressed, the brake switch opens, and in response to this signal the BCM deactivates the cruise control. Other inputs are the vehicle speed sensor (VSS) and ECM-to-BCM communications on the serial data circuit which the BCM uses for vehicle speed and gear position information. The BCM operates the vent and vacuum solenoids in the cruise servo to control the servo

vacuum and the cruise set speed. Information is also supplied from the servo to the BCM to monitor cruise control operation. The cruise control circuit is shown in Figure 9-42.

Instrument Panel Cluster

General Information. The instrument panel cluster (IPC) contains a microprocessor that communicates with the rest of the system through the serial data circuit. Both vacuum fluorescent displays and incandescent telltale bulbs are used in the IPC. The IPC vacuum fluorescent displays are used for these functions:

1. Digital speedometer
2. Odometer information
3. Bar-graph display of system voltage
4. Bar-graph display of engine coolant temperature
5. Bar-graph display of engine oil pressure
6. Bar-graph display of fuel level
7. PRNDL indicator for transaxle gear range selection. (This was not available on early 1986 production vehicles.)

FIGURE 9-42. Cruise Control Circuit. (Courtesy of GM Product Service Training, General Motors Corporation)

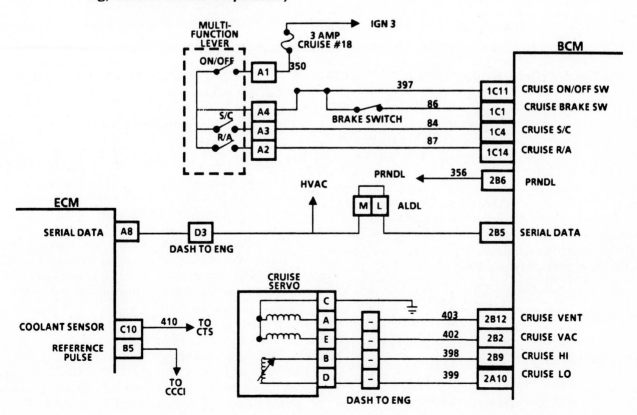

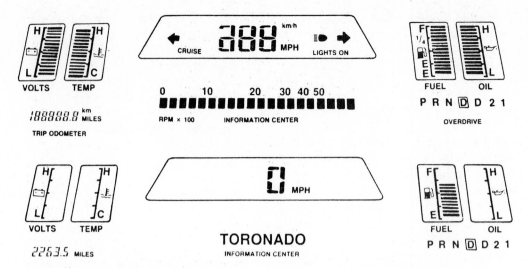

FIGURE 9-43. Instrument Panel Cluster. (Courtesy of GM Product Service Training, General Motors Corporation)

Vacuum fluorescent indicators are used for these items:

1. Turn signal indicators
2. High-beam indicator
3. Lights-on indicator
4. Cruise-on indicator

Driver requested functions are:

1. Bar-graph tachometer display
2. Fuel and trip information

When the ignition switch is turned on, all segments of the IPC are illuminated for a few seconds, as indicated in top diagram in Figure 9-43.

After the initial IPC display, each bar graph displays brightly illuminated bars to provide specific indications. For example, if the fuel tank is half full, the bars on the fuel gauge will be brightly illuminated to the center position on the fuel gauge. The IPC information center display has a 20-character vacuum fluorescent dot matrix display with these messages available:

1. Fuel level
2. Engine hot
3. Low coolant
4. Low refrigerant
5. Electrical problem
6. Generator problem
7. Low brake fluid
8. Park brake
9. Low washer fluid
10. Lighting problem
11. A/C system problem
12. Low oil pressure
13. Right door open
14. Left door open
15. Both doors open
16. Both doors open
17. Tail lamp out
18. Park lamp out
19. Stop lamp out
20. Lamp fuse out
21. Cruise set speed

The following messages are provided by incandescent telltale lamps:

1. Security
2. Fasten belts
3. Service engine soon
4. Brake
5. Lights on

Switch panels on each side of the IPC supply driver-initiated inputs to the IPC. The left switch panel is shown in Figure 9-44, and the right switch panel is pictured in Figure 9-45.

Left Switch Panel Operation. The "odo/trip" button switches the mileage display from season odometer to trip odometer. If this button is pressed a second time, the mileage display changes back to season odometer.

When the "trip reset" button is pressed, the trip odometer is reset to zero.

The "English/metric" button is used to switch the IPC and electronic climate control (ECC) displays for speed distance, fuel consumption, and temperature from English to metric units.

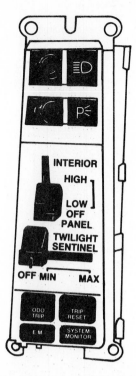

FIGURE 9-44. Left Switch Panel. (Courtesy of GM Product Service Training, General Motors Corporation)

A vacuum fluorescent display check is initiated if the driver presses the "system monitor" button. This button is also used to acknowledge messages in the information center and redisplay these messages.

The left switch panel also contains the headlight and park light switches, the panel dimming switch, and the twilight sentinel control switch.

Right Switch Panel Operation. If the driver presses the "tach" button, the tachometer display appears in the information center.

The "gauge scale" button switches the fuel level gauge from full scale to an expanded 1/4-tank scale. After a specific length of time, the display changes back to full scale.

When the "fuel used" button is pressed, the information center displays the amount of fuel used since the last driver-requested reset.

Pressing the "fuel economy" button causes fuel consumption information to be displayed in the information center. Either instant fuel economy or average fuel economy since the last reset may be displayed.

The "range" button allows the driver to request fuel range information. This value is based on the amount of fuel in the tank and the average fuel economy during the last 25 miles (40 km) of driving.

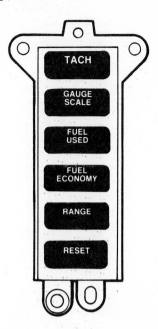

FIGURE 9-45. Right Switch Panel. (Courtesy of GM Product Service Training, General Motors Corporation)

The "reset" button allows the driver to reset the fuel-used and average fuel economy readings to zero, depending on which reading is displayed in the information center.

Specific Instrument Panel Cluster Functions

Data Communicating. Serial data from the BCM is the most important input to the IPC. The IPC controls its displays on the basis of the information received from the BCM on the serial data line. If this input information from the BCM is not available, the IPC will not function. Under this condition "Electrical Problem" will be displayed in the information center. An IPC wiring diagram which includes the serial data line to the BCM is shown in Figure 9-46.

Speedometer and Odometer Operation. The permanent-magnet type of vehicle speed sensor (VSS) is located in the transaxle. This sensor generates an AC voltage pulse that is proportional to vehicle speed. A buffer circuit in the BCM changes this signal into a DC square-wave digital signal. The BCM communicates the vehicle speed information to the IPC on the serial data circuit, and a decoder in the IPC decodes this information before the IPC microprocessor illuminates the speedometer display.

The BCM counts the VSS signal pulses to

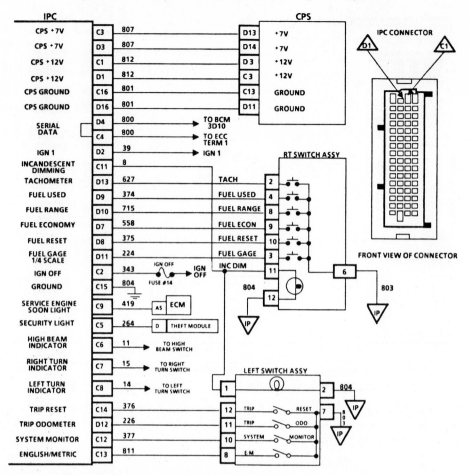

FIGURE 9-46. Instrument Panel Cluster Wiring Diagram. (Courtesy of GM Product Service Training, General Motors Corporation)

calculate mileage for the season odometer display. This total accumulated mileage is continuously written into the BCM's electronically erasable programmable read-only memory (EEPROM). If battery voltage is removed from the BCM, this information is retained in the EEPROM. The BCM also uses VSS input to operate the trip odometer, but this information is stored in the BCM's random access memory (RAM). Removal of battery voltage from the BCM results in the loss of trip odometer information. The IPC only displays either season odometer information or trip information at one time. As mentioned previously, this selection is made by pressing the "odo-trip" button.

The IPC and BCM circuits related to speedometer and odometer operation are illustrated in Figure 9-47.

Park/Reverse/Neutral/Drive/Low Display. The park/reverse/neutral/drive/low (PRNDL) display in the IPC is activated from the ignition-off circuit to

the IPC. This circuit supplies voltage to the IPC when the ignition switch is in the off or run position. Service and towing considerations require this display with the ignition off. In this ignition switch position, all other IPC displays are not illuminated. The gear selector connections to the BCM and ECM are shown in Figure 9-48.

Voltage Display. The BCM monitors voltage directly from the ignition 1 circuit, and compares this voltage to stored value limits. A digital signal from the BCM communicates this voltage information to the IPC, where the signal is decoded. On the basis of this information, the IPC microprocessor decides how many segments to illuminate on the voltage bar graph.

Coolant Temperature and Oil Pressure Displays. The coolant temperature sensor sends an input signal to the ECM. This input affects many ECM outputs. The ECM transmits this information to the

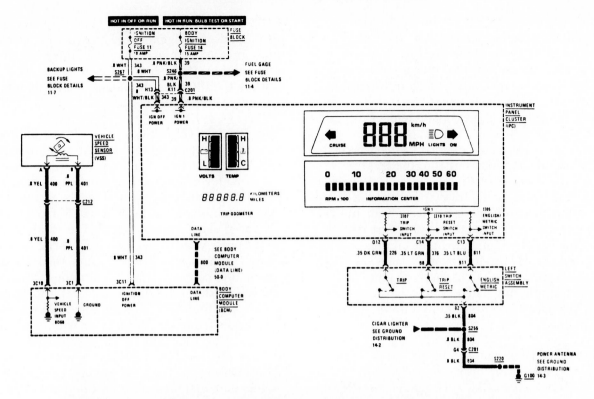

FIGURE 9-47. Instrument Panel Cluster and Body Computer Module Circuits Related to Speedometer and Odometer Operation. (Courtesy of GM Product Service Training, General Motors Corporation)

FIGURE 9-48. Gear Selector Switch Circuit to Body Computer Module and Electronic Control Module. (Courtesy of GM Product Service Training, General Motors Corporation)

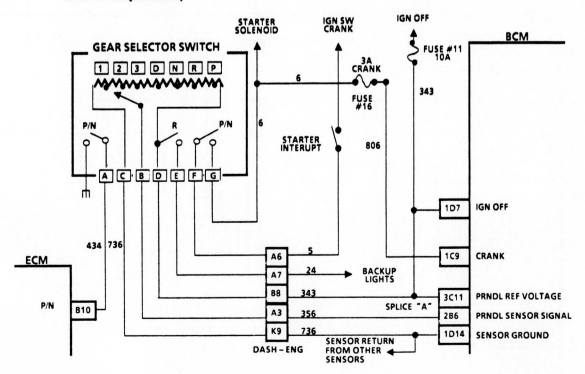

BCM on the serial data line, and the BCM repeats this signal and sends it to the IPC. The BCM also uses coolant temperature information for A/C system control. When the IPC microprocessor receives this coolant temperature information, it compares this value to its stored program to decide how many segments of the coolant temperature bar graph should be illuminated.

An oil pressure sensor in the oil gallery near the oil filter sends an input signal to the BCM in relation to oil pressure. This information is converted to a digital signal by the BCM and then sent to the IPC. When this information is received, the IPC compares it to the stored program before deciding how many bars to illuminate in the oil pressure bar graph. Some early production models had an off/on type of oil pressure switch. On these applications, the IPC illuminates half of the oil pressure bar graph segments as long as minimum oil pressure is available.

Tachometer Display. An engine speed signal is sent from the distributorless direct ignition system (DIS) module to the ECM. This signal is used by the

ECM to control some of its output functions. The ECM sends this signal to the BCM on the serial data line, and the IPC "eavesdrops" on this ECM to BCM information. If the driver has pressed the "tach" button, the IPC uses this engine speed signal to accurately illuminate the tach display.

Fuel Data Calculation. The fuel tank sending unit sends an input signal to the BCM in relation to the fuel level in the tank. Sending unit resistance varies with fuel level, and the BCM measures the voltage drop across the sending unit. A distance-travelled calculation is completed by the BCM from the vehicle speed sensor (VSS) signal. The ECM sends injector pulse width and flow rate information to the BCM. On the basis of these inputs, the BCM determines fuel consumption and supplies this data to the IPC on the serial data line. When the driver requests any fuel data function, the IPC uses this information from the BCM to provide the correct display. The fuel tank sending unit signal to the BCM is also used by the IPC to illuminate the fuel gauge bar graph. The BCM and IPC circuits related to fuel data displays are illustrated in Figure 9-49.

FIGURE 9-49. Body Computer Module and Instrument Panel Cluster Circuits Related to Fuel Data Display. (Courtesy of GM Product Service Training, General Motors Corporation)

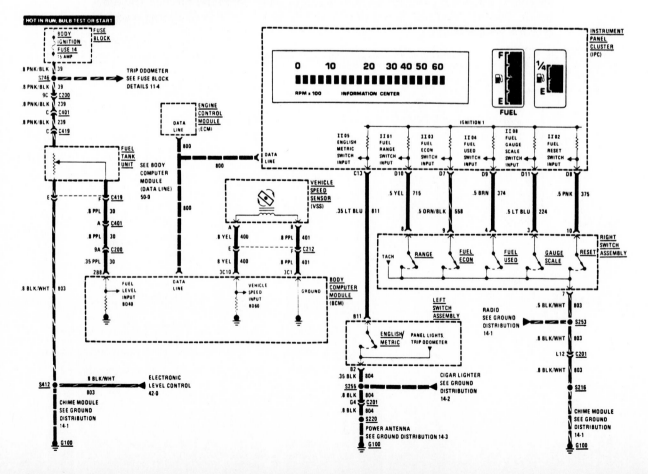

Lamp Monitor Operation. The lamp monitor circuits provide "lamp out" messages on the information center. Four circuits are connected between the BCM and the lamp monitor module. These circuits are "head lamp out," "tail lamp out," "stop lamp out," and "park lamp out." When all the lamps are working normally, the lamp monitor module connects these circuits to ground and causes a low circuit voltage. The input circuit from each lamp switch is connected through two equal resistance wires to the lamp monitor module. Output wires from these same lamp monitor module terminals are connected to the appropriate lamps. The lamp monitor circuit for the front and rear exterior lamps is shown in Figure 9-50.

If a lamp burns out, an open circuit is created and voltage increases at the lamp monitor module terminal to which the burned out lamp is connected. This voltage increase causes the lamp monitor module to open the appropriate lamp-out circuit from the BCM. This action results in the appropriate lamp-out communication from the BCM to the IPC, so that the IPC microprocessor displays the message in the information center.

Twilight Sentinel Operation. The twilight sentinel system keeps the exterior lights on for an adjustable length of time after the ignition switch is turned off. This system is operated by the twilight delay control on the left switch panel. When this switch is off, the lights and head lamp doors operate manually through the head lamp switch. If the twilight sentinel delay control is moved away from the off position, a variable resistor in the control is connected in series with the twilight photocell, and both of these components are connected to the BCM. The twilight photocell is a photoresistor that senses ambient light, and its resistance increases in dark conditions. The BCM senses the voltage drop across the photocell and the delay control. If the delay control is on and the photocell resistance is above a specific value because of dark conditions, the BCM turns the lights on. The BCM performs this function by grounding twilight head lamp relay and twilight park lamp relay windings. When this occurs, the head lamps, park lamps, and head lamp doors are turned on just as though the driver-controlled head lamp switch had been turned on. When the ignition switch is turned off, the BCM keeps the lights on

FIGURE 9-50. Lamp Monitor Circuit for Front and Rear Exterior Lamps. (Courtesy of GM Product Service Training, General Motors Corporation)

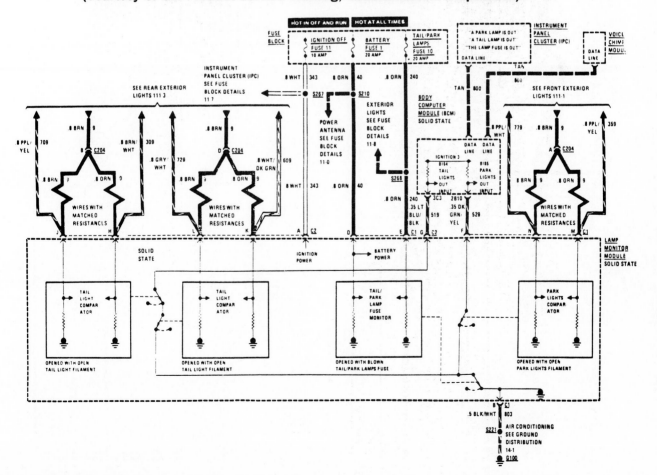

for a specific length of time, which is adjustable with the twilight delay control. The twilight sentinel circuit is illustrated in Figure 9-51.

Illuminated Entry. Battery voltage is applied directly through a fuse to the courtesy light bulbs, and the circuit is completed from each bulb through the courtesy light relay contacts to ground. These contacts are normally open. Voltage is also applied through the same fuse to the courtesy light relay winding, and the circuit from this winding is completed to ground through the BCM. Normally open switches are located in the door locks and door-jambs. If an exterior door handle is pushed, the door lock switch closes and signals the BCM to ground the courtesy light relay winding. This BCM action turns on the interior courtesy lights, and the lights remain on for 20 seconds. If the door is opened, the doorjamb switch signals the BCM to keep the courtesy lights on. When the door is closed, the courtesy lights go out immediately if the ignition switch is turned to the run position.

The twilight photocell input signal to the BCM affects the operation of the courtesy lights. On a bright, sunny day this signal informs the BCM that the courtesy lights are not required. Under this condition the BCM will not ground the courtesy light relay winding. The courtesy light circuit is illustrated in Figures 9-52 and 9-53.

Panel Dimming. The BCM controls the illumination of the left and right switch panels, and the radio, electronic climate control (ECC), and ash tray. Dimming of these lights is also controlled by the BCM in response to the input signal from the panel dimming control in the left switch panel. When the ignition switch is turned on, a 5-V signal is supplied from the BCM to the panel dimming control. The potentiometer in the panel dimming switch is grounded through the BCM. As the panel dimming control is moved toward the dim position, the voltage signal decreases from the switch to the BCM. When this signal is received, the BCM dims the switch panel and displays panel illumination.

A pulse width modulated (PWM) signal is sent from the BCM to the ECC panel and the radio-on circuit 724 to control the intensity of the vacuum fluorescent displays (VFDs) in these components. When the BCM increases the "on time" of the PWM signal, VFD intensity increases. The 724 circuit

FIGURE 9-51. Twilight Sentinel Circuit. (Courtesy of GM Product Service Training, General Motors Corporation)

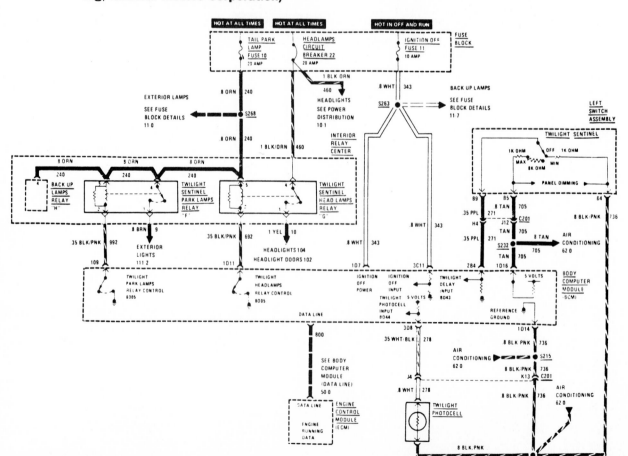

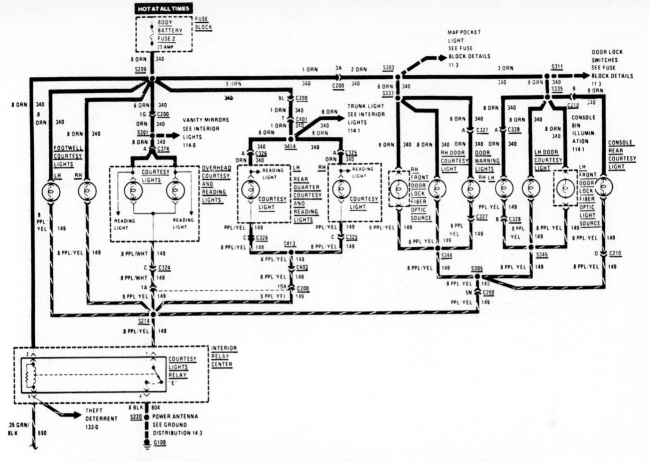

FIGURE 9-52. Illuminated Entry Courtesy Light Circuit. (Courtesy of GM Product Service Training, General Motors Corporation)

FIGURE 9-53. Illuminated Entry Courtesy Light Circuit. (Courtesy of GM Product Service Training, General Motors Corporation)

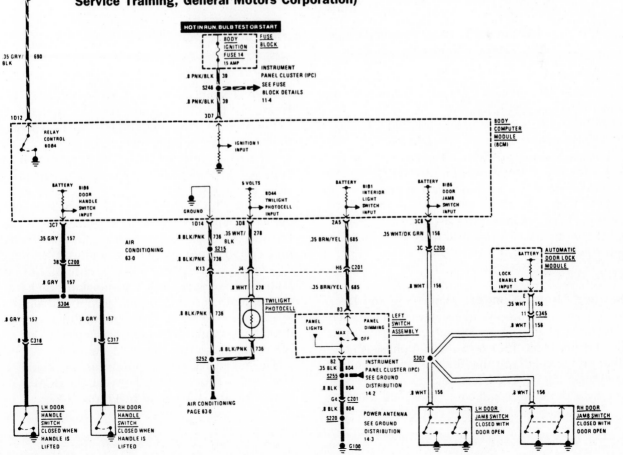

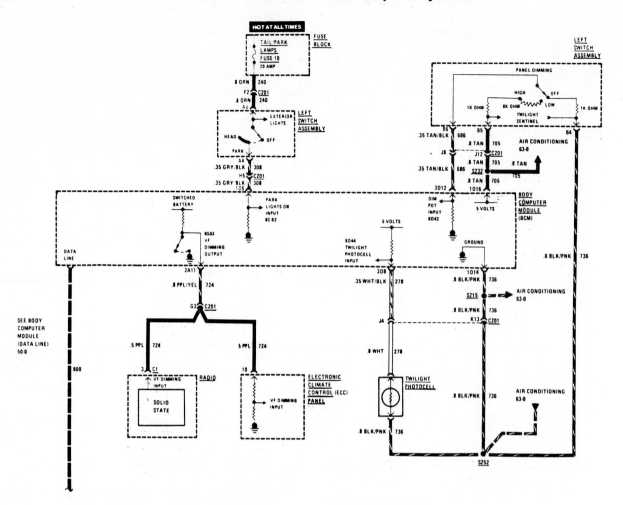

FIGURE 9-54. Panel Dimming Circuit. (Courtesy of GM Product Service Training, General Motors Corporation)

from the BCM to the ECC panel and the radio is shown in Figure 9-54.

The BCM also sends a panel dimming signal on the serial data line to the IPC. The IPC microprocessor uses this signal to control the intensity of the VFD in the IPC. A PWM signal is also sent from the IPC to the incandescent illumination bulbs in the left and right switch panels, ash tray, and face plates in the ECC panel, radio, and tape deck to control the intensity of these bulbs, as illustrated in Figure 9-55.

Charging Circuit Monitor. Many 1986 and later model General Motor vehicles have a new Delco Remy integral alternator. The integral regulator in this alternator has four terminals. A phase (P) terminal on the regulator may be used as a speed or tachometer signal. The L terminal may also be connected to the charge indicator bulb and parallel resistor. This terminal may also be connected to the BCM. The field (F) monitor terminal is connected to the BCM, and the sense (S) terminal is connected to the positive battery terminal.

When the ignition switch is turned on, voltage is supplied through the charge indicator bulb and the BCM to the L terminal and the lamp driver in the regulator. This action causes the transistor to turn on, which allows current to flow from the battery terminal through the transistor and the alternator field coil to ground. When the engine is started, the regulator controls the field current and limits the alternator voltage. While the alternator is charging, the regulator cycling signal is applied from the F terminal to the BCM. This signal informs the BCM if a defect occurs in the charging circuit, and trouble codes are set in the BCM memory.

The integral alternator circuit is shown in Figure 9-56, and the connections between the BCM and the regulator are illustrated in Figure 9-57.

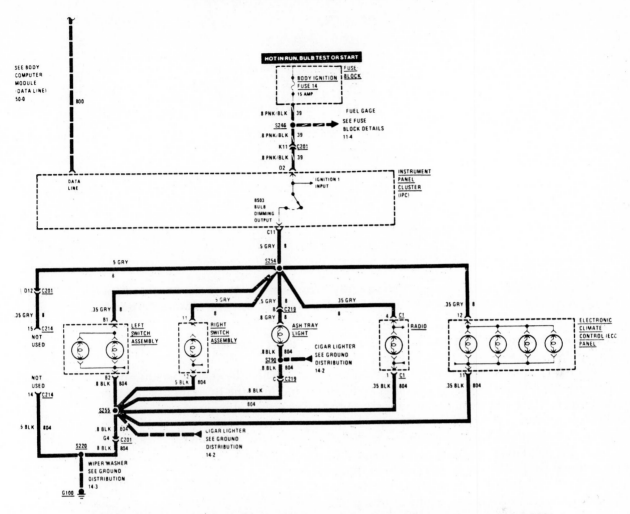

FIGURE 9-55. Panel Dimming Circuit to Incandescent Bulbs. (Courtesy of GM Product Service Training, General Motors Corporation)

FIGURE 9-56. Integral Charging Circuit. (Courtesy of GM Product Service Training, General Motors Corporation)

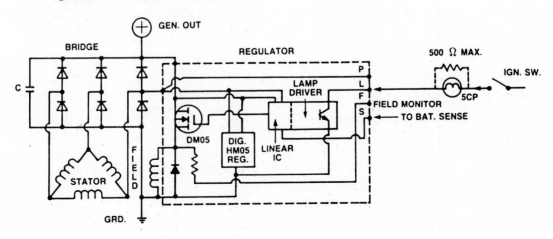

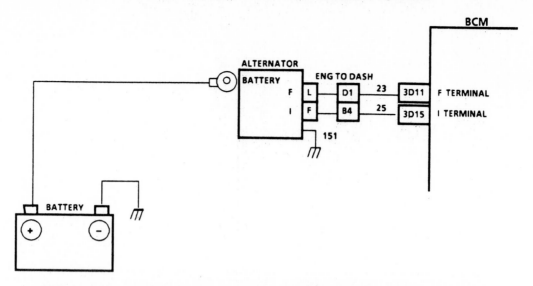

FIGURE 9-57. Integral Voltage Regulator to Body Computer Module Connections. (Courtesy of GM Product Service Training, General Motors Corporation)

Voice/Chime Warning System. Some General Motors vehicles are equipped with a chime module, while others have an optional voice/chime module. The chime module provides chime warnings and the voice/chime module gives voice warnings through the left-front radio speaker along with the chime warnings. When the BCM receives an input signal that indicates a driver warning is necessary, it sends a signal to the IPC on the serial data circuit and the IPC displays a visual warning on the information center. At the same time, the BCM signals the chime or voice/chime warning. The following chime warnings and information center displays are available:

1. Left door open—continuous slow chime.

2. Right door open—continuous slow chime.

3. Both doors open—continuous slow chime.

4. Head lamp out—5-second medium chime.

5. Tail lamp out—5-second medium chime.

6. Park lamp out—5-second medium chime.

7. Stop lamp out—5-second medium chime.

8. Lamp fuse out—5-second medium chime.

9. Low washer—5-second slow chime.

10. A/C system problem—5-second fast chime.

11. Generator problem—5-second fast chime.

12. Electrical problem—5-second fast chime.

13. Fasten seat belt—5-second slow chime.

The chime module circuit is illustrated in Figure 9-58.

When the vehicle is equipped with a voice/chime module, the BCM inputs and voice/chime warnings are the following:

1. Lights on—voice warning with a continuous fast chime.

2. Key in ignition—voice warning with a continuous fast chime.

3. Engine hot
   a) First warning—voice warning to turn off A/C and continuous fast chime.
   b) Second warning—voice warning to idle engine in park with continuous fast chime and information center display.
   c) Third warning—voice warning to turn off ignition switch with continuous fast chime and information center display.
   d) Fourth message—voice message that engine temperature has returned to normal.

4. Parking brake—voice warning with a continuous medium chime and information center display.

5. Brake fluid level—voice warning with a continuous fast chime and an information center display.

6. Engine oil pressure—voice warning with a continuous medium chime and an information center display.

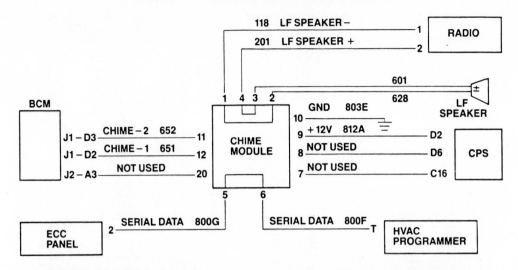

FIGURE 9-58. Chime Module Circuit. (Courtesy of GM Product Service Training, General Motors Corporation)

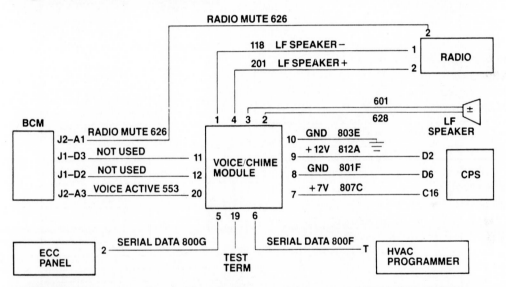

FIGURE 9-59. Voice/Chime Module Circuit. (Courtesy of GM Product Service Training, General Motors Corporation)

The same wiring harness connector will fit both the chime module and voice/chime module. A voice/chime module circuit is shown in Figure 9-59.

Relay Centers. Many General Motors vehicles have two relay centers, one usually located under the hood and the other positioned under the dash or in the console. Each relay is identified on a decal in the relay cover. A fuse panel may be located with the relay center, as pictured in Figure 9-60.

The BCM wiring varies, depending on the make of the car. A typical BCM wiring diagram is illustrated in Figure 9-61.

FIGURE 9-60. Relay Center and Fuse Panel. (Courtesy of GM Product Service Training, General Motors Corporation)

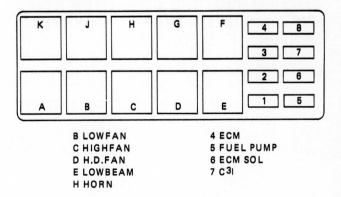

B LOWFAN
C HIGHFAN
D H.D.FAN
E LOWBEAM
H HORN

4 ECM
5 FUEL PUMP
6 ECM SOL
7 C31

Connector #1 / BLACK (D1–D16)

CIR DESC	WIRE COL	CKT NO		CKT NO	WIRE COL	CKT DESC
CRUISE BRAKE SW	BRN	86		801	BLK/RED	CPS GROUND
BRAKE FLD LEV SW	TAN/WHT	33		651	PPL/YEL	CHIME 1
STOP LAMP OUT	PPL/WHT	549		652	PPL/WHT	CHIME 2
C.C. SET/COAST SW	DK BLU	84		733	LT BLU	MIX DR. WIPER
PARK BRAKE SW	TAN/WHT	233		801	BLK/RED	CPS GROUND
SPARE				308	GRA/BLK	PARKLAMPS ON
SPARE				343	WHT	IGNITION OFF
HEADLAMPS ON	YEL	10				SPARE
CRANK INPUT	LT GRN	80		992	BLK/PNK	TWILIGHT PK REL
CPS 12 VOLT	RED/WHT	812A		760	PPL/WHT	BLOWER CONTROL
CRUISE ON/OFF	GRA/WHT	397		692	BLK/PNK	TWILIGHT H/L REL
CPS WAKE-UP	DK BLU/WHT	555		690	GRA/BLK	COURTESY LT REL
SPARE						SPARE
C.C. RESUME/ACCEL	GRA/BLK	87		736	BLK/PNK	5V RETURN (GND)
CPS 7 VOLT	PPL/WHT	807		801	BLK/RED	CPS GROUND
CPS 7 VOLT	PPL/WHT	807		705	TAN	5V REFERENCE

Connector #2 / BLACK (B1–B12)

CIR DESC	WIRE COL	CKT NO		CKT NO	WIRE COL	CKT DESC
RADIO MUTE	DK GRN/ORN	626				SPARE
IN CAR TEMP	DK GRN/WHT	734		402	LT GRN	CRUISE VAC ON
VOICE ACTIVE	TAN/WHT	553				SPARE
WASHER FLUID LEV	BLK/WHT	99		271	PPL	TWILIGHT DELAY
COURTESY LT SW	BRN/YEL	685		800A	TAN	SERIAL DATA
SPARE				356	GRA/BLK	PRNDL
HEADLAMP OUT	DK BLU/YEL	539				SPARE
CPS GROUND	BLK/RED	801		30	PPL	FUEL LEVEL WIPER
SPARE				398	TAN	CRUISE SERVO HI
CRUISE SERVO LO	LT BLU/BLK	399		529	DK GRN/YEL	PARK LAMP OUT
VF DIMMING	PPL/YEL	724				SPARE
SPARE				403	DK BLU/WHT	CRUISE VENT ON

Connector #3 / RED (D1–D16)

CIR DESC	WIRE COL	CKT NO		CKT NO	WIRE COL	CKT DESC
VSS RETURN	PPL	401				SPARE
SPARE						SPARE
TAILLAMP OUT	LT BLU/BLK	519				SPARE
VSS TO ECM	BRN	437		731	GRA/RED	A/C LO SIDE TEMP
SPARE				732	DK BLU	A/C HI SIDE TEMP
SUNLOAD SENSOR	LT BLU/YEL	590				SPARE
DOOR HANDLE SW	GRA	157		39	PNK/BLK	IGNITION REF
DOOR JAMB SW	WHT/DK GRN	156		278	WHT/BLK	TWI PHOTOCELL
SPARE				313	LT GRN	OIL PRESSURE
VSS FEED	YEL	400		800	TAN	SERIAL DATA
IGNITION OFF	WHT	343		23	GRA/WHT	GEN F TERMINAL
DRIVER DOOR AJAR	GRA/BLK	147		686	TAN/BLK	DIM CONTROL
KEY IN IGNITION	LT GRN	80		69	GRA	COOLANT LEVEL
PASS. DOOR AJAR	BLK/ORN	158		735	LT GRN/BLK	OUTSIDE TEMP
SEAT BELT SW	BLK/PNK	238		25	BRN	GEN I TERMINAL
PARK SIGNAL	LT GRN/BLK	275		721	WHT	LO REF PRESSURE

6E OLDS

FIGURE 9-61. Body Computer Module Wiring. (Courtesy of GM Product Service Training, General Motors Corporation)

GENERAL MOTORS COMPUTER SYSTEMS WITH GRAPHIC CONTROL CENTER

General Functions

Buick Rivieras manufactured in 1986 and later have a computer system in which many of the displays are shown on a graphic control center (GCC). These systems also have a BCM, ECM, IPC, and a HVAC programmer. The GCC contains a cathode ray tube (CRT), and a CRT controller (CRTC). This controller is interconnected with the other microprocessors through the 800 circuit serial data line, as indicated in Figure 9-62.

The ECM and BCM inputs and outputs are similar to those explained previously in this chapter, and the serial data line and central power supply (CPS) also operate in the same way as they do in other systems. A chime module is controlled from the IPC, as indicated in Figure 9-63.

The IPC contains a microprocessor that controls the vacuum fluorescent displays (VFDs). A switch pod is located on each side of the IPC. The IPC, switch pods, and GCC are shown in Figure 9-64.

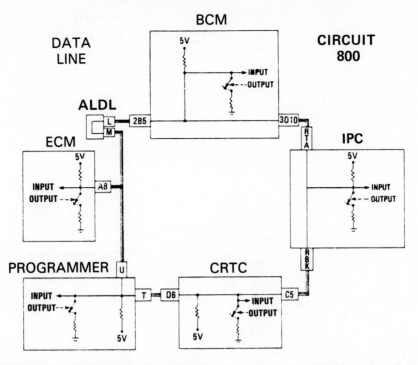

FIGURE 9-62. Serial Data Line 800 Circuit. (Courtesy of GM Product Service Training, General Motors Corporation)

FIGURE 9-63. Buick Riviera Electrical System Components. (Courtesy of GM Product Service Training, General Motors Corporation)

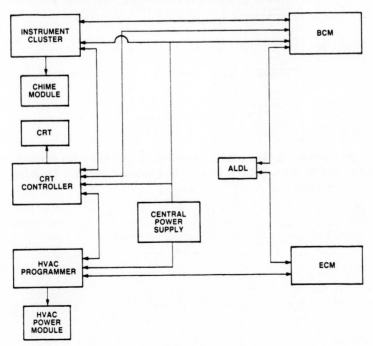

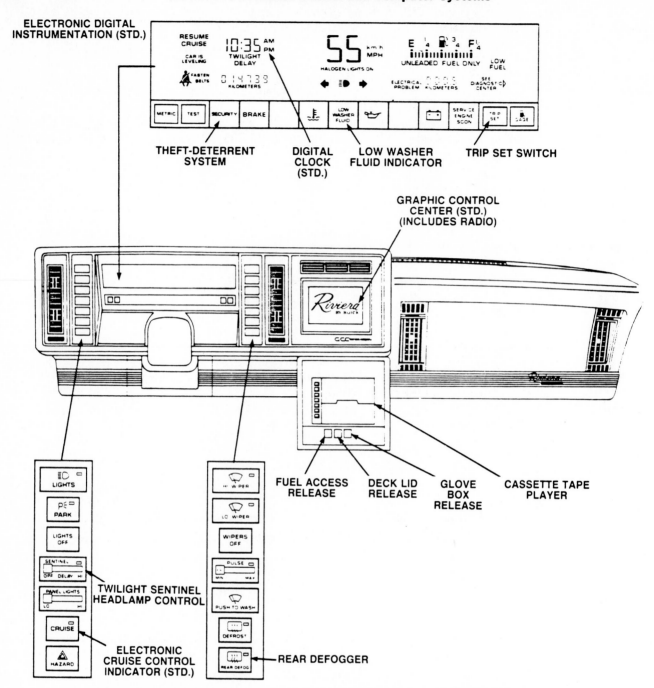

ELECTRONIC DIGITAL
INSTRUMENTATION (STD.)

THEFT-DETERRENT
SYSTEM

DIGITAL
CLOCK
(STD.)

LOW WASHER
FLUID INDICATOR

TRIP SET SWITCH

GRAPHIC CONTROL
CENTER (STD.)
(INCLUDES RADIO)

FUEL ACCESS
RELEASE

DECK LID
RELEASE

GLOVE
BOX
RELEASE

CASSETTE TAPE
PLAYER

TWILIGHT SENTINEL
HEADLAMP CONTROL

ELECTRONIC
CRUISE CONTROL
INDICATOR (STD.)

REAR DEFOGGER

**FIGURE 9-64. Buick Riviera Instrument Panel Cluster with Switch Pods and
Graphic Control Center. (Courtesy of GM Product Service Training, General
Motors Corporation)**

Graphic Control Center Displays

Display Operation. When a vehicle door is opened, the CRT in the GCC are illuminated as illustrated in Figure 9-65.

An invisible mylar switch panel is placed over the CRT screen. This switch panel contains ultrathin wires which are row and column encoded. The titles of the six available display pages are illuminated around the CRT screen. If the driver touches a display page title, the ultrathin wires make contact in that area, and the CRT controller displays the requested page in response to this input signal.

Summary Display Page. After the driver's door has been closed for 30 seconds, the GCC switches from the initial display to the summary page if the driver has not made any GCC requests. The summary page provides partial radio controls, partial A/C controls, partial diagnostic display, and the clock, as pictured in Figure 9-66.

Climate Control Page. If the driver selects the climate control page, all the A/C control switches appear on the CRT, with inside and outside temperature displays. The fan symbol rotates on the display, and its speed increases in relation to the blower speed. The climate control page is illustrated in Figure 9-67.

Trip Monitor Page. Three pages are available in the trip monitor function. If the driver selects "trip monitor," page 1, which indicates current fuel economy, range, and average fuel economy, is displayed, as shown in Figure 9-68.

When the driver touches "trip computer" on page 1 of the trip monitor, page 2 of the trip monitor

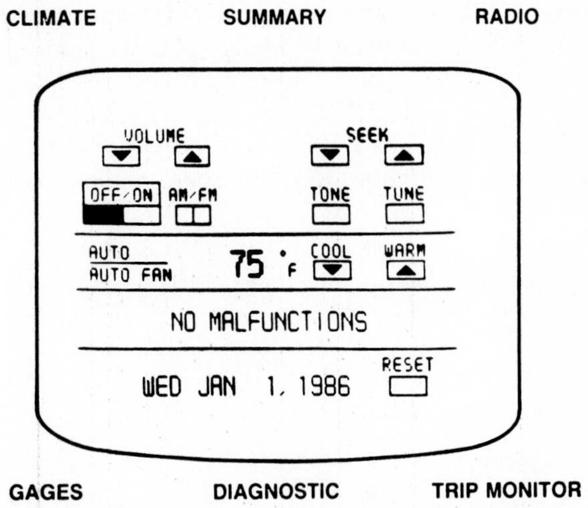

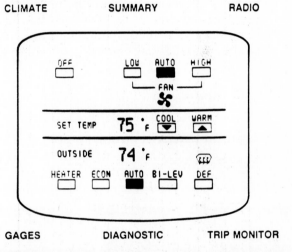

FIGURE 9-66. Graphic Control Center Summary Page. (Courtesy of GM Product Service Training, General Motors Corporation)

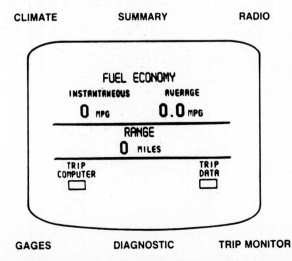

FIGURE 9-67. Graphic Control Center Climate Control Page. (Courtesy of GM Product Service Training, General Motors Corporation)

FIGURE 9-65. Initial Graphic Control Center Display. (Courtesy of GM Product Service Training, General Motors Corporation)

FIGURE 9-68. Graphic Control Center Trip Monitor Page. (Courtesy of GM Product Service Training, General Motors Corporation)

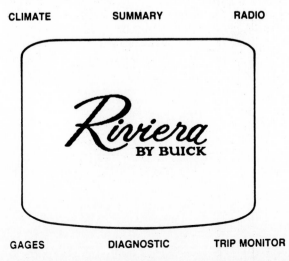

appears. This page may be used to enter the distance to a destination, and the GCC will calculate the estimated time of arrival based on average speed. The third trip monitor page is selected when "trip data" is touched on the first page. This page displays fuel consumption, average speed, and elapsed time.

Diagnostic Page. The diagnostic page monitors these five main areas:

1. Electrical
2. Power train
3. Brakes
4. Vehicle
5. Lamps

If there are no defects in these areas, the diagnostic page appears as shown in Figure 9-69.

The driver may touch any of these five main areas on the CRT for additional diagnosis. If the electrical category is selected, these systems are checked:

1. Engine controls
2. ECM and related circuits
3. Instrument cluster controls
4. Charging system

When the power train diagnosis is requested, engine temperature and oil pressure are diagnosed. The brake diagnosis displays the status of the brake

fluid level and the parking brake system. If the vehicle diagnosis is selected, these parameters are checked:

1. Climate control system
2. Driver and passenger doors ajar
3. Fuel level
4. Washer fluid level (optional)

An optional lamp diagnosis may be available. If this selection is made, the head lamps, tail lamps, and parking lamps are monitored.

Radio Page. The driver may preset five AM and FM stations with the radio page display. These five stations are displayed above the five preset switches. Adjustments for tone, fade, and five-band graphic equalizer may be adjusted on additional pages, depending on the radio option. The radio page is pictured in Figure 9-70.

Gauges Page. The gauges page provides bargraph displays of the following information:

1. Engine speed
2. Coolant temperature
3. Battery voltage
4. Oil pressure (optional)

The gauges page is shown in Figure 9-71.

FIGURE 9-69. Graphic Control Center Diagnostic Page. (Courtesy of GM Product Service Training, General Motors Corporation)

CLIMATE   SUMMARY   RADIO

```
        DIAGNOSTIC SYSTEM
             MONITOR
        ─────────────────────
        ELECTRICAL [OK] VEHICLE [OK]

        POWERTRAIN [OK]

            BRAKES [OK]
```

GAGES   DIAGNOSTIC   TRIP MONITOR

FIGURE 9-70. Graphic Control Center Radio Page. (Courtesy of GM Product Service Training, General Motors Corporation)

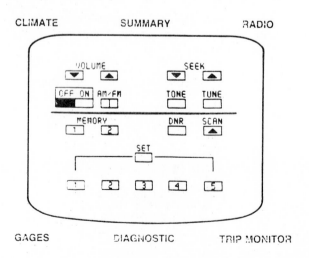

GAGES   DIAGNOSTIC   TRIP MONITOR

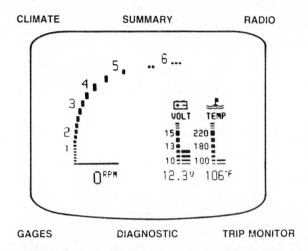

CLIMATE SUMMARY RADIO

GAGES DIAGNOSTIC TRIP MONITOR

FIGURE 9-71. Graphic Control Center Gauges Page. (Courtesy of GM Product Service Training, General Motors Corporation)

DIAGNOSIS OF GENERAL MOTORS THROTTLE BODY INJECTION SYSTEMS

Fuel Pump Circuit

A defective fuel pump relay may cause a slightly longer cranking time when the engine is started because the oil pressure switch contacts must close before the battery voltage can be supplied to the fuel pump. The fuel pump electrical circuit is shown in Figure 9-72.

When the ignition switch is turned on, it should be possible to hear the in-tank fuel pump run for approximately 2 seconds. If the fuel pump does not run, the circuit from the relay through the fuel pump may be tested by supplying battery voltage to terminal G in the 12-terminal assembly line diagnostic link (ALDL), as shown in Figure 9-73.

If the fuel pump does not operate when battery voltage is supplied to terminal G, the fuel pump or the circuit from the fuel pump through the upper fuel pump relay contacts is defective. The fuel pump must be replaced if it fails to run when battery voltage is supplied to the fuel pump connector at the fuel tank. If the fuel pump runs when battery voltage is supplied to the connector at the tank, but it did not run when battery voltage was sup-

plied to terminal G, the defect exists in the circuit from the tank to the relay. If the fuel pump operates when battery voltage is supplied to terminal G but fails to operate when the ignition switch is turned on, the fuel pump relay, connecting wires, or the ECM is defective.

In some vehicles, the fuel pump test connector is located in the engine compartment on the left fender shield rather than in the ALDL. A similar ALDL is used in TBI systems and 3C systems.

Trouble Codes

When performance or economy complaints are received on a TBI system, the ignition system and the engine compression should be tested before the TBI system is diagnosed. A check-engine light in the instrument panel comes on if a defect occurs in the TBI system. When the ignition switch is turned on, the check-engine light should be illuminated and it should remain on for a few seconds after the engine is started. The check-engine light will be on all the time if a continuous defect exists in the TBI system. When an intermittent defect occurs, the check-engine light will be on only when the defect is present. However, the trouble code from the defect is stored in the computer memory until the vehicle has been started 50 times. The check-engine light is illustrated in Figure 9-74.

When the test terminal is connected to the ground terminal in the ALDL with the ignition switch on, the system enters the diagnostic mode. In this mode, the check-engine light flashes out any trouble codes that are stored in the ECM memory. One flash followed by a pause and four flashes in quick sequence indicates trouble code 14. Trouble code 12, which indicates the diagnostic system is working, will be flashed first. The trouble codes are given in numerical order and are repeated three times. The check-engine light keeps repeating the trouble code sequence as long as the test connector is grounded. When the system is in the diagnostic mode, the ECM also energizes all the relays that are controlled by the ECM.

A trouble code indicates a defect in a specific circuit. For example, code 14 indicates that the engine coolant sensor or the connecting wires are defective, or that the ECM may not be able to receive the engine coolant sensor signal. The possible trouble codes and the operating requirements to set each code in the ECM memory are provided in Figure 9-75.

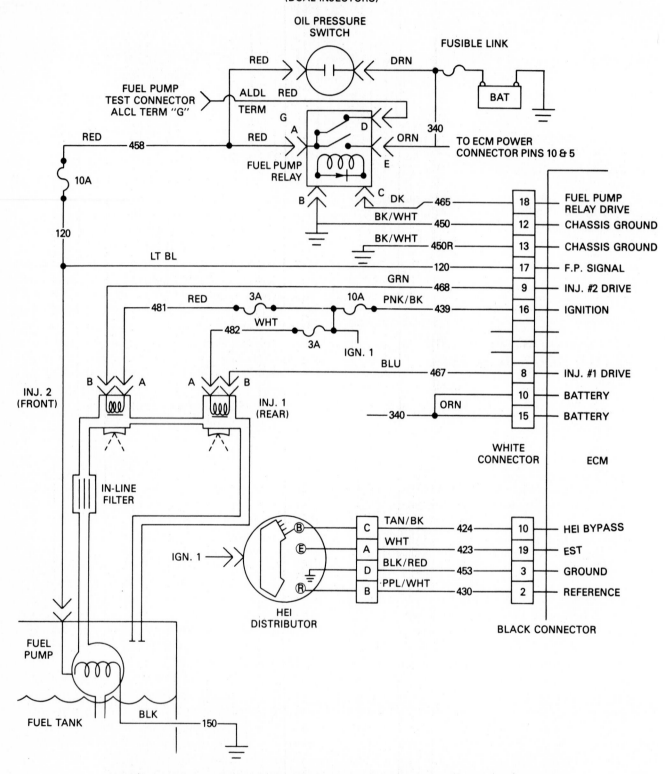

FIGURE 9-72. Electric Fuel Pump Circuit. (Courtesy of GM Product Service Training, General Motors Corporation)

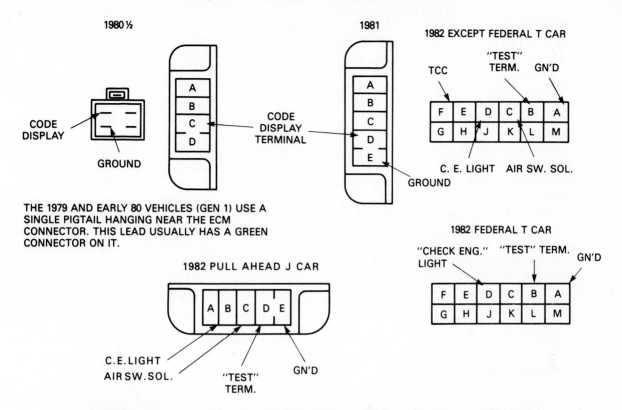

FIGURE 9-73. Assembly Line Diagnostic Link (ALDL). (Courtesy of GM Product Service Training, General Motors Corporation)

FIGURE 9-74. Check-Engine Light. (Courtesy of GM Product Service Training, General Motors Corporation)

DIAGNOSTIC CODE DISPLAY

CHECK ENGINE

CHECK ENGINE   PAUSE   CHECK ENGINE   CHECK ENGINE

FLASH
1

FLASH + FLASH
1 + 1 = 2

1 and 2 = CODE 12

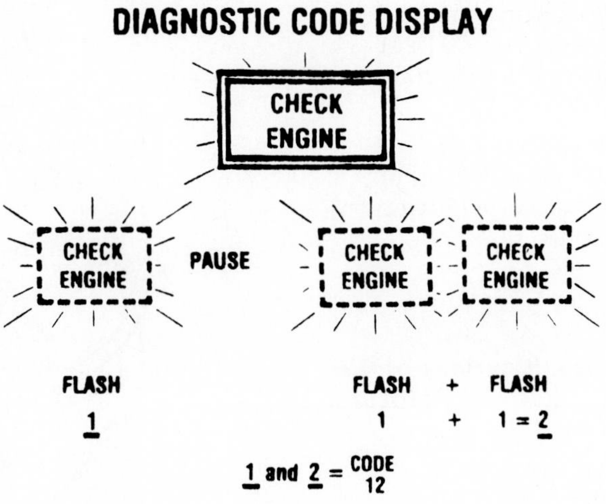

DIAGNOSTIC CODE IDENTIFICATION

Code 12—Will flash three times to indicate that system diagnostics are working. If there are any stored fault codes, they will flash following Code 12. After flashing stored code(s), Code 12 will again flash to indicate that all codes have been displayed.

Code 13—Oxygen Sensor Circuit—Engine running at normal operating temperature for about one minute (1200-1400 RPM in "neutral"). Oxygen sensor signal missing for 6 seconds.

Code 14—High Coolant Temperature—Engine running 2 seconds with no coolant sensor signal voltage.

Code 15—Low Coolant Temperature—Engine running for one minute. Coolant sensor signal too high for 2 seconds.

Code 21—High Throttle Position—Engine running below 1600 RPM, MAP less than 47 KPA (2.5 volts) and TPS is above 50% (2.5 volts) for 2 seconds.

Code 22—Low Throttle Position—Engine running with TPS signal voltage zero for 2 seconds.

Code 24—Vehicle Speed—Vehicle speed about 40-45 MPH steady throttle (decelerating 2.5 L) with no VSS signal for one minute.

Code 33—MAP Too High—Engine idling and MAP signal is above 6t KPA (3 volts) for 5 seconds.

Code 34—MAP Too Low—Engine running .025 seconds (250 m. seconds) with MAP signal voltage too low.

Code 42—EST—Open or grounded EST line, open or grounded bypass line, and engine speed above 500 RPM'.

Code 43—ESC—Engine running and ESC circuit 485 is less than 6 volts for 4 seconds.

Code 44—Lean Exhaust Indication—Engine running 2 minutes steady throttle above 2000 RPM with oxygen sensor signal less than 200 mv. (0.2 volts) for 60 seconds.

Code 45—Rich Exhaust Indication—Engine running 3 minutes steady throttle above 2000 RPM with oxygen sensor signal above 750 mv. (.075 volts) for 60 seconds.

Code 51—PROM Error

Code 55—A/D Error

FIGURE 9-75. TBI Trouble Codes. (Courtesy of GM Product Service Training, General Motors Corporation)

Code 55 is interpreted as an ECM error. This indicates a defective ECM, oxygen sensor, or 5-V supply wire from the ECM to some of the sensors. If the test terminal in the ALDL is grounded with the engine running, the system enters the field service mode, where the frequency of check-engine light flashes indicates whether the system is in the open-loop mode or closed-loop mode. The check-engine light flashes twice per second in the open-loop mode and once per second in the closed-loop mode.

When the necessary repairs have been completed, the trouble codes can be cleared from the computer memory by disconnecting the wire attached to the positive battery cable, as shown in Figure 9-76.

A complete wiring diagram of the ECM used on various engines is provided in Figure 9-77.

Idle Air Control (IAC) Motor Service

The IAC motor may be tested by removing the motor and reconnecting the wiring harness. With the ignition switch on and the test terminal in the ALDL grounded, the motor pintle should pulse in and out. Erratic idle operation may be caused by an IAC motor that is sticking. When the IAC motor is reinstalled, the measurement from the pintle tip to the motor housing should not exceed 1.25 in (32 mm), as shown in Figure 9-78.

FIGURE 9-76. Trouble Code Clearing. (Courtesy of GM Product Service Training, General Motors Corporation)

① BATTERY   ③ FUSIBLE LINK
② TO E.C.M.   ④ TO E.C.M. HARNESS

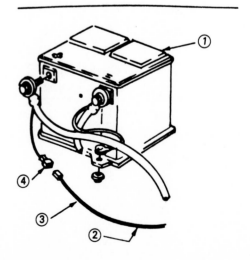

WHITE POWER CONNECTOR

1.8L 2.5L	2.0L	5.0L 5.7L		#	#		1.8L 2.5L	2.0L	5.0L 5.7L
	—	—	SPARE	1	24	SPARE	—	—	—
BRN	→	→	VEHICLE SPEED SENSOR	2	23	E-CELL		—	DK.BLU/ WHT.
WHT/BLK	→	→	DIAGNOSTIC TEST ALCL	3	22	4TH GEAR SWITCH	—	—	5.7L M.T. DK.GRN 5.0L M.T.
—	—	BLK	ELECTRONIC SPARK CONTROL		21	A/C CLUTCH	DK.GRN/ WHT	DK.GRN	→
ORN/BLK	→	→	PARK/NEUTRAL SWITCH	5	20	CHECK ENGINE LIGHT	BRN/WHT	→	→
—	—	LT.BLU	DUAL INJECTOR SELECT	6	19	CONVERTER CLUTCH	TAN/BLK	→	→
ORN	→	→	SERIAL DATA	7	18	FUEL PUMP RELAY DRIVE	DK.GRN/ WHT	→	→
BLU	2	2	INJECTOR #1	8	17	VOLTAGE MONITOR	TAN/WHT	LT.BLU	→
—	—	LT.GRN	INJECTOR #2	9	16	SWITCHED IGNITION	PNK/BLK	→	→
ORN	→	→	BATTERY	10	15	BATTERY	ORN	→	→
GRY	→	→	5 VOLT REFERENCE	11	14	MAP GROUND	BLK	→	1
BLK/WHT	→	→	ECM GROUND	12	13	ECM GROUND	BLK/WHT	→	→

1.8L 2.5L	2.0L	5.0L 5.7L		#	#		1.8L 2.5L	2.0L	5.0L 5.7L
PPL/WHT	PPL	PPL	CRANK SIGNAL	1	22	EGR	—	GRY	→
PPL/WHT	→	→	HEI REFERENCE	2	21	AMBIENT TEMP. SENSOR 1.8L	TAN	—	—
BLK/RED	→	→	HEI DIST. GROUND	3	20	MANIFOLD ABSOLUTE PRESSURE SIGNAL	LT.GRN	→	→
YEL	→	→	COOLANT SENSOR SIGNAL	4	19	EST SIGNAL	WHT	→	→
DK. BLU	→	→	T.P.S. SIGNAL	5	18	I.A.C. COIL "A" LO	LT.BLU/ BLK	LT.BLU RED	→
1-8L DK.GRN/ WHT.	—	5.7L M.T. GRY/RED	3RD GEAR SIGNAL	6	17	I.A.C. COIL "A" HI	LT.BLU/ BLK	LT.BLU/ RED	→
DK. BLU	GRY/RED	5.0L M.T. GRY/RED	A/C RELAY OR HOOD LOUVRE CONTROL	7	16	AIR DIVERT SO.	—	DK.GRN/ YEL	5.7L M.T. BLK/PNK 5.0l M.T.
PPL	→	→	OXYGEN SENSOR SIGNAL	8	15	OXYGEN SENSOR GROUND	TAN	→	→
1-8L LT.BLU/ BLK	DK.GRN/ YEL	→	COOLING FAN OR CANNISTER PURGE	9	14	AIR SWITCH SOLENOID	—	BRN	→
TAN/BLK	→	→	EST BYPASS	10	13	I.A.C. COIL "B" LO	LT.GRN/ BLK	→	→
BLK	→	→	COOLANT & TPS GROUND	11	12	I.A.C. COIL "B" HI	LT.GRN/ WHT	LT.GRN/ RED	→

BLACK I/O CONNECTOR

———▶ = THE SAME

1. BLK/WHT - 5.7L
   - 5.0L
2. DARK BLUE AT INJECTOR. LIGHT BLUE AT ECM.

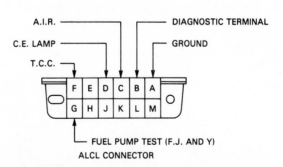

A.I.R. — DIAGNOSTIC TERMINAL
C.E. LAMP — GROUND
T.C.C.

F E D C B A
G H J K L M

FUEL PUMP TEST (F.J. AND Y)
ALCL CONNECTOR

FIGURE 9-77. Electronic Control Module (ECM) Terminals. (Courtesy of GM Product Service Training, General Motors Corporation)

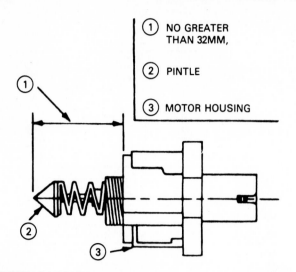

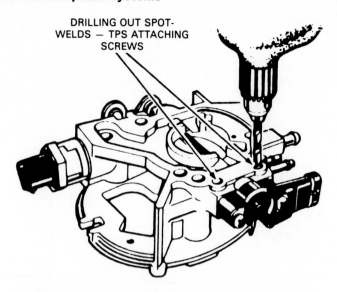

FIGURE 9-78. Idle Air Control Motor Installation. (Courtesy of GM Product Service Training, General Motors Corporation)

FIGURE 9-79. Throttle Position Sensor (TPS) Screw Removal. (Courtesy of GM Product Service Training, General Motors Corporation)

When an IAC motor has been removed, the ECM will not know the position of the motor pintle until the vehicle speed reaches 35 MPH (56 KPH) and the ECM drives the motor to the fully extended position. Until the vehicle reaches this speed, idling may be erratic. This problem can be overcome by reinstalling the IAC motor and disconnecting the speedometer cable from the speedometer head to the cruise control. With the engine idling, spin the speedometer cable by hand until the speedometer indicates more than 35 MPH (56 KPH). The ECM

will then fully extend the IAC motor and begin controlling the motor in the normal manner.

Throttle Position Sensor (TPS) Adjustment

The TPS sensor can be adjusted in some applications. In other systems, no means of adjusting the TPS is provided. Before the rear TPS screw can be loosened, the spot weld on the screw must be drilled out as illustrated in Figure 9-79.

FIGURE 9-80. Throttle Position Sensor (TPS) Adjustment. (Courtesy of GM Product Service Training, General Motors Corporation)

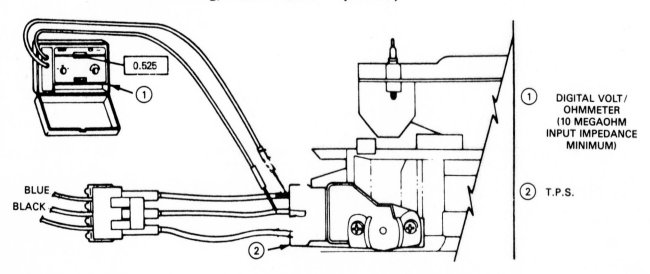

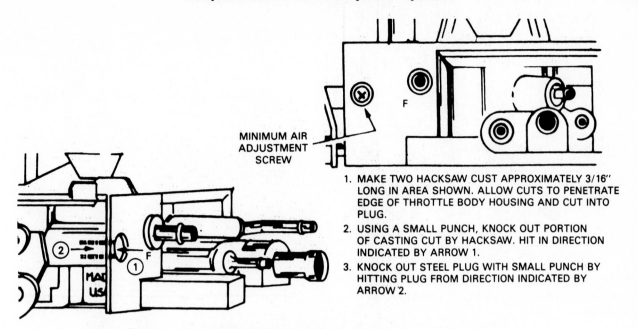

MINIMUM AIR
ADJUSTMENT
SCREW

1. MAKE TWO HACKSAW CUST APPROXIMATELY 3/16"
   LONG IN AREA SHOWN. ALLOW CUTS TO PENETRATE
   EDGE OF THROTTLE BODY HOUSING AND CUT INTO
   PLUG.
2. USING A SMALL PUNCH, KNOCK OUT PORTION
   OF CASTING CUT BY HACKSAW. HIT IN DIRECTION
   INDICATED BY ARROW 1.
3. KNOCK OUT STEEL PLUG WITH SMALL PUNCH BY
   HITTING PLUG FROM DIRECTION INDICATED BY
   ARROW 2.

FIGURE 9-81. Tamper-Resistant Plug Removal. (Courtesy of GM Product Service Training, General Motors Corporation)

With the ignition switch on and the engine stopped, connect a digital voltmeter to the TPS as illustrated in Figure 9-80. The voltmeter should read 0.450–0.600 V. If the correct reading is not obtained on the voltmeter, loosen the TPS screws and rotate the TPS until the voltage is within specifications. Hold the TPS in this position and tighten the mounting screws.

Minimum Air Rate Adjustment

The minimum air rate should not have to be adjusted unless some throttle body parts have been replaced. It may be necessary to remove the throttle body unit and then remove the tamper-resistant plug on the minimum air adjustment screw as outlined in Figure 9-81.

With the engine at normal operating temperature, plug the air passage to the IAC motor as shown in Figure 9-82. With the transmission in park, adjust the minimum air rate screw to obtain an engine speed of 500 RPM. The injectors may be replaced in the throttle body assembly as shown in Figure 9-83.

The injector winding may be tested for continuity with an ohmmeter. Do not connect a 12-V source to the injector winding for more than 5 seconds. The pressure regulator should not be disassembled.

FIGURE 9-82. Minimum Air Rate Adjustment. (Courtesy of GM Product Service Training, General Motors Corporation)

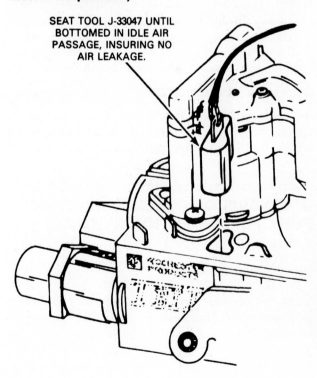

SEAT TOOL J-33047 UNTIL
BOTTOMED IN IDLE AIR
PASSAGE, INSURING NO
AIR LEAKAGE.

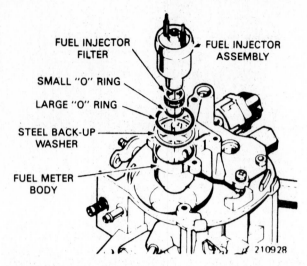

FIGURE 9-83. Injector Removal. (Courtesy of GM Product Service Training, General Motors Corporation)

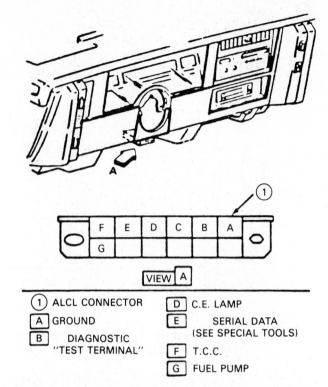

FIGURE 9-84. Assembly Line Diagnostic Link (ALDL). (Courtesy of GM Product Service Training, General Motors Corporation)

Fuel Pump Pressure Test

The fuel pump pressure gauge should be connected in the fuel inlet at the throttle body assembly. Before the gauge is connected, the pressure in the system should be relieved. This may be accomplished by grounding one injector terminal and connecting the other terminal to a 12-V source momentarily. Never loosen a fuel line without relieving the system pressure. The pressure gauge may be adapted to fit TBI systems with single or dual throttle body assemblies. With the ignition switch on, fuel pump pressure should equal manufacturer's specifications. Never turn on the ignition switch with a fuel line disconnected.

DIAGNOSIS OF GENERAL MOTORS PORT FUEL INJECTION SYSTEMS

Diagnostic Codes

Many PFI systems use a "service engine soon" light in place of the check-engine light that is used on General Motors 3C and throttle body injection systems. The diagnostic code diagnosis of PFI and SFI systems is very similar to the diagnosis of 3C and throttle body injection systems. When the diagnostic test terminal is connected to the ground terminal in the assembly line diagnostic link (ALDL)

with the ignition switch on, the "service engine soon" light flashes out any diagnostic codes stored in the ECM. The ALDL is pictured in Figure 9-84.

When the diagnostic test terminal is grounded, the ECM activates all the relays in the system and fully extends the IAC motor. The spark advance also remains in a fixed position when the diagnostic test terminal is grounded. When the initial timing is being checked, the instructions on the underhood emission label should be followed. The diagnostic codes for PFI systems are shown in Figure 9-85.

If the diagnostic test terminal is grounded with the engine running, the ECM enters the field service mode in which the frequency of "service engine soon" light flashes indicates whether the system is in open-loop or closed-loop mode.

Checking Minimum Air Rate

The following procedure should be followed to check the minimum air rate in the throttle body.

12 NO REFERENCE PULSES TO ECM	33 *MASS FLOW SENSOR READING TOO HIGH
13 OXYGEN SENSOR CIRCUIT FAILED	33 MAP SENSOR HIGH
14 COOLANT READING TOO HIGH	34 *MASS FLOW SENSOR READING TOO LOW OR NO SIGNAL FROM SENSOR
15 COOLANT READING TOO LOW	34 MAP SENSOR LOW
21 THROTTLE POSITION SENSOR READING TOO HIGH	41 *CAM SENSOR FAILED
22 THROTTLE POSITION SENSOR READING TOO LOW	41**CYLINDER SELECT ERROR
23**MANIFOLD AIR TEMPERATURE LOW	42 *ERROR IN DISTRIBUTOR OR C³ SYSTEM
24 VEHICLE SPEED SENSOR FAILED	43 ELECTRONIC SPARK CONTROL FAILURE
25**MANIFOLD AIR TEMPERATURE HIGH	44 OXYGEN SENSOR LEAN TOO LONG
31 *WASTEGATE ELECTRICAL SIGNAL OPEN OR GROUNDED	45 OXYGEN SENSOR RICH TOO LONG
	51 CALIBRATION PROM ERROR
32 EGR ELECTRICAL OR VACUUM CIRCUIT MALFUNCTION	52 *CALPAK MISSING
	53**OVER VOLTAGE CONDITION
	54**LOW FUEL PUMP VOLTAGE
	55 INTERNAL ECM ERROR (A/D CONVERTER)

NOTES
* = NEW CODES 1984
** = NEW CODES FOR 1985
NOT ALL CODES USED ON ALL APPLICATIONS

FIGURE 9-85. PFI Diagnostic Codes. (Courtesy of GM Product Service Training, General Motors Corporation)

1. With the engine at normal operating temperature and the ignition switch on, connect a jumper wire across terminals A and B in the ALDL.

2. Wait 30 seconds until the ECM extends the IAC motor.

3. Disconnect the IAC motor connector.

4. Connect a tachometer from the coil "tach" terminal to ground and start the engine.

5. Idle speed should be 475 to 525 RPM. If the idle speed is not within specifications, the minimum air rate screw in the throttle body must be adjusted. (This procedure was explained earlier in this chapter.)

Injector Testing

The injectors may be tested as outlined in Figure 9-86.

The various sensors and circuits may be tested by measuring the voltage at the ECM terminals with a digital voltmeter. Voltage specifications at each ECM terminal on an SFI turbocharged engine are provided in Figure 9-87.

Fuel Pump Pressure Test

The fuel pump pressure gauge must be connected to the test point on the fuel rail. Before the pressure gauge is connected, the fuel system pressure must be relieved. This may be done by disconnecting the injector terminals and connecting one injector terminal to ground while 12 V is supplied momentarily to the other terminal. Never turn on the ignition switch while a fuel line is disconnected. Do not attempt to loosen a fuel line or connect the pressure gauge until the fuel system pressure is relieved. With the ignition switch on, the fuel pump pressure should be 26–46 PSI (179–317 kPa).

Basic Timing

If the basic timing is being checked, the diagnostic test terminal must be connected to the ground terminal in the ALDL. On some older models, an underhood timing connector has to be grounded while checking basic timing. A single wire underhood in-line timing connector must be disconnected on some models. The underhood emission label should provide the correct basic timing procedure.

BEFORE PERFORMING THIS TEST, THE ITEMS LISTED BELOW MUST BE DONE.

- CHECK SPARK PLUGS AND WIRES.
- CHECK COMPRESSION.
- CHECK FUEL INJECTION HARNESS FOR BEING OPEN OR SHORTED.

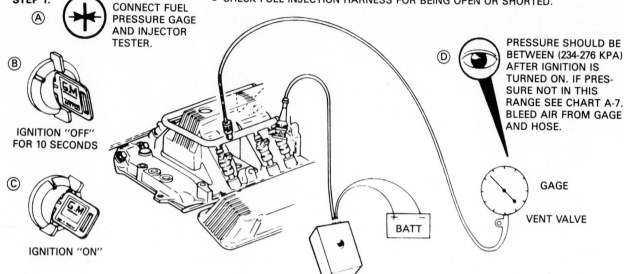

STEP 1.

Ⓐ CONNECT FUEL PRESSURE GAGE AND INJECTOR TESTER.

Ⓑ IGNITION "OFF" FOR 10 SECONDS

Ⓒ IGNITION "ON"

Ⓓ PRESSURE SHOULD BE BETWEEN (234-276 KPA) AFTER IGNITION IS TURNED ON. IF PRESSURE NOT IN THIS RANGE SEE CHART A-7. BLEED AIR FROM GAGE AND HOSE.

GAGE

VENT VALVE

BATT

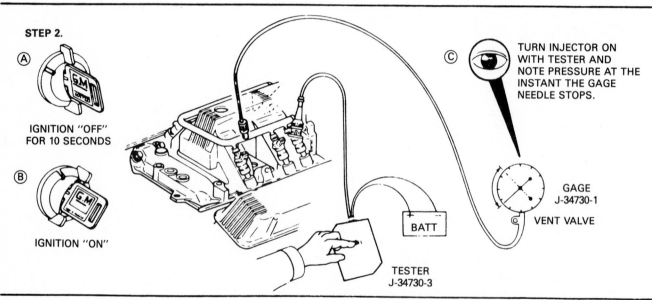

STEP 2.

Ⓐ IGNITION "OFF" FOR 10 SECONDS

Ⓑ IGNITION "ON"

Ⓒ TURN INJECTOR ON WITH TESTER AND NOTE PRESSURE AT THE INSTANT THE GAGE NEEDLE STOPS.

GAGE J-34730-1

VENT VALVE

BATT

TESTER J-34730-3

STEP 3.

REPEAT TEST AS IN STEP 2 ON ALL INJECTORS AND RECORD PRESSURE DROP ON EACH.

RETEST INJECTORS THAT APPEAR FAULTY. REPLACE ANY INJECTORS THAT HAVE A 10 KPA DIFFERENCE EITHER (MORE OR LESS) IN PRESSURE.

— EXAMPLE —

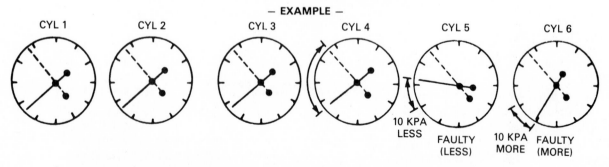

CYL 1   CYL 2   CYL 3   CYL 4   CYL 5   CYL 6

10 KPA LESS

FAULTY (LESS)

10 KPA MORE

FAULTY (MORE)

FIGURE 9-86. Injector Balance Test. (Courtesy of Buick Motor Division, General Motors Corporation)

FUEL INJECTION ECM CONNECTOR IDENTIFICATION

THIS ECM VOLTAGE CHART IS FOR USE WITH A DIGITAL VOLTMETER TO FURTHER AID IN DIAGNOSIS. THE VOLTAGES YOU GET MAY VARY DUE TO LOW BATTERY CHARGE OR OTHER REASONS, BUT THEY SHOULD BE VERY CLOSE.
THE FOLLOWING CONDITIONS MUST BE MET BEFORE TESTING.
● ENGINE AT OPERATING TEMPERATURE ● ENGINE IDLING IN CLOSED LOOP (FOR "ENGINE RUN" COLUMN) ●
● TEST TERMINAL NOT GROUNDED ● ALCL TOOL NOT INSTALLED ●

	VOLTAGE			CIRCUIT	PIN
	KEY "ON"	ENG. RUN	OPEN CRT.		
④	.13	13.48	0	FUEL PUMP RELAY	A1
⑥	12.46	ON.1 OFF13.8	12.21	A/C CLUTCH CONTROL	A2
	12.45	13.8	12.21	CANISTER PURGE CONTROL	A3
	12.45	13.8	12.21	EGR CONTROL	A4
	.14	13.77	12.19	"CHECK ENGINE" CONTROL (ALCL)	A5
	12.34	13.6	12.15	IGN.—ECM FUSE	A6
	12.40	13.78	12.20	TCC CONTROL ALCL	A7
③	4.30	2.5 4.5	.03	SERIAL DATA ALCL	A8
	4.95	4.95	.02	DIAG. TERM ALCL	A9
①	.49	.49	.55	SPEED SENSOR SIGNAL	A10
⑤	.11 10.75	11.3	.01 11.40	CAM HI	A11
	.07	.07	0	GRN'D	A12

PIN	CIRCUIT	VOLTAGE			
		KEY "ON"	ENG. RUN	OPEN CRT.	
B1	NOT USED				
B2	NOT USED				
B3	CRANK REF LO	.14	.14	0	
B4	EST CONTROL	.08	1.10	0	
B5	CRANK REF HI	.09 10.02	5.56	.03 10.43	⑤
B6	MASS AIRFLOW SENSOR SIGNAL	2.50	2.48	.01	
B7	ESC SIGNAL	8.71	8.75	9.60	
B8	A/C SIGNAL	ON 11.97 OFF .02	13.34 .02	12.2 .01	⑥
B9	NOT USED				
B10	PARK/NEUTRAL SW. SIGNAL	.06	.07	.04	
B11	NOT USED				
B12	INJ. 5	12.06	13.2		

A1   B1

BACK

24 PIN A-B CONNECTOR

				CIRCUIT	PIN
				NOT USED	C1
				NOT USED	C2
	.82	.83	0	IAC-B-LO	C3
	10.59	12.35	0	IAC-B-HI	C4
	10.59	12.35	0	IAC-A-HI	C5
	.82	.83	0	IAC-A-LOW	C6
	.13	.13	0	3RD GEAR SIGNAL	C7
	.13	.14	0	4TG GEAR SIGNAL	C8
		13.5			C9
②	2.04	2.18	.02	COOLANT TEMP SIGNAL	C10
				NOT USED	C11
	11.99	13.15	12.16	INJ. 6	C12
	.43	.42	.03	TPS SIGNAL	C13
	4.96	4.95	.02	TPS 5V REF	C14
	11.98	13.15	12.15	INJ. 2	C15
	11.89	13.73	12.16	BATT. 12 VOLTS	C16

D-PIN	CIRCUIT	KEY "ON"	ENG. RUN	OPEN CRT.	
D1	GRN'D	.0	.01	.04	
D2	NOT USED				
D3	WASTEGATE CONTROL	11.94	13.70	12.14	
D4	NOT USED				
D5	EST-BYPASS	4.56	4.55	0	
D6	GRN'D. (O_2)	.14	.14	0	
D7	O_2 SENSOR SIGNAL	.26	.51	0	③
D8	NOT USED				
D9	EGR DIAG	11.77	13.38	0	
D10	GRN'D	.07	.07		
D11	NOT USED				
D12	TPS/CTS GRN'd	.02	.03	.20	
D13	NOT USED				
D14	INJ. 1	11.90	13.15	12.12	
D15	INJ. 3	11.90	13.15	12.12	
D16	INJ. 4	11.90	13.15	12.12	

C1   D1

BACK

32 PIN C-D CONNECTOR

① Varies from .45 to battery voltage depending on position of drive wheels.

② Normal operating temperature.

③ Varies.

④ 12V first two seconds.

⑤ Depends on position of vane in relation to "hall-effect" switch. Voltage will be low when vane is passing through switch.

⑥ Engine running voltage will be high or low depending whether A/C is on or off.

FIGURE 9-87. ECM Terminal Identification and Voltage Specifications, SFI System with Turbocharger. (Courtesy of GM Product Service Training, General Motors Corporation)

DIAGNOSIS OF GENERAL MOTORS COMPUTER SYSTEM WITH BODY COMPUTER MODULE

General Diagnostic Procedure

The diagnostic procedure is entered by pressing the "off" and "warm" buttons in the electronic climate control panel simultaneously for 3 seconds with the ignition switch on. An exit from the diagnostic mode occurs if the "bi-level" button is pressed. If the "off" button is pressed, the diagnostic procedure returns to the next selection in the previous test mode. The "hi" and "lo" buttons are used to select level, test type, or device in the diagnosis. The "hi" button could be considered a "yes" command and the "lo" button a "no" command. For example, if a specific test mode appears in the instrument panel cluster (IPC) information center, the technician can select this test mode by pressing the "hi" button. When the technician does not want a certain test mode indicated in the information center, the "lo" button is pressed. The "hi" button is used to move on to the next parameter in a specific test mode, while the "lo" button returns the diagnostic procedure to previous parameter.

The diagnostic codes or parameters appear in the IPC information center and certain override values appear in the electronic climate control (ECC) display. For example, an override value of 0 to 99 is given for blower speed. The "warm" button increases the override value and blower speed, whereas the "cool" button may be used to decrease these values.

All of the diagnostic functions performed by the ECC buttons are illustrated in Figure 9-88.

ECM and BCM Diagnostic Codes

When the diagnostic mode is entered, all the segments in the IPC and ECC panel are illuminated for a short time followed by a display of ECM and BCM codes in the IPC information center. The ECM codes begin with the letter E, while the BCM codes begin with a B. The ECM codes are listed in Figure 9-89, the BCM codes in Figure 9-90, and diagnostic code comments in Figure 9-91.

FIGURE 9-88. Diagnostic Functions of Electronic Climate Control Buttons. (Courtesy of GM Product Service Training, General Motors Corporation)

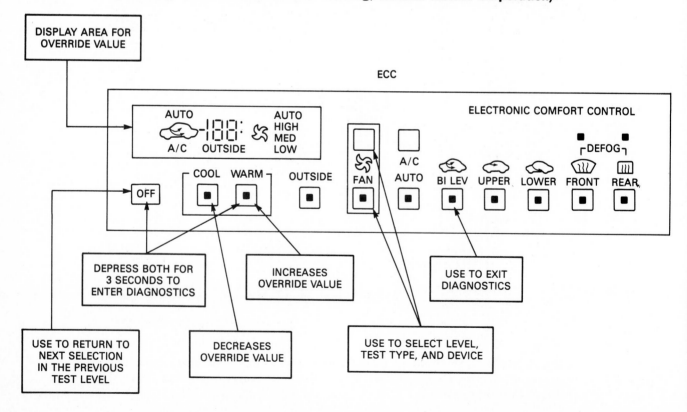

ECM DIAGNOSTIC CODES					
CODE	DESCRIPTION	COMMENTS	CODE	DESCRIPTION	COMMENTS
EQ13	OPEN OXYGEN SENSOR CIRCUIT (CANISTER PURGE)	Ⓐ	EQ37	MAT (ATS) SENSOR TEMPERATURE TOO HIGH	Ⓐ
EQ14	COOLANT SENSOR TEMPERATURE TOO HIGH	Ⓐ/Ⓕ/Ⓜ	EQ38	MAT (ATS) SENSOR TEMPERATURE TOO LOW	Ⓐ
EQ15	COOLANT SENSOR TEMPERATURE TOO LOW	Ⓐ/Ⓕ/Ⓜ	EQ40	OPEN POWER STEERING PRESSURE SWITCH CIRCUIT (A/C CLUTCH)	Ⓐ
EC16	SYSTEM VOLTAGE OUT OF RANGE (ALL SOLENOIDS)	Ⓐ	EQ41	CAM SENSOR CIRCUIT - (C³I)	Ⓐ
EQ21	THROTTLE POSITION SENSOR VOLTAGE TOO HIGH (TCC)	Ⓐ	EQ42	IGNITION SYSTEM - (C³I)	Ⓐ/Ⓙ
EQ22	THROTTLE POSITION SENSOR VOLTAGE TOO LOW (TCC)	Ⓐ	EQ43	ESC SYSTEM	Ⓐ
EQ24	SPEED SENSOR CIRCUIT (TCC)	Ⓐ	EQ44	LEAN EXHAUST SIGNAL	Ⓐ/Ⓘ
EQ29	OPEN FOURTH GEAR CIRCUIT	Ⓐ	EQ45	RICH EXHAUST SIGNAL	Ⓐ/Ⓘ
EQ32	EGR SYSTEM FAULT	Ⓐ	EQ47	BCM-ECM COMMUNICATION (A/C CLUTCH & CRUISE)	Ⓐ
EQ33	MAF SENSOR VOLTAGE HIGH	Ⓐ	EQ51	ECM PROM ERROR	Ⓐ/Ⓙ/Ⓚ
EQ34	MAF SENSOR VOLTAGE LOW	Ⓐ	EQ52	CALPAK ERROR	Ⓐ
			EQ55	ECM ERROR	Ⓐ

FIGURE 9-89. Electronic Control Module (ECM) Codes. (Courtesy of Oldsmobile Division, General Motors Corporation)

BCM DIAGNOSTIC CODES					
CODE	DESCRIPTION	COMMENTS	CODE	DESCRIPTION	COMMENTS
B110	OUTSIDE AIR TEMPERATURE CIRCUIT	Ⓑ/Ⓗ	B409	GENERATOR DETECTED CONDITION	Ⓑ
B111	A/C HIGH SIDE TEMPERATURE CIRCUIT	Ⓑ	B411	BATTERY VOLTS TOO LOW (CRUISE)	Ⓑ
B112	A/C LOW SIDE TEMPERATURE CIRCUIT (A/C CLUTCH)	Ⓑ/Ⓔ	B412	BATTERY VOLTS TOO HIGH (CRUISE)	Ⓑ
B113	IN-CAR TEMPERATURE CIRCUIT	Ⓑ	B440	AIR MIX DOOR	Ⓑ
B115	SUNLOAD TEMPERATURE CIRCUIT	Ⓑ	B445	COMPRESSOR CLUTCH ENGAGEMENT (A/C CLUTCH)	Ⓑ/Ⓔ
B118	DOOR JAM/AJAR CIRCUIT	Ⓒ	B446	LOW A/C REFRIGERANT CONDITION WARNING	Ⓑ
B119	TWILIGHT SENTINEL PHOTOSENSOR CIRCUIT	Ⓑ	B447	VERY LOW A/C REFRIGERANT (A/C CLUTCH)	Ⓑ/Ⓔ
B120	TWILIGHT SENTINEL DELAY POT CIRCUIT	Ⓑ	B448	VERY LOW A/C REFRIGERANT PRESSURE CONDITION (A/C CLUTCH)	Ⓑ/Ⓔ
B122	PANEL LAMP DIMMING POT CIRCUIT	Ⓑ	B449	A/C HIGH SIDE TEMPERATURE TOO HIGH (A/C CLUTCH)	Ⓒ/Ⓔ
B123	COURTESY LAMPS ON CIRCUIT	Ⓑ	B450	COOLANT TEMPERATURE TOO HIGH (A/C CLUTCH)	Ⓒ/Ⓔ
B124	SPEED SENSOR CIRCUIT (CRUISE)	Ⓗ	B552	BCM MEMORY RESET INDICATOR	
B127	PRNDL SENSOR CIRCUIT (CRUISE)	Ⓜ	B556	BCM EEPROM ERROR	Ⓓ
B131	OIL PRESSURE SENSOR CIRCUIT	Ⓜ	B660	CRUISE - TRANSMISSION NOT IN DRIVE (CRUISE)	Ⓒ
B132	OIL PRESSURE SENSOR CIRCUIT	Ⓜ	B663	CRUISE - CAR SPEED AND SET SPEED DIFFERENCE TOO HIGH (CRUISE)	Ⓒ
B334	LOSS OF ECM SERIAL DATA (CRUISE AND A/C CLUTCH)	Ⓐ/Ⓑ/Ⓔ/Ⓖ	B664	CRUISE - CAR ACCELERATION TOO HIGH (CRUISE)	Ⓒ
B335	LOSS OF ECC SERIAL	Ⓑ/Ⓖ/Ⓛ	B667	CRUISE - CRUISE SWITCH SHORTED (CRUISE)	Ⓑ
B336	LOSS OF IPC SERIAL DATA	Ⓑ/Ⓖ	B671	CRUISE - SERVO POSITION SENSOR CIRCUIT (CRUISE)	Ⓑ
B337	LOSS OF PROGRAMMER SERIAL DATA (A/C CLUTCH)	Ⓑ/Ⓔ/Ⓖ	B672	CRUISE - VENT SOLENOID CIRCUIT (CRUISE)	Ⓑ
B338	LOSS OF VOICE SERIAL DATA	Ⓑ/Ⓖ	B673	CRUISE - VACUUM SOLENOID CIRCUIT (CRUISE)	Ⓑ

FIGURE 9-90. Body Computer Module (BCM) Codes. (Courtesy of Oldsmobile Division, General Motors Corporation)

FIGURE 9-91. Diagnostic Code Comments. (Courtesy of Oldsmobile Division, General Motors Corporation)

DIAGNOSTIC CODE COMMENTS	
Ⓐ "SERVICE ENGINE SOON" INDICATOR LIGHTS	Ⓘ FORCES OL OPERATION
Ⓑ DISPLAYS DIAGNOSTIC MESSAGE ON IPC	Ⓙ CAUSES SYSTEM TO OPERATE ON BYPASS SPARK
Ⓒ NO INDICATOR LIGHT OR MESSAGE	Ⓚ CAUSES SYSTEM TO OPERATE ON BACK-UP FUEL
Ⓓ DISPLAYS "ERROR" IN SEASON ODOMETER	Ⓛ ECC DISPLAYS 3 DASHES
Ⓔ SWITCHES A/C COMPRESSOR "OFF", IF IN AUTO	Ⓜ APPROPRIATE SEGMENTS FLASH ON IPC
Ⓕ FORCES COOLING FANS ON	[ ] FUNCTIONS WITHIN BRACKET ARE DISENGAGED WHILE SPECIFIED MALFUNCTIONS REMAINS CURRENT
Ⓖ DISPLAYS "ELECTRICAL PROBLEM" ON IPC	
Ⓗ DISABLES ECI	

ECM Diagnostic Functions

When the ECM and BCM codes have been displayed, specific ECM diagnostic functions are displayed in the information center. These functions include data, inputs, and outputs. If the technician wants to enter one of these functions, the "hi" button must be pressed. When the "lo" button is pressed, the next function choice appears. Once a specific func-

tion is selected, the "hi" button advances the diagnosis to the next parameter in that function. The diagnosis of ECM data, inputs, and outputs is shown in Figure 9-92.

After the ECM outputs display, "clear ECM codes" appears in the information center. If ECM codes are in the computer memory, they may be erased by pressing the "hi" button for 3 seconds.

FIGURE 9-92. Electronic Control Module (ECM) Diagnostic Functions. (Courtesy of Oldsmobile Division, General Motors Corporation)

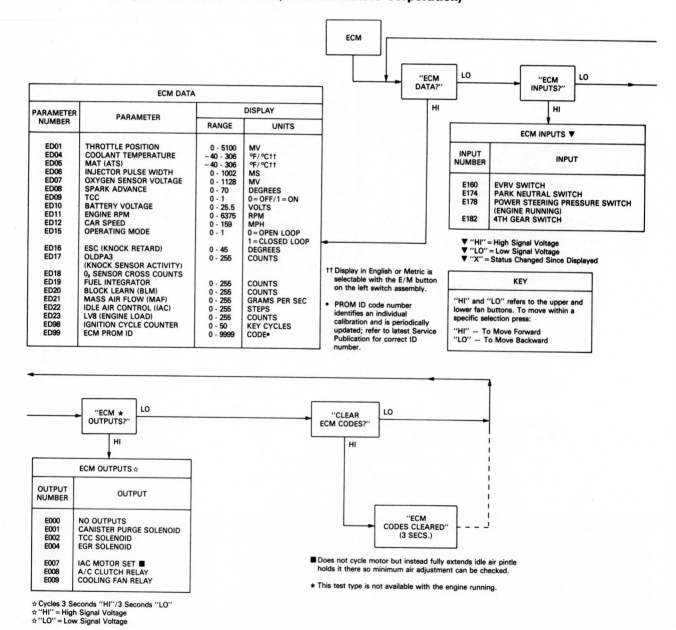

ECM DATA

PARAMETER NUMBER	PARAMETER	DISPLAY RANGE	DISPLAY UNITS
ED01	THROTTLE POSITION	0 - 5100	MV
ED04	COOLANT TEMPERATURE	−40 - 306	°F/°C††
ED05	MAT (ATS)	−40 - 306	°F/°C††
ED06	INJECTOR PULSE WIDTH	0 - 1002	MS
ED07	OXYGEN SENSOR VOLTAGE	0 - 1128	MV
ED08	SPARK ADVANCE	0 - 70	DEGREES
ED09	TCC	0 - 1	0 = OFF/1 = ON
ED10	BATTERY VOLTAGE	0 - 25.5	VOLTS
ED11	ENGINE RPM	0 - 6375	RPM
ED12	CAR SPEED	0 - 159	MPH
ED15	OPERATING MODE	0 - 1	0 = OPEN LOOP 1 = CLOSED LOOP
ED16	ESC (KNOCK RETARD)	0 - 45	DEGREES
ED17	OLDPA3 (KNOCK SENSOR ACTIVITY)	0 - 255	COUNTS
ED18	O₂ SENSOR CROSS COUNTS		
ED19	FUEL INTEGRATOR	0 - 255	COUNTS
ED20	BLOCK LEARN (BLM)	0 - 255	COUNTS
ED21	MASS AIR FLOW (MAF)	0 - 255	GRAMS PER SEC
ED22	IDLE AIR CONTROL (IAC)	0 - 255	STEPS
ED23	LV8 (ENGINE LOAD)	0 - 255	COUNTS
ED98	IGNITION CYCLE COUNTER	0 - 50	KEY CYCLES
ED99	ECM PROM ID	0 - 9999	CODE•

†† Display in English or Metric is selectable with the E/M button on the left switch assembly.

• PROM ID code number identifies an individual calibration and is periodically updated; refer to latest Service Publication for correct ID number.

ECM INPUTS ▼

INPUT NUMBER	INPUT
E160	EVRV SWITCH
E174	PARK NEUTRAL SWITCH
E178	POWER STEERING PRESSURE SWITCH (ENGINE RUNNING)
E182	4TH GEAR SWITCH

▼ "HI" = High Signal Voltage
▼ "LO" = Low Signal Voltage
▼ "X" = Status Changed Since Displayed

KEY

"HI" and "LO" refers to the upper and lower fan buttons. To move within a specific selection press:

"HI" — To Move Forward
"LO" — To Move Backward

ECM OUTPUTS ☆

OUTPUT NUMBER	OUTPUT
E000	NO OUTPUTS
E001	CANISTER PURGE SOLENOID
E002	TCC SOLENOID
E004	EGR SOLENOID
E007	IAC MOTOR SET ■
E008	A/C CLUTCH RELAY
E009	COOLING FAN RELAY

■ Does not cycle motor but instead fully extends idle air pintle holds it there so minimum air adjustment can be checked.

★ This test type is not available with the engine running.

☆ Cycles 3 Seconds "HI"/3 Seconds "LO"
☆ "HI" = High Signal Voltage
☆ "LO" = Low Signal Voltage

BCM Diagnostic Functions

The BCM diagnostic functions are available following the ECM diagnostic functions. BCM diagnostic functions include data, inputs, outputs and overrides, and these functions are selected with the same procedure as the ECM diagnostic functions. The BCM diagnostic functions are illustrated in Figure 9-93.

At the conclusion of the BCM diagnostic functions, "Clear BCM Codes" appears in the information center. These codes are erased in the same way as ECM codes. The IPC inputs shown in Figure 9-94 are available after the BCM diagnostic functions.

The BCM and ECM codes and diagnostic functions vary, depending on the model and year of vehicle. The codes and functions shown in this chapter are from a 1986 Toronado.

FIGURE 9-93. Body Computer Module (BCM) Diagnostic Functions. (Courtesy of Oldsmobile Division, General Motors Corporation)

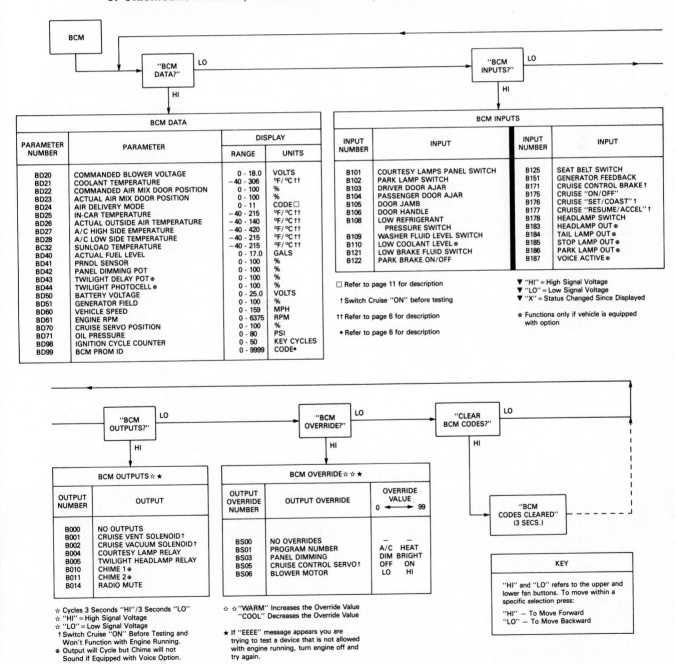

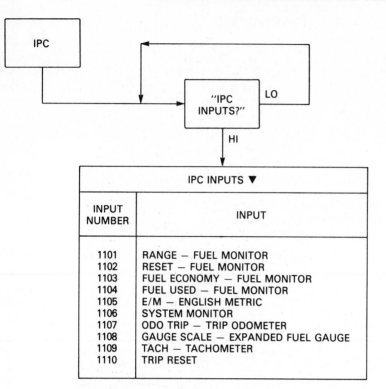

▼ "HI" = High Signal Voltage
▼ "LO" = Low Signal Voltage
▼ "X" = Status Changed Since Displayed

FIGURE 9-94. Instrument Panel Cluster (IPC) Input Diagnosis. (Courtesy of Oldsmobile Division, General Motors Corporation)

CHRYSLER SINGLE-POINT ELECTRONIC FUEL INJECTION (EFI) USED ON 2.2L ENGINE

Modules

Power Module. The power module is located in the left front fender well, behind the battery. Adequate cooling for the electronic components in the module is supplied by intake air flowing through the module before it enters the air cleaner. The power module supplies an 8-V signal to the logic module and the distributor pickup. A ground circuit for the automatic shutdown (ASD) relay is provided in the power module. When this ground circuit is completed, ASD relay supplies voltage to the electric fuel pump, logic module, ignition coil positive terminal, and the injector and ignition coil drive circuits in the power module.

The power module controls the operation of the fuel injector by opening and closing the injector

ground circuit. Another function of the power module is to open and close the circuit from the coil negative terminal to ground, which operates the ignition system. Commands from the logic module are used to control the power module illustrated in Figure 9-95.

Logic Module. The logic module is located inside the vehicle behind the right front kick pad. This module supplies a 5-V signal to the sensors in the system, and it also receives input signals from the

FIGURE 9-95. Power Module. (Courtesy of Chrysler Canada Ltd.)

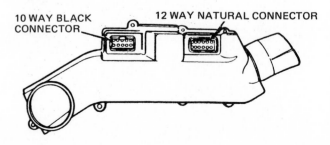

10 WAY BLACK CONNECTOR

12 WAY NATURAL CONNECTOR

sensors and the distributor pickup. On the basis of all the input signals received, the logic module sends the appropriate spark advance schedule to the power module under all engine operating conditions, and the power module opens the primary ignition circuit at the right instant to provide the correct spark advance. The logic module also commands the power module to supply the right injector pulse width, or "on time," to maintain engine performance, fuel economy, and emission levels. Other functions of the logic module include the operation of the exhaust gas recirculation (EGR) and canister purge solenoids and the automatic idle speed (AIS) motor.

The logic module has the capability to test many of its own input and output circuits. If a fault is found in a major system, the information is stored in the logic module for future reference. This fault code can be displayed to the service technician by a flashing power-loss light on the instrument panel, or by a digital reading on a tester which can be connected to the system.

The logic module places the system in a "limp-in" mode if an unacceptable signal is received from the manifold absolute pressure sensor, throttle position sensor, or coolant sensor. In this mode, the logic module ignores some of the sensor signals and maintains the spark advance and injector pulse width to keep the engine running, but fuel economy and engine performance decrease. In the "limp-in" mode, the logic module illuminates the power-loss light. The logic module is shown in Figure 9-96.

Switch Inputs

Park/Neutral Safety Switch. This switch supplies information to the logic module which is used by the module to control the automatic idle speed (AIS) motor and provide the correct idle speed in all transmission selector positions. The park/neutral safety switch signal to the logic module may affect spark advance to some extent.

Electric Backlite (EBL) Switch. When the EBL is turned on, the logic module operates the AIS motor to increase throttle opening slightly to compensate for the additional alternator load on the engine.

Brake Switch. In the event that the logic module does not receive a signal from the electric idle switch, the brake light switch is used to sense idle throttle position.

Air Conditioning Switch. If this switch is activated, the logic module operates the AIS motor to increase idle speed.

Air Conditioning Clutch Switch. The logic module activates the AIS motor to give a one-time kick, which maintains engine speed when the clutch engages and prevents variations in idle speed. The switch inputs to the logic module are indicated in Figure 9-97.

Sensor Inputs

Manifold Absolute Pressure (MAP) Sensor. The MAP sensor sends a signal to the logic module in relation to manifold vacuum and barometric pressure. A reference voltage of 5 V is applied from the logic module to the sensors. The MAP sensor sends a voltage signal of 0.3–4.9 V to the logic module. This voltage is 4.9 V at zero vacuum, and it can be as low as 0.3 V at maximum vacuum.

FIGURE 9-96. Logic Module. (Courtesy of Chrysler Canada Ltd.)

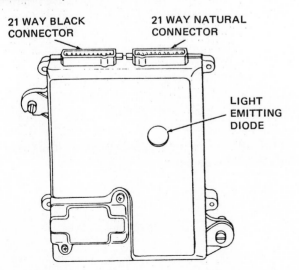

21 WAY BLACK CONNECTOR   21 WAY NATURAL CONNECTOR

LIGHT EMITTING DIODE

FIGURE 9-97. Switch Inputs to Logic Module. (Courtesy of Chrysler Canada Ltd.)

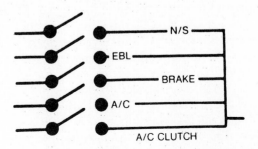

N/S

EBL

BRAKE

A/C

A/C CLUTCH

WIRING CONNECTOR

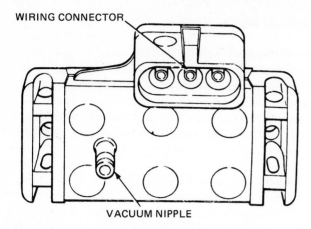

VACUUM NIPPLE

FIGURE 9-98. Manifold Absolute Pressure (MAP) Sensor. (Courtesy of Chrysler Canada Ltd.)

When engine load increases, manifold vacuum decreases, and this signal is sent from the MAP sensor to the logic module. When this signal is received, the logic module commands the power module to increase the injector pulse width, which supplies the additional fuel requirements of the engine. As manifold vacuum increases, the MAP sensor signal causes the logic and power modules to shorten the injector pulse width and supply less fuel to the engine.

The logic module uses the information from the MAP sensor and other inputs to determine the correct spark advance schedule under all engine operating conditions. A MAP sensor is pictured in Figure 9-98.

Throttle Position Sensor. The throttle position sensor is a variable resistor connected to the throttle shaft on the throttle body assembly. As the throttle is opened, a signal of 0.16–4.7 V is sent from the sensor to the logic module. This signal and other sensor information is used by the logic module to adjust the air-fuel ratio to meet various conditions during acceleration, deceleration, wide-open throttle, and idle.

Oxygen Sensor. The oxygen sensor is very similar to the oxygen sensor in the 3C system. (Refer to Chapter 8 for a description of this sensor.) This sensor generates a signal from 0 to 1 V as the air-fuel ratio becomes richer. The oxygen sensor signal is used by the logic module along with other sensor data to provide the air-fuel ratio for optimum engine performance, fuel economy, and emission levels. When there is a need for additional fuel enrichment, such as on sudden acceleration, the MAP sensor and throttle position sensor signals may override the oxygen sensor signal.

Coolant Temperature Sensor. The coolant temperature sensor is mounted in the thermostat housing. The resistance values of the resistive element vary from 11,000 Ω at −4°F (−15°C) to 800 Ω at 195°F (90.5°C). Along with other input data, the coolant sensor signal is used by the logic module in the scheduling of idle speeds, air-fuel ratio, and spark advance curves for all engine operating conditions. When the engine is cold, the idle speed increases, the air-fuel ratio is enriched, and the spark advance curve is altered to improve cold-engine performance.

Vehicle Speed Sensor. The vehicle speed sensor is located in the speedometer cable. This sensor contains an on/off microswitch that generates eight pulses per speedometer cable revolution. The vehicle speed sensor signal and the throttle position sensor signal are interpreted by the logic module to tell the difference between closed-throttle deceleration and normal idle speed with the vehicle not in motion. During deceleration the logic module operates the AIS motor to maintain a slightly higher idle speed to reduce emissions. The vehicle speed sensor is illustrated in Figure 9-99.

Distributor Pickup Signal. A Hall Effect switch unit is used in the distributor pickup assembly. Four metal shutter blades, one for each cylinder, are attached to the bottom of the distributor rotor. Each time one of these shutter blades rotates through the magnetic field of the Hall Effect switch, a pulse signal is generated by the switch. This pulse signal is sent to the logic module and the power module. When the distributor pickup signal is received, the power module grounds the automatic shutdown (ASD) relay winding, which supplies voltage to the fuel pump, ignition coil positive terminal, and the injector drive and coil drive circuits of the power module. If the distributor pickup signal is not present or is not correct, the ASD relay is not energized by the power module.

The logic module uses the distributor pulse signals as speed information to help determine the spark advance. Piston position is also determined

FIGURE 9-99. Vehicle Speed Sensor. (Courtesy of Chrysler Canada Ltd.)

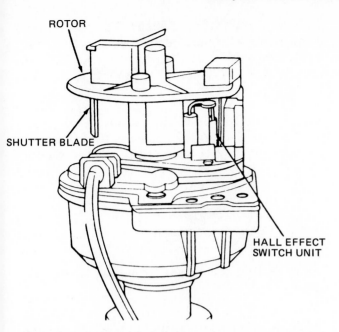

FIGURE 9-100. Hall Effect Distributor Pickup. (Courtesy of Chrysler Canada Ltd.)

by the logic module from the distributor pulse signal. The logic module signals the power module to operate the injector when the piston position signal is received from the distributor pickup.

A distributor with the Hall Effect switch and shutter blade is shown in Figure 9-100. All the sensor inputs to the logic module are shown in Figure 9-101. (A complete explanation of Chrysler electronic ignition systems is provided in Chapter 6.)

FIGURE 9-101. Sensor Inputs to Logic Module. (Courtesy of Chrysler Canada Ltd.)

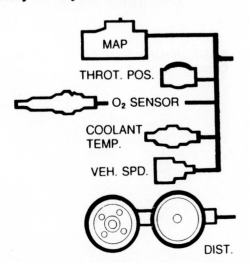

Automatic Shutdown (ASD) Relay

Operation. When the ignition switch is turned on, voltage is supplied through the J2 circuit to the power module, and the power module supplies 8 V to the logic module and the distributor pickup. The power module also supplies voltage through the fused J2 (FJ2) circuit to the ASD relay. When this occurs, the power module grounds the ASD relay winding and voltage is supplied through the relay contacts to the electric fuel pump, logic module, and coil positive terminal.

When the ASD relay closes, the injector drive and ignition coil drive circuits are activated in the power module. The logic module commands the power module to send a single "prime shot" of fuel to the engine when the ASD relay closes. This is accomplished by the power module grounding the injector winding for a brief instant. If the engine is not cranked within 1/2 second, the power module opens the circuit from the ASD relay winding to ground and the relay contacts open the circuit to the fuel pump, logic module, and the coil positive terminal. If the power module does not sense battery voltage and distributor pulses at the rate of 60 revolutions per minute (RPM) within 1/2 second the power module opens the circuit from the ASD relay winding to ground. This causes the relay contacts to open, and the voltage supplied to the electric fuel pump, logic module, and coil positive terminal is shut off. The power module also opens the circuit to the injector and deactivates the ignition coil drive circuit in the module.

The ASD relay is a safety feature that shuts down the fuel pump and the ignition system if the vehicle is involved in an accident and the ignition switch is left on. The power module grounds the ASD relay winding continuously while the module receives distributor pulses at a rate above 60 RPM. This keeps the ASD relay contacts closed so that voltage is supplied to the circuits mentioned previously.

The ASD relay is located under the right front kick pad with the logic module and MAP sensor. The ASD relay and the entire EFI system are illustrated in Figure 9-102.

Fuel System Components

Electric Fuel Pump. The roller-vane type fuel pump is driven by a permanent magnet electric motor. A check valve on the inlet side of the pump limits the maximum pump pressure to 120 PSI (827 kPa) if the fuel system becomes completely plugged

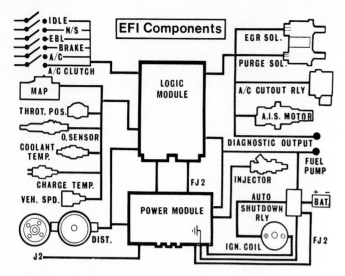

FIGURE 9-102. EFI System with ASD Relay. (Courtesy of Chrysler Canada Ltd.)

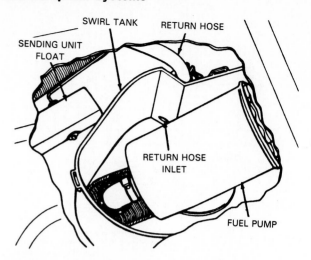

FIGURE 9-104. Swirl Tank. (Courtesy of Chrysler Canada Ltd.)

or restricted. Another check valve in the pump outlet prevents any movement of fuel in either direction when the pump is not in operation. A 70-micron filter is provided by the fuel inlet sock on the pump inlet, which prevents water and other foreign particles from entering the fuel system. The fuel pump is shown in Figure 9-103.

The electric fuel pump and the fuel gauge sending unit are located in separate openings in the fuel tank. A swirl tank surrounds the fuel pump in the fuel tank. The return fuel line is connected from the throttle body assembly to the swirl tank. Excess fuel is being returned continuously from the throttle body assembly to the swirl tank while the engine is running. The swirl tank pictured in Figure 9-104 provides a supply of fuel at the pump inlet during all driving conditions.

The fuel flows through a low-pressure orifice in the end of the return hose, which creates a slight low pressure area at the end of the hose so that fuel flow from the main fuel tank to the swirl tank is increased. A return line check valve prevents fuel flow from the tank into the return line if the vehicle is rolled over in an accident.

FIGURE 9-103. Electric Fuel Pump. (Courtesy of Chrysler Canada Ltd.)

In-Line Fuel Filter. Besides the filter sock on the pump inlet, a 50-micron in-line fuel filter, as shown in Figure 9-105, is located near the fuel tank under the vehicle.

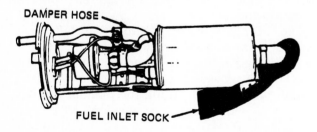

FIGURE 9-105. In-Line Fuel Filter. (Courtesy of Chrysler Canada Ltd.)

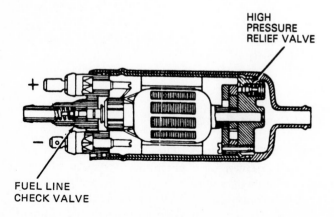

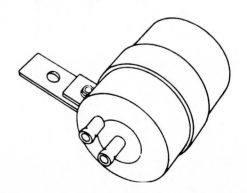

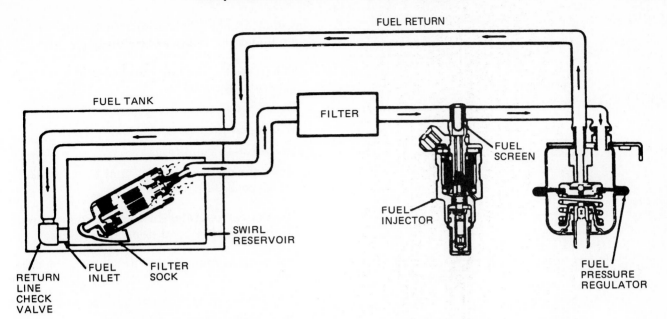

FIGURE 9-106. Fuel System with Injector and Pressure Regulator. (Courtesy of Chrysler Canada Ltd.)

Throttle Body Assembly. The throttle body assembly contains the injector, pressure regulator, automatic idle speed (AIS) motor, throttle position sensor, and throttle valve. The fuel pump supplies fuel through the in-line filter to the injector and pressure regulator. When pump pressure reaches 36 PSI (248 kPa), the pressure regulator diaphragm is forced downward by the fuel pressure and excess fuel is returned from the pressure regulator through the return line to the fuel tank, as illustrated in Figure 9-106.

A special high-pressure fuel hose and clamps are used in the EFI system. A vacuum hose is connected from the area above the throttle in the throttle body to the pressure regulator spring chamber. Since the injector tip is mounted above the throttle, this tip is subjected to venturi vacuum, which increases with engine speed. Therefore, a pressure decrease occurs at the injector tip at wide throttle openings. When this occurs, the pressure regulator vacuum signal assists the fuel pressure to move the regulator diaphragm downward. This action moves the diaphragm downward at a lower pressure than at idle speed, when no vacuum is applied to the pressure regulator. Thus, fuel pressure to the injector is decreased when pressure at the tip of the injector is decreased at high speed, and a constant pressure drop of 36 PSI (248 kPa) is maintained across the injector.

The injector has a winding surrounding the movable plunger. Voltage is supplied to the injector

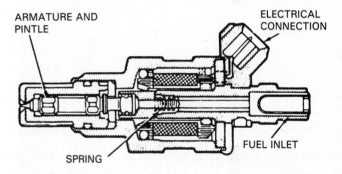

FIGURE 9-107. Injector Design. (Courtesy of Chrysler Canada Ltd.)

winding from the power module and the module also grounds the other end of the injector winding to energize the injector. The internal design of the injector is shown in Figure 9-107.

Air enters the throttle body assembly from the air cleaner. When the injector is energized, fuel is sprayed from the injector into the airstream above the throttle.

Output Control Functions

Fuel Injection Control. The power module energizes the injector once for each piston intake stroke, and the amount of fuel delivered by the injector is determined by the pulse width, or "on time," of the injector. The pulse width is measured

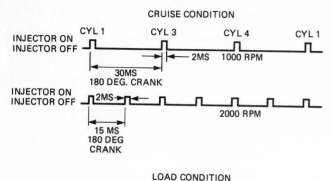

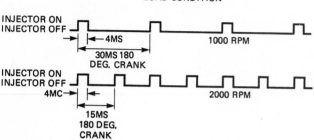

FIGURE 9-108. Injector Pulse Width. (Courtesy of Chrysler Canada Ltd.)

in milliseconds (ms), and the fuel flow from the injector increases in relation to the pulse width. Input data received by the logic module indicates the engine fuel requirements. The logic module commands the power module to supply the precise injector pulse width to meet these fuel requirements. As indicated in Figure 9-108, the pulse width will be 2 ms at 1,000 RPM cruise conditions and the entire injector on/off time is 30 ms.

If the engine is operating at a 2,000 RPM cruise condition, the entire injector on/off time is only 15 ms because the crankshaft is turning much faster. Under this condition, the injector pulse width is still 2 ms, but more fuel is delivered by the injector because the injector is being energized every 15 ms. When the engine is operating under heavy load conditions at 1,000 RPM to 2,000 RPM, the injector pulse width is increased to 4 ms. The logic module and the power module supply the correct pulse width to maintain engine performance, fuel economy, and emission levels under all operating conditions. When the engine is being started, the injector is energized twice for each piston intake stroke. The injector pulse width is increased when starting a cold engine to provide the necessary mixture enrichment.

Under certain operating conditions, the system operates in open-loop mode. In this mode, the oxygen sensor signal is ignored by the logic module and the module maintains the air-fuel ratio at a pre-

determined value. The system remains in open-loop mode under the following conditions:

1. Cold engine operation, until the oxygen sensor generates a signal.
2. Park/neutral idle operation.
3. Wide-open throttle operation.
4. Deceleration conditions.
5. When the oxygen sensor signal is not available for a specified period of time.

The system operates in the closed-loop mode when the following conditions are present:

1. The coolant temperature is above a specified value.
2. The start-up delay timer in the logic module has timed out.
3. The oxygen sensor is generating a valid signal to the logic module.
4. The vehicle is being operated under drive/idle or cruise conditions.

Ignition Spark Advance Control. The logic module determines the precise spark advance requirements by interpreting data from the distributor RPM signal, MAP sensor, and coolant temperature sensor. When the engine is cold, the logic module increases the spark advance for improved engine performance. The logic module commands the power module to open the primary ignition circuit at the right instant to provide the precise spark advance required by the engine.

Idle Speed Control. While the engine is being started, the logic module positions the automatic idle speed (AIS) motor to provide easy starting without touching the accelerator pedal. When the engine is cold, the logic module positions the AIS motor to provide the correct cold fast idle speed. The AIS motor allows more air to flow past the motor plunger into the intake manifold to increase the idle speed. This airflow bypasses the throttle, as indicated in Figure 9-109.

The AIS motor provides the correct idle speed when the air conditioner is on and the correct throttle opening when the engine is decelerating.

Exhaust Gas Recirculation (EGR) Control. The module energizes a vacuum solenoid in the EGR vacuum system. If the solenoid is energized, it shuts off vacuum to the EGR valve, and a de-energized

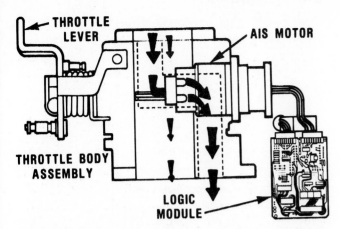

FIGURE 9-109. AIS Motor. (Courtesy of Chrysler Canada Ltd.)

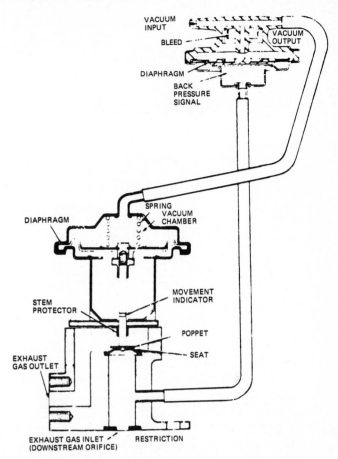

FIGURE 9-110. EGR Valve with Backpressure Transducer. (Courtesy of Chrysler Canada Ltd.)

solenoid allows vacuum to pass through the solenoid to the EGR system. The solenoid is energized at speeds below 1,200 RPM, during wide-open throttle operation, or when the coolant temperature is below 70°F (21°C). During all other engine operating conditions, the solenoid is de-energized and vacuum is supplied through the solenoid to the backpressure transducer. If the vehicle speed is below 30 to 35 MPH (48 KPH), the backpressure transducer vents the vacuum in the EGR system. Above this speed, the exhaust pressure closes the vacuum bleed in the transducer and vacuum is supplied to open the EGR valve. The EGR valve and the backpressure transducer are shown in Figure 9-110.

Canister Purge Control. The logic module operates a solenoid connected in the vacuum hose to the canister purge valve. When engine temperature is below 180°F (82°C), the logic module energizes the solenoid, which shuts off vacuum to the canister purge valve. Above this coolant temperature, the solenoid is de-energized, which supplies vacuum through the solenoid to the purge control valve. Under this condition, the canister is purged through a port in the throttle body.

The canister purge solenoid and the EGR solenoid are located with the diagnostic connector under a cover on the fender well, as illustrated in Figure 9-111.

Air Conditioning Control. When the throttle approaches the wide-open position and the throttle position sensor voltage is above a specific value, the logic module de-energizes the air conditioning (A/C) wide-open throttle (WOT) cutout relay and the relay contacts open the circuit to the A/C compressor

FIGURE 9-111. Canister Purge and EGR Solenoids with Diagnostic Connector. (Courtesy of Chrysler Canada Ltd.)

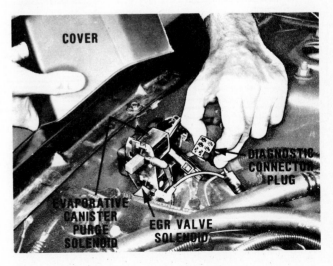

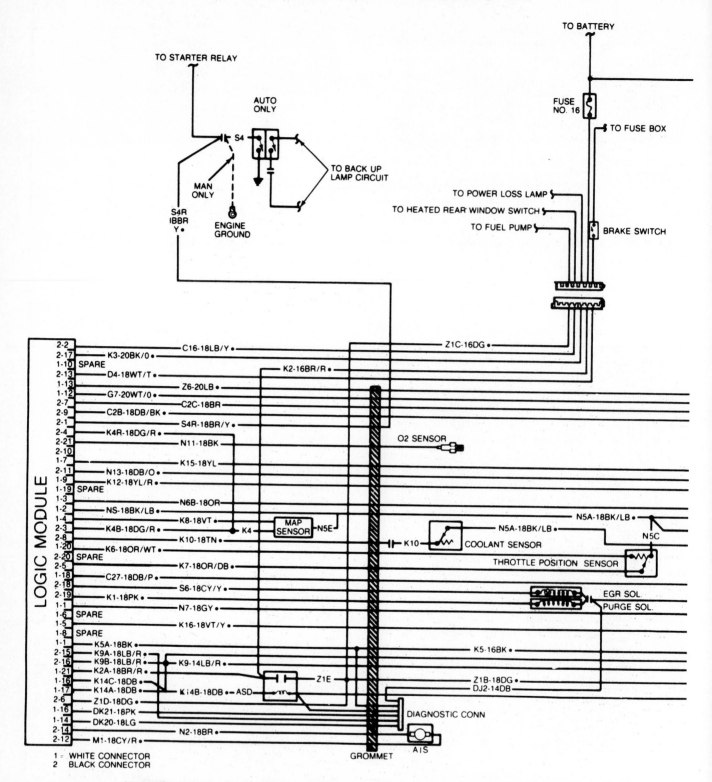

FIGURE 9-112. Logic Module Wiring Diagram. (Courtesy of Chrysler Canada Ltd.)

clutch. If the throttle is in the idle or cruising speed range, the logic module energizes the A/C WOT cutout relay and the relay contacts close, which supplies voltage to the A/C compressor clutch. If the engine speed drops below 500 RPM, the logic module de-energizes the A/C WOT cutout relay and opens the circuit to the compressor clutch. This prevents the engine from stalling under unusual conditions. If the engine is being cranked and the air conditioning is turned on, the logic module does not engage the compressor clutch until the engine speed exceeds 500 RPM.

The A/C WOT cutout relay is located beside the starter relay near the battery. The complete wiring diagrams for the logic module and power module are shown in Figures 9-112 and 9-113.

FIGURE 9-113. Power Module Wiring Diagram. (Courtesy of Chrysler Canada Ltd.)

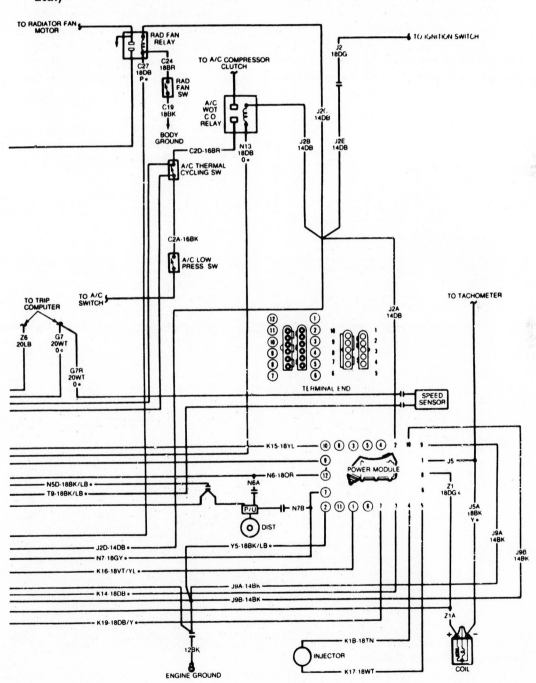

Later Model Chrysler Electronic Fuel Injection Systems

System Changes. The wiring diagrams in Figure 9-112 and 9-113 are for 1984 models. In the 1985 model year, several changes were made to the EFI system. For example, the automatic shutdown (ASD) relay is contained in the power module, and the logic module provides necessary signals to the power module for ASD relay control and for wastegate control on turbocharged engines. An external alternator voltage regulator is no longer required. A battery temperature sensor is located in the power module, and a battery charge signal is also sent to the power module. The power module operates a radiator fan relay and a solenoid for turbo wastegate control. Some new fault codes which apply to the battery and charging system have been added.

In the 1986 model year, a redesigned throttle body assembly is used. This assembly contains an improved fuel injector and a throttle body temperature sensor. Since the pressure regulator limits fuel pressure to 14.5 PSI (100 kPa), vapors may form in the fuel system. When this occurs, the air-fuel mixture becomes leaner. If the throttle body temperature increases to the point at which fuel vapors occur, the throttle body temperature sensor signals the logic module to supply a slightly richer air-fuel ratio. The throttle body assembly with the temperature sensor is illustrated in Figure 9-114, and the improved injector design is shown in Figure 9-115.

Wiring diagrams for 1986 single-point EFI systems are provided in Figures 9-116 through 9-120.

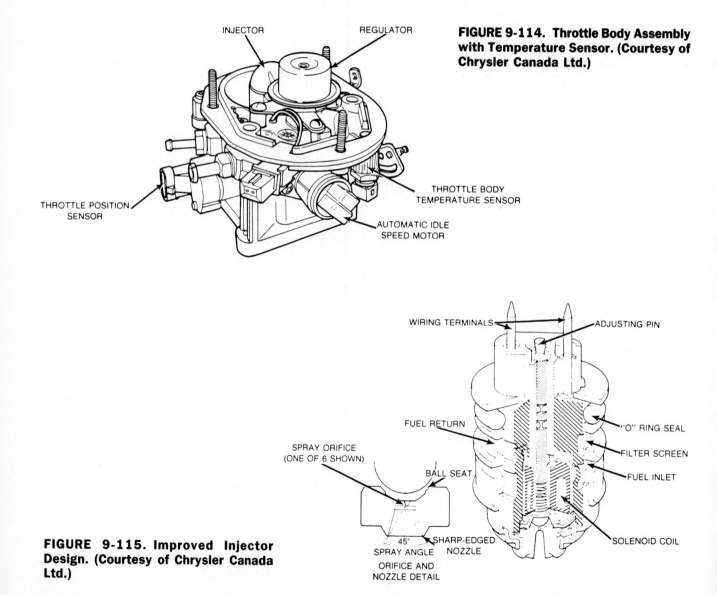

FIGURE 9-114. Throttle Body Assembly with Temperature Sensor. (Courtesy of Chrysler Canada Ltd.)

FIGURE 9-115. Improved Injector Design. (Courtesy of Chrysler Canada Ltd.)

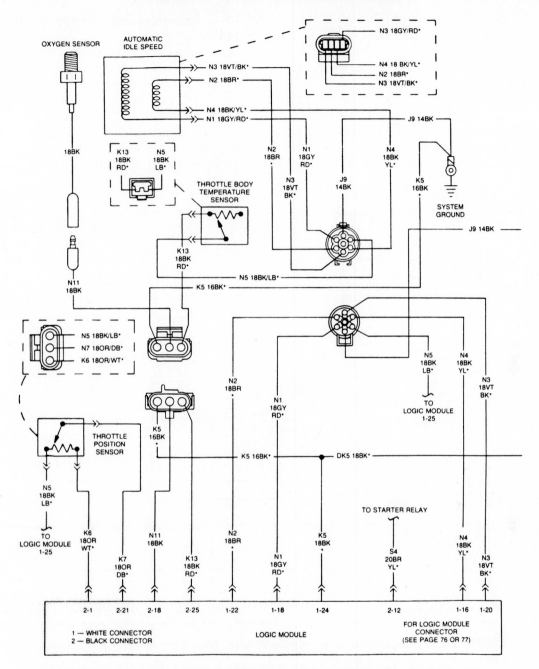

FIGURE 9-116. Logic Module Wiring. (Courtesy of Chrysler Canada Ltd.)

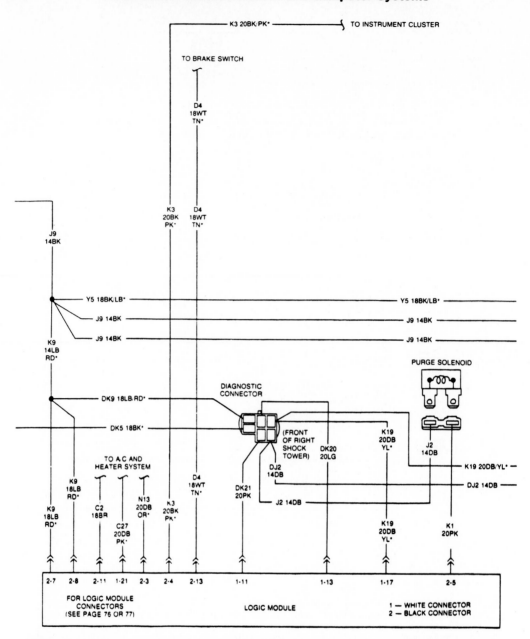

FIGURE 9-117. Logic Module Wiring Continued. (Courtesy of Chrysler Canada Ltd.)

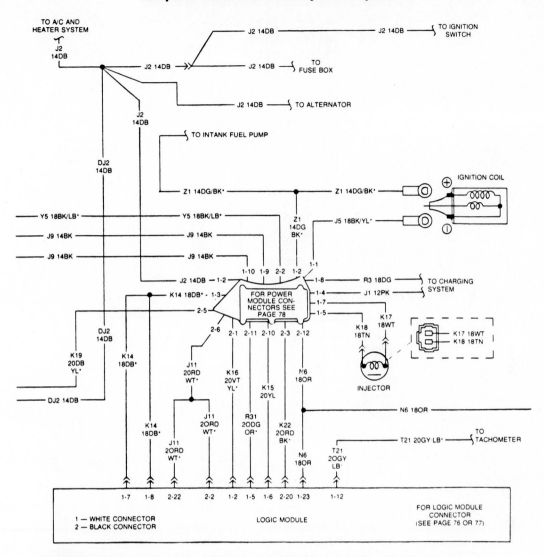

FIGURE 9-118. Power Module and Logic Module Wiring. (Courtesy of Chrysler Canada Ltd.)

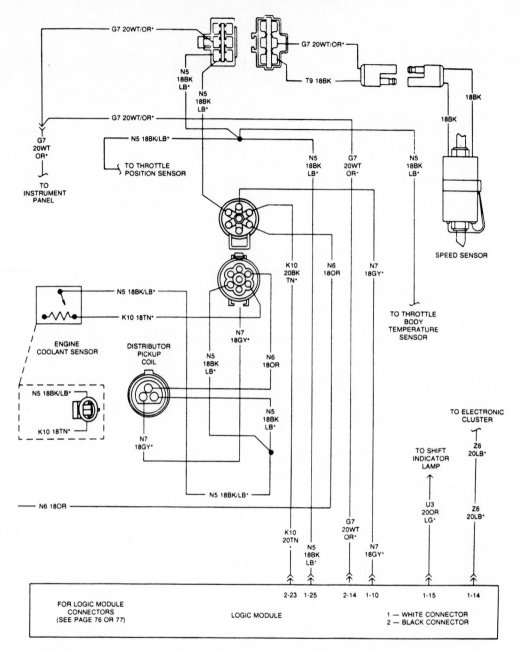

FIGURE 9-119. Logic Module Wiring Continued. (Courtesy of Chrysler Canada Ltd.)

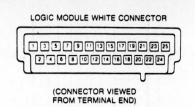

LOGIC MODULE WHITE CONNECTOR

(CONNECTOR VIEWED
FROM TERMINAL END)

CAV	CIRCUIT	GAUGE	COLOR	FUNCTION
1				
2	K16	20	VT/YL*	INJECTOR CONTROL TO POWER MODULE
3				
4				
5	R31	20	DG/OR*	ALTERNATOR FIELD CONTROL TO POWER MODULE
6	K15	20	YL	IGNITION CONTROL TO POWER MODULE
7	K14	18	DB*	FUSED J2 FROM POWER MODULE
8	K14	18	DB*	FUSED J2 FROM POWER MODULE
9				
10	N7	18	GY*	DISTRIBUTOR PICKUP SIGNAL
11	DK21	20	PK	DIAGNOSTIC CONNECTOR
12	T21	20	GY/LB*	TACHOMETER SIGNAL
13	DK20	20	LG	DIAGNOSTIC CONNECTOR
14	Z6	20	LB*	FUEL MONITOR OUTPUT SIGNAL
15	U3	20	OR/LG*	SHIFT INDICATOR LIGHT
16	N4	18	BK/YL*	AUTOMATIC IDLE SPEED MOTOR
17	K19	20	DB/YL*	AUTO SHUTDOWN RELAY CONTROL
18	N1	18	GY/RD*	AUTOMATIC IDLE SPEED MOTOR
19				
20	N3	18	VT/BK*	AUTOMATIC IDLE SPEED MOTOR
21	C27	20	DB/PK*	RADIATOR FAN RELAY CONTROL
22	N2	18	BR*	AUTOMATIC IDLE SPEED MOTOR
23	N6	18	OR	8 VOLT SUPPLY FROM POWER MODULE
24	K5	18	BK*	SENSOR GROUND
25	N5	18	BK/LB*	SENSOR GROUND

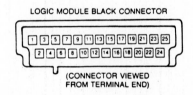

LOGIC MODULE BLACK CONNECTOR

(CONNECTOR VIEWED
FROM TERMINAL END)

CAV	CIRCUIT	GAUGE	COLOR	FUNCTION
1	K6	18	OR/WT*	5 VOLT SUPPLY FOR THROTTLE POSITION SENSOR
2	J11	20	RD/WT*	DIRECT BATTERY FEED FROM POWER MODULE
3	N13	20	DB/OR*	A/C CUTOUT RELAY CONTROL
4	K3	20	BK/OR*	POWER LOSS LAMP CONTROL
5	K1	20	PK*	PURGE SOLENOID CONTROL
6				
7	K9	18	LB/RD*	POWER GROUND
8	K9	18	LB/RD*	POWER GROUND
9				
10				
11	C2	18	BR	AIR CONDITIONING CLUTCH SIGNAL
12	S4	20	BR/YL*	PARK/NEUTRAL SWITCH SIGNAL
13	D4	18	WT/TN*	BRAKE SWITCH SIGNAL
14	G7	20	WT/OR*	SPEED SENSOR SIGNAL
15				
16				
17				
18	N11	18	BK	OXYGEN SENSOR SIGNAL
19				
20	K22	20	RD/BK*	BATTERY TEMPERATURE SIGNAL FROM POWER MODULE
21	K7	18	OR/DB*	THROTTLE POSITION SIGNAL
22	J11	20	RD/WT*	DIRECT BATTERY SENSE VOLTAGE FROM POWER MODULE
23	K10	20	TN*	ENGINE COOLANT TEMPERATURE SIGNAL
24				
25	K13	18	BK/RD*	THROTTLE BODY TEMPERATURE SIGNAL

FIGURE 9-120. Logic Module Terminal Identification. (Courtesy of Chrysler Canada Ltd.)

CHRYSLER MULTI-POINT ELECTRONIC FUEL INJECTION (EFI) USED ON 2.2L TURBOCHARGED ENGINE

Modules and Inputs

Operation. The logic module and the power module are very similar to the modules used in the single-point throttle body injection system, but they are not interchangeable. The modules perform basically the same functions as they did in the single-point EFI system; therefore, we describe here only the differences in the multi-point system.

Inputs. The same inputs are used in the multi-point EFI system with the addition of a detonation sensor and a charge temperature sensor. If the engine detonates, the detonation sensor signals the power module to retard the spark advance until the detonation stops. The charge temperature sensor sends a signal to the logic module in relation to the temperature of the air-fuel mixture in the intake manifold. This signal acts as a backup for the coolant temperature sensor if the coolant temperature sensor fails.

All the inputs and output control functions in the multi-point EFI systems are illustrated in Figure 9-121.

The distributor in the multi-point EFI system has a reference pickup and a synchronizer ("sync") pickup. The reference pickup is the same as the Hall Effect switch used in the single-point EFI system.

FIGURE 9-121. Multi-Point EFI System Components. (Courtesy of Chrysler Canada Ltd.)

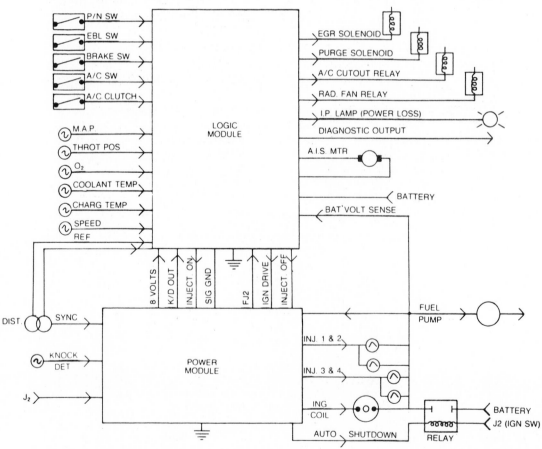

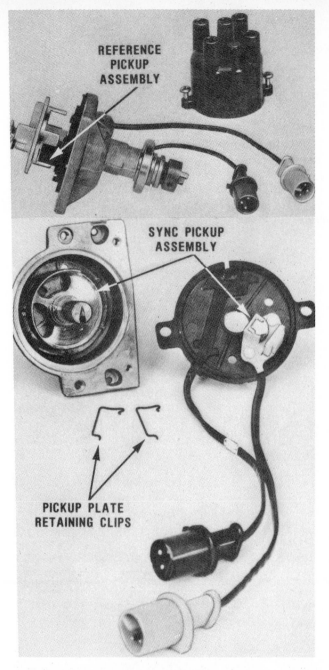

FIGURE 9-122. Reference and "Sync" Pickup Assemblies. (Courtesy of Chrysler Canada Ltd.)

Information data regarding engine speed and crankshaft position is supplied to the logic module by the reference pickup signal. The logic module uses this information with the other input data to determine the correct spark advance and injector pulse width.

The "sync" pickup is mounted under the pickup plate, and a notched "sync" ring attached to the distributor shaft rotates through the pickup. This signal from the "sync" pickup tells the logic and power modules which pair of injectors to turn on and which pair of injectors to turn off. If the engine speed exceeds 6,600 RPM, the logic module shuts off the fuel until the speed drops to a safe 6,100 RPM.

The reference pickup and "sync" pickup are pictured in Figure 9-122.

Output Control Functions

Fuel Injection Control. All the output control functions in the multi-point EFI system are the same as the single-point EFI system, except the fuel injection control. In the multi-point system, a single injector is placed in each intake port in the intake manifold rather than having one injector in the throttle body assembly. The injectors are similar to the injector in the single-point system. The power module energizes injectors 1 and 2 and injectors 3 and 4 in pairs. Each pair of injectors is energized once for every two crankshaft revolutions. The four injectors located in the intake manifold are shown in Figure 9-123.

The logic module commands the power module to provide the precise injector pulse width, which supplies the exact air-fuel ratio required by the engine. The pressure regulator maintains a constant pressure drop of 55 PSI (379 kPa) across the

FIGURE 9-123. Multi-Point EFI Fuel System. (Courtesy of Chrysler Canada Ltd.)

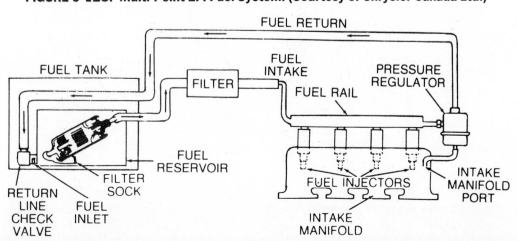

injectors in the multi-point system. Many of the other components in the multi-point EFI system are similar to the components in the single-point system. The changes on the 1985 single-point system also apply to the multi-point system. These changes were explained under the single-point system, and they pertain to the alternator voltage

regulator, battery charge and temperature signals, and the automatic shutdown relay. Fault codes and wiring diagrams vary, depending on the year of vehicle.

Wiring diagrams for 1986 multi-point EFI systems are provided in Figures 9-124 through 9-129.

FIGURE 9-124. Logic Module Wiring, Chrysler Multi-Point EFI Systems. (Courtesy of Chrysler Canada Ltd.)

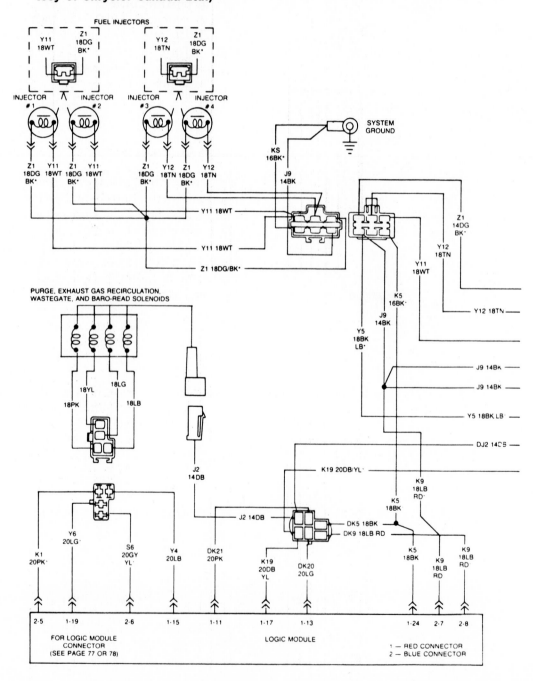

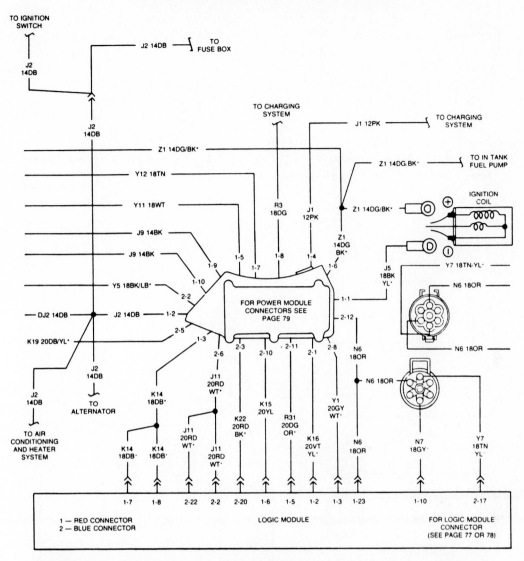

FIGURE 9-125. Logic Module and Power Module Wiring. (Courtesy of Chrysler Canada Ltd.)

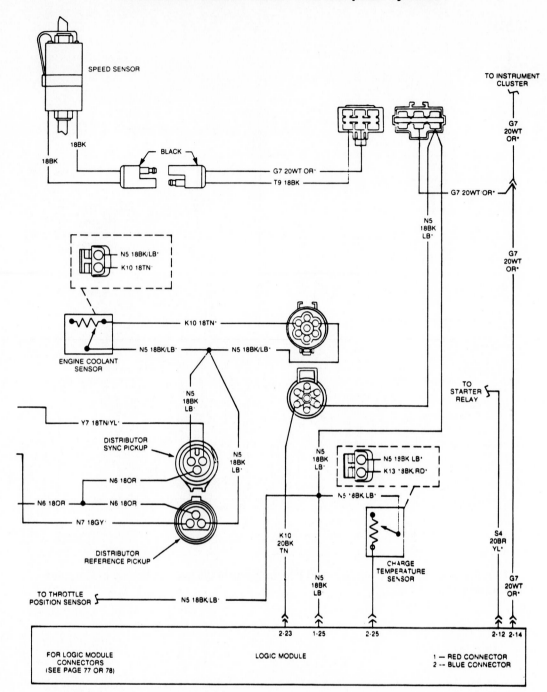

FIGURE 9-126. Logic Module Wiring Continued. (Courtesy of Chrysler Canada Ltd.)

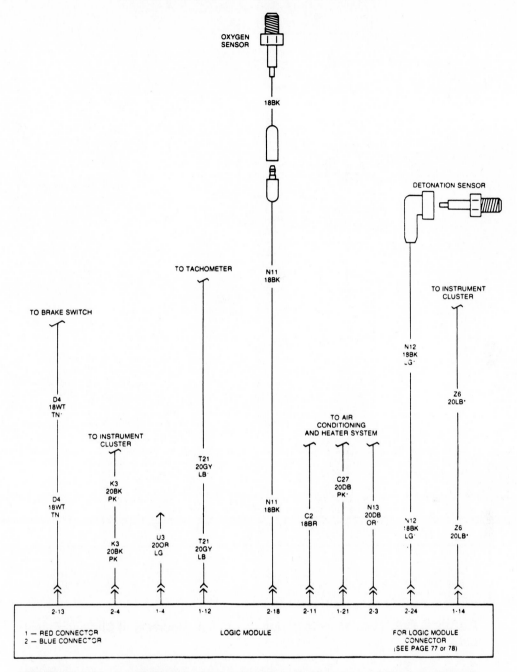

FIGURE 9-127. Logic Module Wiring Continued. (Courtesy of Chrysler Canada Ltd.)

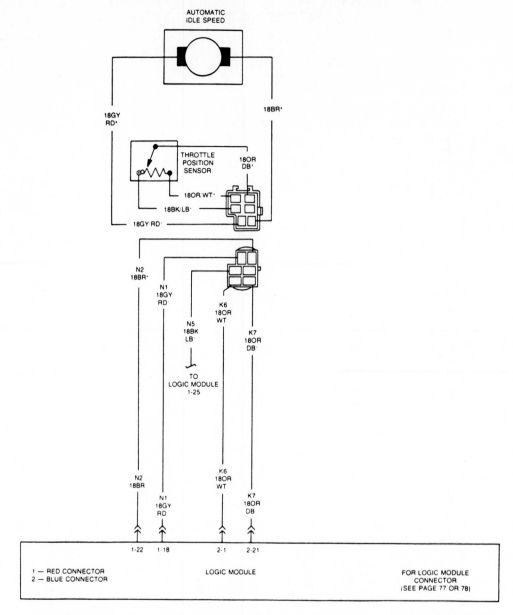

FIGURE 9-128. Logic Module Wiring Continued. (Courtesy of Chrysler Canada Ltd.)

LOGIC MODULE RED CONNECTOR

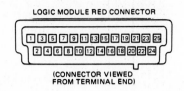

(CONNECTOR VIEWED
FROM TERMINAL END)

CAV	CIRCUIT	GAUGE	COLOR	FUNCTION
1	—			
2	K16	20	VT·YL·	INJECTOR CONTROL (1-2) TO POWER MODULE
3	Y1	20	GY WT·	INJECTOR CONTROL (3-4) TO POWER MODULE
4	—			
5	R31	20	DG OR	ALTERNATOR FIELD CONTROL TO POWER MODULE
6	K15	20	YL	IGNITION CONTROL TO POWER MODULE
7	K14	18	DB·	FUSED J2 FROM POWER MODULE
8	K14	18	DB	FUSED J2 FROM POWER MODULE
9	—			
10	N7	18	GY	DISTRIBUTOR REFERENCE PICKUP SIGNAL
11	DK21	20	PK	DIAGNOSTIC CONNECTOR
12	T21	20	GY LB	TACHOMETER SIGNAL
13	DK20	20	LG	DIAGNOSTIC CONNECTOR
14	Z6	20	LB	FUEL MONITOR OUTPUT SIGNAL
15	Y4	20	LB	BARO READ SOLENOID CONTROL
16	N4	18	BK YL	AUTOMATIC IDLE SPEED MOTOR
17	K19	20	DB YL	AUTO SHUTDOWN RELAY CONTROL
18	N1	18	GY RD	AUTOMATIC IDLE SPEED MOTOR
19	Y6	20	LG	WASTEGATE SOLENOID CONTROL
20	N3	18	VT BK	AUTOMATIC IDLE SPEED MOTOR
21	C27	20	DB PK	RADIATOR FAN RELAY CONTROL
22	N2	18	BR	AUTOMATIC IDLE SPEED MOTOR
23	N6	18	OR	8 VOLT SUPPLY FROM POWER MODULE
24	K5	18	BK	SENSOR GROUND
25	N5	18	BK LB	SENSOR GROUND

LOGIC MODULE RED CONNECTOR

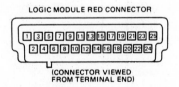

(CONNECTOR VIEWED
FROM TERMINAL END)

CAV	CIRCUIT	GAUGE	COLOR	FUNCTION
1	K6	18	OR/WT·	5 VOLT SUPPLY FOR THROTTLE POSITION SENSOR
2	J11	20	RD/WT·	DIRECT BATTERY FEED FROM POWER MODULE
3	N13	20	DB/OR·	A/C CUTOUT RELAY CONTROL
4	K3	20	BK/PK·	POWER LOSS LAMP CONTROL
5	K1	20	PK·	PURGE SOLENOID CONTROL
6	S6	20	GY/YL·	EGR SOLENOID CONTROL
7	K9	18	LB/RD·	POWER GROUND
8	K9	18	LB/RD·	POWER GROUND
9	—			
10	—			
11	C2	18	BR	AIR CONDITIONING CLUTCH SIGNAL
12	S4	20	BR/YL·	PARK/NEUTRAL SWITCH SIGNAL
13	D4	18	WT/TN·	BRAKE SWITCH SIGNAL
14	G7	20	WT/OR·	SPEED SENSOR SIGNAL
15	—			
16	—			
17	Y7	18	TN/YL·	DISTRIBUTOR SYNC PICKUP SIGNAL
18	N11	18	BK	OXYGEN SENSOR SIGNAL
19	—			
20	K22	20	RD/BK·	BATTERY TEMPERATURE SIGNAL FROM POWER MODULE
21	K7	18	OR/DB·	THROTTLE POSITION SIGNAL
22	J11	20	RD/WT·	DIRECT BATTERY SENSE VOLTAGE FROM POWER MODULE
23	K10	20	TN·	ENGINE COOLANT TEMPERATURE SIGNAL
24	N12	18	BK/LG·	DETONATION SENSOR SIGNAL
25	K13	18	BK/RD·	CHARGE TEMPERATURE SENSOR SIGNAL

FIGURE 9-129. Logic Module Terminal Identification. (Courtesy of Chrysler Canada Ltd.)

SERVICING AND DIAGNOSIS OF CHRYSLER ELECTRONIC FUEL INJECTION (EFI) SYSTEMS

Service Precautions

Personal injury or damage to the system components could result from improper service procedures. The following service precautions must be observed:

1. Using a 12-V test light to test for continuity in EFI electrical circuits can damage expensive components in the system. Always use a voltmeter, ohmmeter, or diagnostic readout tester for the test procedure.

2. Before any fuel line is loosened, the pressure in the fuel system must be relieved. If a fuel line connection is loosened without relieving the fuel pressure, high pressure in the fuel system could cause personal injury or start a fire. The fuel pressure can be relieved by disconnecting the injector electrical connector and grounding one injector terminal with a jumper wire while the other terminal is connected to the battery positive terminal. In a multipoint system, the fuel pressure can be relieved by activating one injector. The injector must not be activated for more than 10 seconds.

3. Never restrict a fuel line completely, because this causes extremely high pump pressure.

4. Fuel line clamps must be torqued to 10 in lb (Nm).

5. Never substitute conventional gas line hose in place of the EFI hose. The EFI system also has special clamps that must not be replaced with other clamps.

Adjusting Basic Ignition Timing

When the basic timing is being checked, proceed as follows:

1. Operate the engine until normal operating temperature is reached.

2. Connect a timing light and tachometer to the engine.

3. Disconnect and reconnect the coolant temperature sensor connector at the thermostat housing. This will put the system in the "limp-in" mode and the power loss light will be on. If the system is working normally, the logic module varies the timing continuously at idle to provide smooth idle operation. In the "limp-in" mode, the timing is fixed by the logic module, and this condition is essential for checking basic timing.

4. Check the timing marks with the timing light. If the timing light is not set at the specifications shown on the underhood emission label, loosen the distributor clamp and rotate the distributor to correct the timing. Retighten the distributor clamp.

5. Shut the engine off and remove the timing light and tachometer. Disconnect the quick disconnect terminal in the heavy red wire near the battery positive terminal for 10 seconds to clear the fault codes from the logic module's memory.

6. Reconnect the quick-disconnect terminal at the battery and start the engine. Check to make sure the power-loss light goes out.

Checking Spark Advance

The basic timing must be checked before the spark advance, and the power-loss light must be off. When the spark advance is being checked, there must not be any fault codes stored in the logic module's memory. (Fault code diagnosis is discussed later in this chapter.) With an advance-type timing light and a tachometer connected to the engine, and the engine at normal operating temperature, increase the engine speed to 2,000 RPM. Rotate the advance control on the timing light until the timing mark returns to the basic timing light indicator. If the spark advance is not within specifications, the logic module should be replaced.

Testing Fuel Pump Pressure

On single-point systems, the fuel system pressure must be relieved before the pressure is tested. To test fuel pump pressure, the pressure gauge must be connected in the fuel line at the throttle body assembly as illustrated in Figure 9-130.

The fuel pressure gauge can be connected to the Schrader valve on the fuel rail to test fuel pump pressure on the multi-point system.

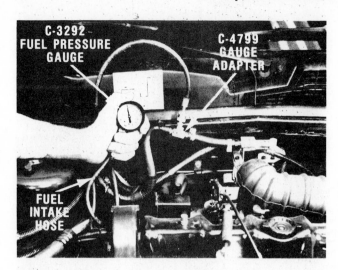

FIGURE 9-130. Testing Fuel Pump Pressure, Single-Point EFI System. (Courtesy of Chrysler Canada Ltd.)

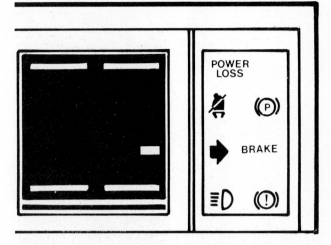

FIGURE 9-131. Power-Loss Light. (Courtesy of Chrysler Canada Ltd.)

With the engine idling, the fuel pressure should be 48–62 PSI (331–427 kPa) on multi-point systems, and 29–43 PSI (200–296 kPa) on single-point systems.

On-Board Diagnostics

The logic module tests many of its input circuits and output control functions continuously. If an input signal is out of range or completely missing for a specific number of times, the logic module considers this a "real" fault and a fault code is placed in the logic module memory. An output control function malfunctioning for a specific number of times also causes a fault code to be stored in the logic module memory. The logic module does not store transient faults in its memory, because they do not occur enough times. If the problem is repaired or disappears, the logic module is programmed to erase the fault from its memory after 30 ignition switch cycles. The fault code may be cleared from the logic module memory by momentarily disconnecting the quick-disconnect terminal in the heavy red wire connected to the battery positive cable.

Limp-In Mode

If a defect occurs in the MAP sensor, throttle position sensor, or coolant temperature sensor, the logic module places the system in a "limp-in" mode. In this mode, the logic module substitutes other information for the missing or incorrect sensor input signal. For example, if the MAP sensor signal is not available, the logic module uses the engine speed signal and the throttle position sensor signal to provide a "modified" MAP signal. In the "limp-in" mode, the logic module keeps the output control functions operational, but engine performance and fuel economy decrease and exhaust emissions could increase. When the system is in the "limp-in" mode, the power-loss light on the instrument panel will be on. Some models use a power-limited light in place of the power-loss light, as indicated in Figure 9-131.

Once the fault is repaired and the fault codes are cleared from the logic module memory, the power-loss light will go out after the engine is started. When a defect that caused a "limp-on" mode corrects itself, the power-loss light will go out after the engine is stopped and restarted.

Diagnostic Readout Tester

A diagnostic tester is available to read the fault codes in the EFI systems, as shown in Figure 9-132. The tester must be connected to the diagnostic connector as illustrated in the figure. When the tester is connected, it will perform six test functions:

1. A display check, code 88, should appear for a few seconds when the ignition switch is turned on, which indicates the tester is operational.

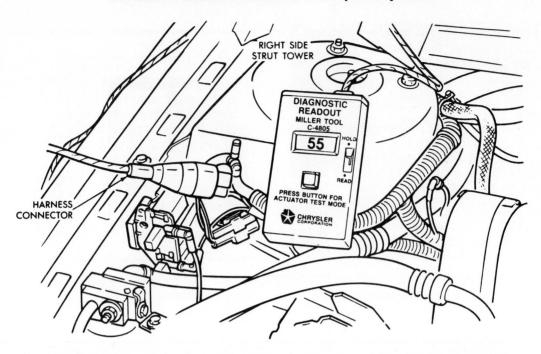

FIGURE 9-132. Diagnostic Readout Tester. (Courtesy of Chrysler Canada Ltd.)

2. All fault codes in the logic module memory will be displayed on the tester.

3. A distributor signal should be seen on the tester when the engine is being cranked. As long as the display changes while the engine is being cranked, a distributor signal is being received.

4. A switch test procedure may be performed by turning each switch input to the logic module on and off. It does not matter what code is on display before a switch is turned on and off, but if the switch is working normally, the display will change when the switch is turned on and off.

5. An oxygen sensor signal test may be performed with the engine running. If the oxygen signal is generating a normal signal, the readout will continuously switch from 0 to 1 as the slight variations from lean to rich mixture occur.

6. An actuator test mode may be selected by depressing the actuator test mode (ATM) button on the tester with the ignition switch on. In this mode, the ASD relay is grounded and the service technician can observe the following test sequence:
   a) Three sparks from the coil wire when the coil wire is held 1/4 in (6.3 mm) from a good engine ground.

   b) Two automatic idle speed motor movements: one open, one closed.
   c) One fuel pulse from the injector.

This ATM test sequence applies to Chrysler cars through 1984.

On 1985 EFI systems, use the following procedure for fault code diagnosis:

1. Connect the digital readout tester to the diagnostic connector and place the read/hold test switch in the read position.

2. Turn the ignition switch on/off, on/off, on within a 5-second interval.

3. Record all the fault codes displayed on the tester.

After the fault code display is completed, 55 will appear on the tester display. When this occurs, the actuation test mode (ATM) may be entered if the tester ATM button is pressed. In this mode, ATM codes will be displayed on the tester. Each ATM code represents a specific solenoid, relay, or component in the circuit that will be cycled on and off if the ATM button is released when the code appears on the display representing the specific component.

Cycling of a component will be stopped after 5 minutes, or when the ignition switch is turned off. The ATM tests allow the technician to listen to, observe, or perform voltage tests on specific components to locate various defects. ATM codes are provided with the fault codes later in this chapter.

After the fault code diagnosis, a switch test procedure may be performed as follows:

1. Turn off all the input switch signals to the logic module.

2. With 55 indicated on the tester display at the end of the fault code diagnosis, turn each input switch on and off. The display reading should change if the switch signal is received by the logic module.

On the 1985 EFI systems, an oxygen sensor test with the engine running can be performed as outlined previously. Sensor access codes are available on 1986 models. These codes are obtained when the ATM test mode is entered with the ATM button depressed and the read/hold switch in the read position. Sensor access codes and ATM codes have the same numbers. When the desired ATM code appears on the digital readout meter, move the read/hold switch to the hold position. This action changes the ATM code to a sensor access code. In this mode the digital readout tester will indicate the specific sensor voltage. To obtain the actual sensor voltage output, the readings displayed on the tester must be divided by 10, except the coolant sensor voltage, which is multiplied by 10. A listing of all codes is provided later in this chapter.

An engine-running test mode is also available. On 1987 models this test mode has the following capabilities:

1. With the engine running and the digital readout switch in the read position, the tester will switch from 0 to 1 as the oxygen sensor cycles from lean to rich. This feature was available on earlier models, as explained previously.

2. When the engine is running in neutral or park and the oxygen sensor cycling is displayed, a check of the idle air control motor can be performed if the tester switch is moved to the hold position. If this action is taken, engine speed should increase to approximately 1,500 RPM.

3. If the engine speed is increased above 1,500 RPM, an 8 will appear beside the 0 and 1 when the detonation sensor signal is satisfactory.

4. When the engine is running and oxygen sensor cycling is displayed on the tester, with the tester switch in the read position, depress the ATM button. This action will start the engine-running sensor test code sequence. When the desired engine-running test code appears on the tester, release the ATM button and move the switch to the hold position. The selected sensor output will now appear on the tester. All the sensor readings are divided by 10 except the coolant sensor and engine RPM, which are multiplied by 10. Actual readings are provided for battery voltage, manifold vacuum, and vehicle speed.

Power-Loss Light Diagnosis

The power-loss light should come on when the ignition switch is turned on, and it should remain on for a few seconds after the engine is started. Once the light goes off, it should remain off while the engine is running. The power-loss light may be used to perform three test functions:

1. A test of the distributor signal is performed while the engine is being cranked. If the power-loss light is on while the engine is being cranked, a distributor signal is being received.

2. Fault codes in the logic module memory will be displayed by a flashing power-loss light if the ignition switch is turned off/on, off/on, off/on in a 5-second interval. For example, if the light flashes five times, pauses, and then flashes once, code 51 has been indicated.

3. A switch test may be performed after all the fault codes have been displayed. Each time a switch input to the logic module is turned on and off, the light should be illuminated momentarily.

A fault code indicates a problem in a certain area, not necessarily in a specific component. For example, a code 21 indicates that there is a fault in the oxygen sensor or sensor wire, or that the logic module is unable to receive an oxygen sensor signal. Code 12 indicates that the battery voltage has been disconnected from the logic module. This code can only be erased by 30 ignition switch cycles.

Fault codes, ATM codes, and sensor access codes vary, depending on the year and model of car. The fault codes for a 1987 2.5L turbocharged engine are shown in Figure 9-133.

Code	Type	Power Loss/ Limit Lamp	Circuit	When Monitored By The Logic Module	When Put Into Memory	ATM Test Code	Sensor Access Code
11	Fault	No	Distributor Signal	During cranking.	If no distributor signal is present since the battery was disconnected.	None	None
12	Indication	No	Battery Feed to the Logic Module	All the time when the ignition switch is on.	If the battery feed to the logic module has been disconnected within the last 20-40 engine starts.	None	None
13	Fault	Yes	M.A.P. Sensor (Vacuum)	When the throttle is closed during cranking and after the engine starts.	If the M.A.P. sensor vacuum level does not change between cranking and when the engine starts.	None	None
14	Fault	Yes	M.A.P. Sensor (Electrical)	All the time when the ignition switch is on.	If the M.A.P. sensor signal is below .02 or above 4.9 volts.	None	08
15	Fault	No	Vehicle Speed Sensor	Over a 7 second period during decel from highway speeds when the throttle is closed.	If the speed sensor signal indicates less than 2 mph when the vehicle is moving.	None	None
16	Fault	Yes	Battery Voltage Sensing (Charging System)	All the time after one minute from when the engine starts.	If the battery sensing voltage drops below 4 or between 7½ and 8½ volts for more than 20 seconds.	None	07
17	Fault	No	Engine Cooling System	During cranking when engine coolant temperature is between −20 and 212°F.	If engine coolant temperature does not reach 160°F within 20 minutes after the engine is started.	None	None
21	Fault	No	Oxygen Sensor	All the time after 12 minutes from when the engine starts.	If there is no oxygen sensor signal for more than 22 seconds when in closed loop.	None	02
22	Fault	Yes	Engine Coolant Sensor	All the time when the ignition switch is on.	If the coolant sensor voltage is above 4.96 volts when the engine is cold or below .51 volts when the engine is warm.	None	04
23	Fault	Yes	Charge Temperature Sensor	All the time when the ignition switch is on.	If the charge temperature sensor voltage is above 4.98 or below .06.	None	03
24	Fault	Yes	Throttle Position Sensor	All the time when the ignition switch is on.	If the throttle position sensor signal is below .16 or above 4.7 volts.	None	05
25	Fault	No	Automatic Idle Speed Motor (AIS)	Only when the AIS system is required to control the engine speed.	If proper voltage in the AIS system is not present. **NOTE:** Open circuit will not activate code.	03	None
26	Fault	No	Injector 1 and 2	During cranking.	If injectors 1 and 2 do not fire correctly.	02	None
27	Fault	No	Injectors 3 and 4	During cranking.	If injectors 3 and 4 do not fire correctly.	02	None
31	Fault	No	Canister Purge Solenoid	All the time when the ignition switch is on.	If the solenoid does not turn on and off when it should.	07	None
33	Fault	No	A/C Cutout Relay	All the time when the ignition switch is on.	If the relay does not turn on and off when it should.	05	None
34	Fault	No	E.G.R. Solenoid	All the time when the ignition switch is on	If the solenoid does not turn on and off when it should.	08	None
35	Fault	No	Radiator Fan Relay Circuit	All the time when the ignition switch is on.	If the relay does not turn on and off when it should.	04	None

FIGURE 9-133. EFI Fault Codes. (Courtesy of Chrysler Canada Ltd.)

Code	Type	Power Loss/ Limit Lamp	Circuit	When Monitored By The Logic Module	When Put Into Memory	ATM Test Code	Sensor Access Code
36	Fault	Yes	Wastegate Control Solenoid	All the time when the ignition switch is on.	If the solenoid does not turn on and off when it should.	09	None
37	Fault	No	Baro Read Solenoid	All the time when the ignition switch is on.	If the solenoid does not turn on and off when it should.	10	None
41	Fault	No	Alternator Field Control (Charging System)	All the time when the ignition switch is on.	If the field control fails to switch properly.	None	None
42	Fault	No	Auto Shutdown	All the time when the ignition switch is on.	If the control voltage of the relay pull in coil in the power module is not correct.	06	None
43	Fault	No	Spark Control	All the time when the ignition switch is on.	If the spark control interface fails to switch properly.	01	None
44	Fault	No	Battery Temperature Sensor (Charging System)	All the time when the ignition switch is on.	If the battery temperature sensor signal is below .04 or above 4.9 volts.	None	01
45	Fault	Yes	Overboost Monitor	All the time when the engine is running.	When M.A.P. sensor signal exceed a predetermine amount of boost indication.	None	None
46	Fault	Yes	Battery Voltage Sensing (Charging System)	All the time when the engine is running.	If the battery sense voltage is more than 1 volt above the desired control voltage for more than 20 seconds.	None	None
47	Fault	No	Battery Voltage Sensing (Charging System)	When the engine has been running for more than 6 minutes, engine temperature above 160°F and engine rpm above 1,500 rpm.	If the battery sense voltage is less than 1 volt below the desired control voltage for more than 20 seconds.	None	None
51	Fault	No	Oxygen Feedback System	During all closed loop conditions.	If the system stays lean for more than 12 minutes.	None	None
52	Fault	No	Oxygen Feedback System	During all closed loop conditions.	If the system stays rich for more than 12 minutes.	None	None
53	Fault	No	Logic Module	All the time in the diagnostic mode.	If the logic module fails.	None	None
54	Fault	No	Distributor Sync. Pickup	All the time when the engine is running.	If there is no distributor sync. pickup signal.	None	None
55	Indication	No			Indicates end of diagnostic mode.	None	None
88	Indication	No			Indicates start of diagnostic mode. **NOTE:** This code must appear first in the diagnostic mode or fault codes will be inaccurate.		
0	Indication	No			Indicates oxygen feedback system is lean with the engine running.	None	None
1	Indication	No			Indicates oxygen feedback system is rich with the engine running.		
8	Indication	No	Knock Circuit		Indicates knock sensor system is detecting knock.	None	06

FIGURE 9-133 (Continued). EFI Fault Codes. (Courtesy of Chrysler Canada Ltd.)

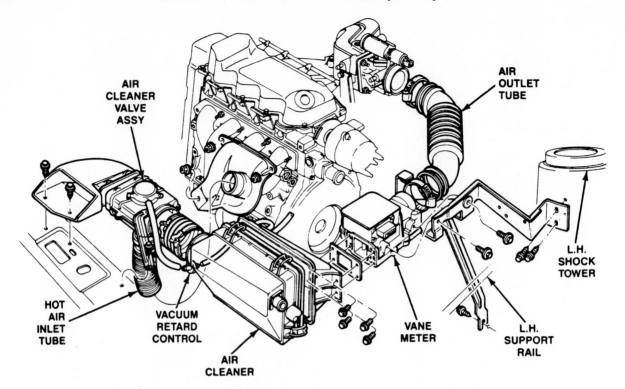

FIGURE 9-134. Vane Meter Location. (Courtesy of Ford Motor Co.)

FORD ELECTRONIC ENGINE CONTROL IV (EEC IV) SYSTEMS USED WITH 2.3L TURBOCHARGED ENGINE

Input Sensors

The 2.3L turbocharged engine has an EEC IV system with electronic fuel injection (EFI). This system uses many of the same input sensors that are used in other EEC IV systems. A vane meter that contains two sensors is used in this system. These sensors are referred to as the vane air flow (VAF) sensor and the vane air temperature (VAT) sensor. The vane meter is located in the engine air intake system, as illustrated in Figure 9-134.

All the airflow into the engine must travel through the vane meter. The airflow rotates a vane that is mounted on a pivot pin in the meter body, as pictured in Figure 9-135.

The movement of the vane is proportional to the volume of airflow through the air intake system. The VAF sensor is a variable resistor which has a sliding contact that is attached to the vane shaft. The electronic control assembly (ECA) applies a 5-V reference voltage to the VAF sensor. As the vane

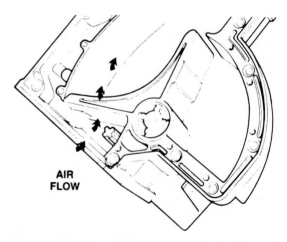

FIGURE 9-135. Vane Meter Air Vane. (Courtesy of Ford Motor Co.)

shaft rotates, it moves the sliding contact on the variable resistor. The voltage output signal from the VAF sensor to the ECA will vary between 0 and 5 V. A higher volume of airflow will produce a higher voltage output signal from the VAF sensor. The VAF sensor in the vane meter is shown in Figure 9-136.

A vane air temperature (VAT) sensor is also located in the vane meter. The ECA calculates a mass airflow value from the VAT and VAF sensor signals. This value is used by the ECA to provide

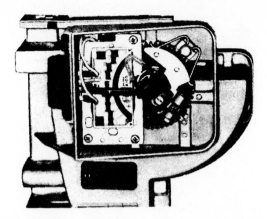

FIGURE 9-136. Vane Airflow (VAF) Sensor. (Courtesy of Ford Motor Co.)

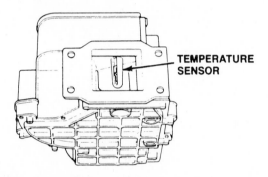

FIGURE 9-137. Vane Air Temperature Sensor. (Courtesy of Ford Motor Co.)

The throttle body assembly contains the throttle valve, throttle position sensor (TPS), and the throttle air bypass valve solenoid, as illustrated in Figure 9-139.

The fuel injectors are mounted in the intake ports and there is no fuel delivered to the throttle body assembly. The throttle air bypass valve solenoid is operated by the ECA. When the ECA increases the voltage that is supplied to the solenoid, the throttle air bypass valve opening increases and allows additional air to bypass the throttle valve, which increases the engine idle speed. The throttle air bypass valve controls both the cold fast idle speed and the idle speed at normal engine temperatures. There are no curb idle or fast idle adjustments on the throttle air bypass valve solenoid. While the engine is being cranked, the ECA moves the throttle air bypass valve to the fully open position. This allows "no-touch starting," which means the engine can be started at any temperature without touching the throttle. Each time the engine is started, the throttle air bypass valve provides fast idle speed for a period of time. The time period increases as engine coolant temperature decreases.

The throttle air bypass valve solenoid provides the same functions as the throttle kicker or idle speed control motor on other systems. Airflow through the throttle air bypass valve is illustrated in Figure 9-140.

the correct air-fuel ratio of 14.7:1. The VAT sensor in the vane meter is shown in Figure 9-137.

Some EEC IV systems have a three-wire exhaust gas oxygen (EGO) sensor in place of the single-wire sensor. This three-wire sensor has an internal electric heater to improve sensor response time. An air charge temperature sensor is used in some systems on 2.3L engines. The input sensors and output controls in the EEC IV system on the 2.3L turbocharged engine are pictured in Figure 9-138.

Output Controls

An exhaust gas recirculation (EGR) solenoid controls the vacuum that is applied to the EGR valve. The EGR solenoid shuts off the vacuum to the EGR valve if the engine coolant is cold, or when the engine is operating at idle speed or wide-open throttle. At all other operating conditions, the ECA energizes the EGR solenoid to apply vacuum to the EGR valve.

Fuel System

The EEC IV system with electronic fuel injection (EFI) has a low-pressure electric pump mounted in the fuel tank and a high-pressure electric pump located on the right frame rail, as shown in Figure 9-141.

When the ignition switch is turned on, the ECA grounds the fuel pump relay winding and the relay contacts close to provide power to the fuel pumps, as shown in Figure 9-142.

If the engine is not cranked within one second, the ECA opens the circuit from the fuel pump relay to ground and the relay contacts open the circuit to the fuel pumps. The dual fuel pumps are capable of supplying fuel at 100 PSI (690 kPa). A fuel filter is mounted near the high-pressure pump in the frame rail.

Fuel is delivered from the fuel pumps to the fuel rail and the pressure regulator. The fuel rail, pressure regulator, and the injectors are mounted on the intake manifold, as shown in Figure 9-143.

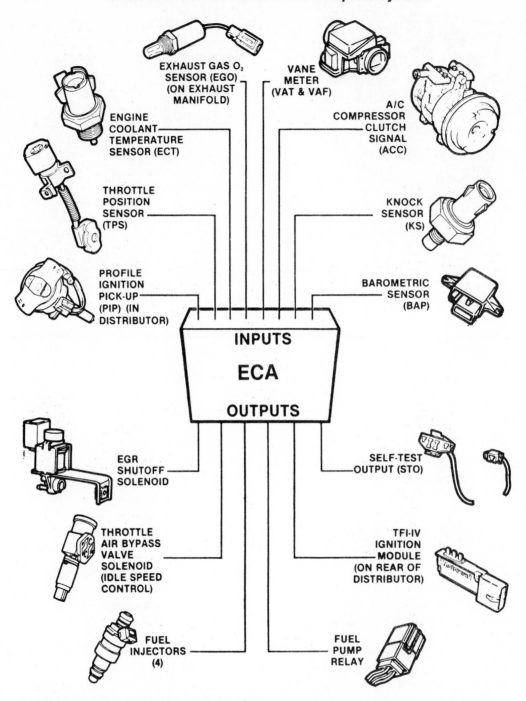

FIGURE 9-138. Wiring Diagram for EEC IV System Used on 2.3L Turbocharged Engine.

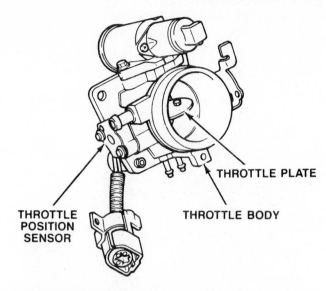

FIGURE 9-139. Throttle Body Assembly. (Courtesy of Ford Motor Co.)

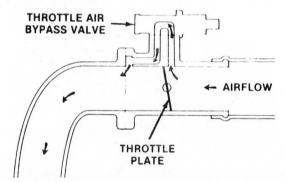

FIGURE 9-140. Throttle Air Bypass Valve Operation. (Courtesy of Ford Motor Co.)

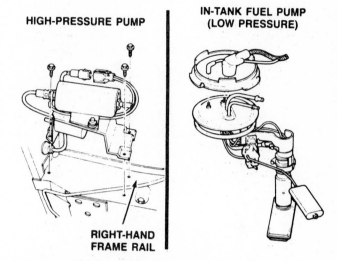

FIGURE 9-141. EEC IV with EFI Fuel Pumps. (Courtesy of Ford Motor Co.)

FIGURE 9-142. Electric Fuel Pump Circuit. (Courtesy of Ford Motor Co.)

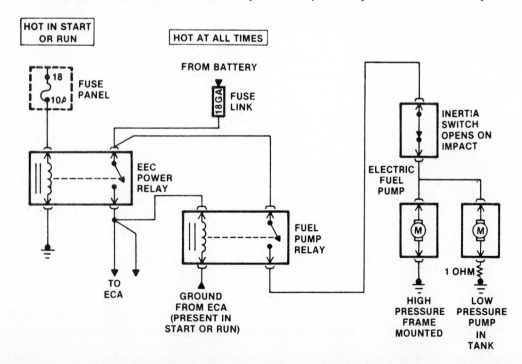

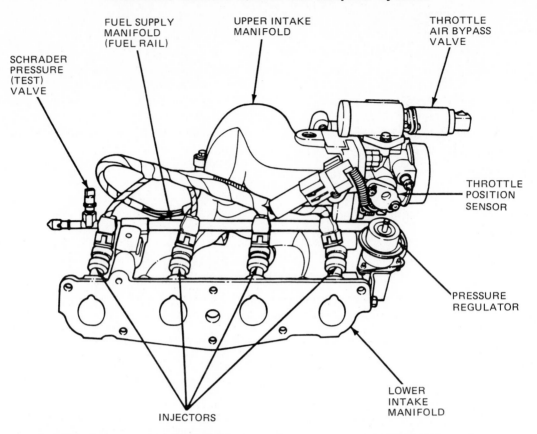

FIGURE 9-143. Fuel Rail with Pressure Regulator and Injectors. (Courtesy of Ford Motor Co.)

When the pressure in the fuel rail reaches 39 PSI (269 kPa), the pressure regulator diaphragm is forced upward, which opens the valve so that excess fuel can return to the fuel tank, as illustrated in Figure 9-144.

A vacuum hose is connected from the intake manifold to the top of the pressure regulator. This

FIGURE 9-144. Pressure Regulator Design. (Courtesy of Ford Motor Co.)

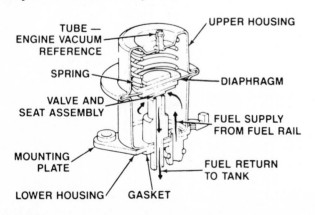

vacuum signal maintains a constant pressure drop across the injectors during all intake manifold conditions. When the engine is idling, high manifold vacuum is applied to the upper side of the pressure regulator diaphragm, and the diaphragm can be forced upward at 39 PSI (269 kPa). At high engine speeds, the turbocharger pressurizes the intake manifold to 8 PSI (55 kPa), and this pressure is applied to upper side of the pressure regulator diaphragm. Under this condition, 50 PSI (345 kPa) is required to move the diaphragm upward, and fuel pressure to the injectors is maintained at this higher value. When the intake manifold pressure increases at high speeds, the increase in fuel pressure maintains the same pressure difference across the injectors.

The injectors are mounted in the intake ports of the intake manifold. Each injector contains an electric solenoid that is energized by the ECA. The amount of fuel that each injector delivers is determined by the "on time" of the injector, because there is a constant pressure across the injector. A cutaway view of an injector is shown in Figure 9-145.

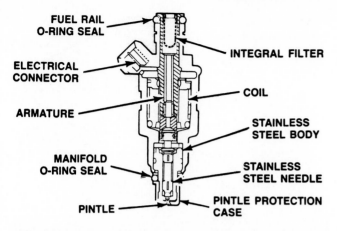

FIGURE 9-145. Injector Design. (Courtesy of Ford Motor Co.)

The ECA controls the "on time" of the injectors to provide the precise air-fuel ratio that is required by the engine under all operating conditions. When the ignition switch is turned on, current flows from the switch through the power relay winding to ground. This action closes the relay contacts, which supplies voltage to the injectors, the solenoids in the system, and ECA terminals 37 and 57. Some power relays have a time-delay feature that causes the relay to remain closed for 10 seconds after the ignition switch is turned off. This feature allows the ECA to position the throttle air bypass valve for no-touch starting. The injectors for cylinders 1 and 2 and the injectors for cylinders 3 and 4 are connected in parallel electrically, as shown in the wiring diagram in Figure 9-146.

FIGURE 9-146. EEC IV with EFI Wiring Diagram. (Courtesy of Ford Motor Co.)

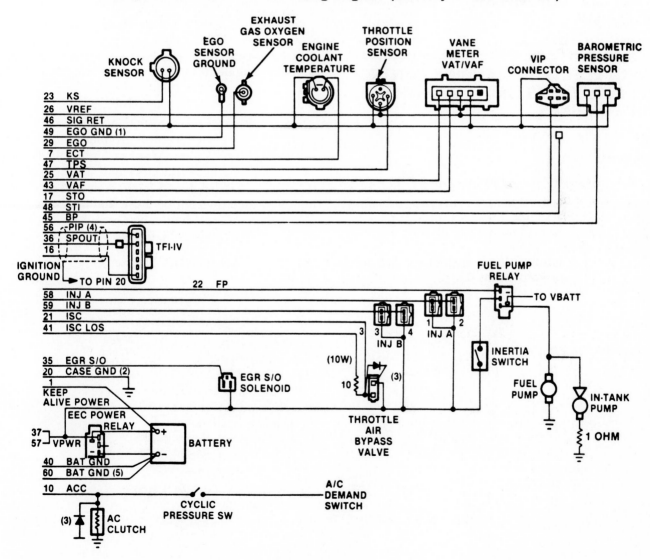

The ECA energizes each pair of injectors when it completes the circuit from the injectors to ground through the ECA. When the engine is decelerated in the closed-throttle mode, the ECA does not energize the injectors. This provides improved emission levels and fuel economy during deceleration. The injectors are energized again when the engine speed drops to a predetermined RPM. Enrichment of the air-fuel ratio on a cold engine is provided by increasing the "on time" of the injectors, which eliminates the need for a conventional choke.

If the engine becomes flooded, the condition can be cleared by depressing the throttle to the wide-open position while cranking the engine. When the ECA receives a wide-open throttle signal from the throttle position sensor (TPS), the ECA stops energizing the injectors while the engine is being cranked. This clears the flooded condition, and as soon as the engine starts on the excess fuel in the intake manifold, the ECA begins energizing the injectors again.

FORD ELECTRONIC ENGINE CONTROL IV (EEC IV) SYSTEMS USED WITH 5.0L ENGINES

Design

The EEC IV systems used with 5.0L engines have injectors located in each intake port. These engines are equipped with high-rise manifolds, as shown in Figures 9-147 and 9-148.

FIGURE 9-148. High-Rise Intake Manifold with EFI 5.0L Car Engine. (Courtesy of Ford Motor Co.)

Some 5.0L engines have sequential fuel injection (SFI), in which each injector is grounded individually through the ECA as shown in Figure 9-149.

On 5.0L engines with electronic fuel injection (EFI), injectors 2, 3, 6, and 7 are connected together and grounded through the ECA, and injectors 1, 4, 5, and 8 also share a common ground through the ECA, as illustrated in Figure 9-150.

FIGURE 9-147. High-Rise Intake Manifold with EFI 5.0L Truck Engine. (Courtesy of Ford Motor Co.)

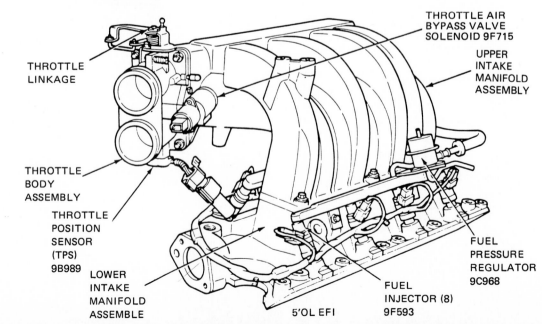

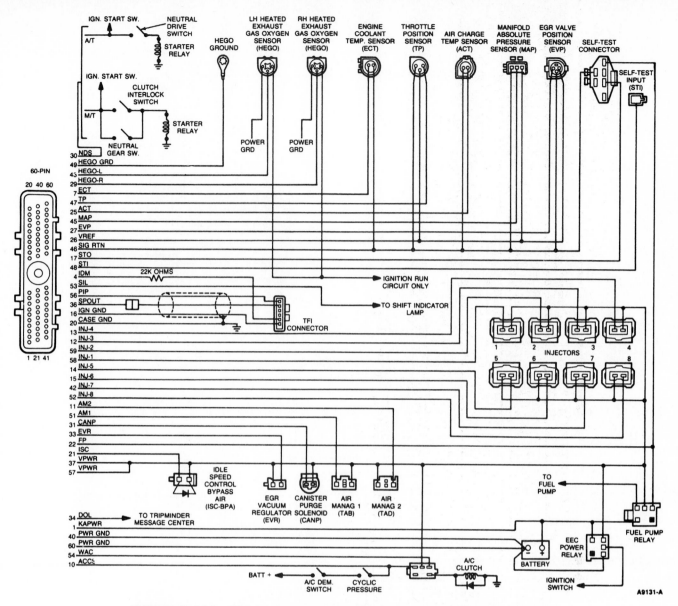

FIGURE 9-149. EEC IV System with SFI. (Courtesy of Ford Motor Co.)

On some EFI systems, injectors 1, 4, 6, and 7 are connected together and injectors 2, 3, 5, and 8 share a common connection. These groups of injectors are grounded through the ECA. On early model systems, the ECA grounded the eight injectors simultaneously, whereas on later systems the ECA grounds the groups of injectors individually.

While many of the same sensors are used on the SFI and EFI systems, there are some differences. The SFI systems have an oxygen (O_2) sensor in each exhaust manifold. These three-wire sensors contain an electric heater to provide faster heating. The EGR valve position (EVP) sensor sends a signal to the ECA in relation to EGR valve position. On the EFI system, the inferred mileage sensor (IMS) sends a signal to the ECA at 22,500 mi (36,209 km). When

this signal is received, the ECA modifies its calibration slightly. The IMS is located under the left side of the instrument panel.

A profile ignition pickup (PIP) signal is sent from the distributor pickup to the ECA. This signal is modified by the ECA and sent to the distributor module on the spark output (SPOUT) circuit to provide the spark advance required by the engine. An ignition diagnostic monitor (IDM) circuit is connected to terminal 4 on the ECA.

If a difference occurs in the PIP and SPOUT signals because of defects in the wires or ECA, the IDM circuit places fault codes 14 and 18 in the ECA.

On the SFI systems, a signature PIP armature is used in the distributor, as shown in Figure 9-151.

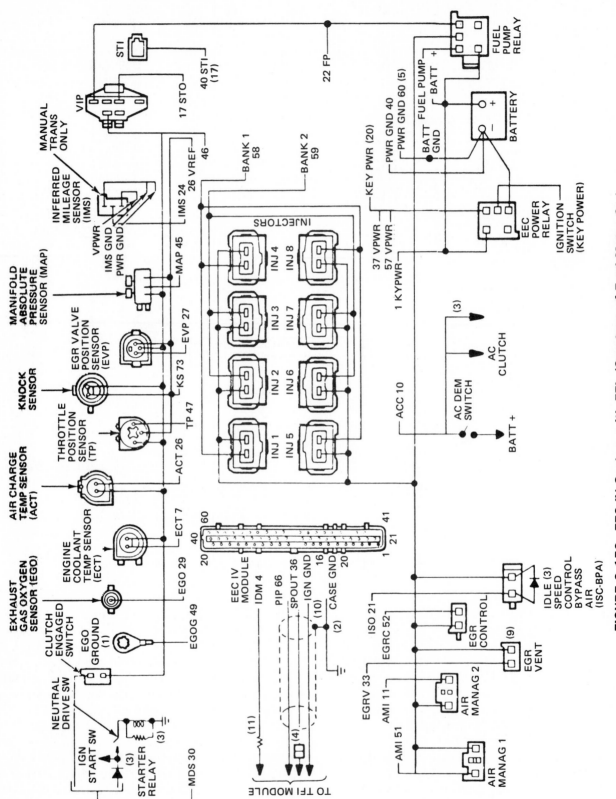

FIGURE 9-150. EEC IV System with EFI. (Courtesy of Ford Motor Co.)

292

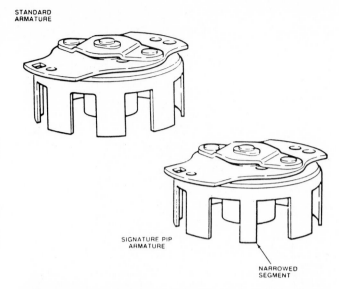

STANDARD
ARMATURE

SIGNATURE PIP
ARMATURE

NARROWED
SEGMENT

FIGURE 9-151. Signature PIP Armature. (Courtesy of Ford Motor Co.)

The signature PIP armature has one narrow blade segment that informs the ECA when number 1 piston is approaching top dead center (TDC). The ECA uses this signal to properly time and fire the injectors. The signature PIP armature must not be interchanged with other armatures.

Some EEC IV systems have a throttle body injection assembly. On a 4-cylinder engine, this assembly contains a single injector, whereas on a V6 or V8 engine dual injectors are used. This type of EEC IV system is referred to as central fuel injection (CFI), as shown in Figure 9-152.

Cruise Control

On some EEC IV systems, such as those used on Taurus and Sable, the cruise control module is integrated into the ECA. The inputs and outputs on the integrated vehicle speed control (IVSC) system are shown in Figure 9-153. Self-diagnostics related to the IVSC system are contained in the ECA.

Integrated Relay Control Module

The integrated relay control module provides control of the cooling fan, A/C clutch, and fuel pump. On previous EEC IV systems, these functions were controlled by separate relays. An integrated relay module is illustrated in Figure 9-154, and a wiring diagram for this module is provided in Figure 9-155.

When the A/C switch is turned on, a signal is sent through the cyclic pressure switch to a solid state relay in the integrated control module. A signal is also sent from the cyclic pressure switch to the ECA. The solid state relay energizes the A/C clutch. If the throttle is wide open, a signal is sent from the ECA through the wide-open throttle A/C cutoff (WAC) circuit to the solid state relay. This signal prevents A/C clutch operation.

If the ECA input signals indicate that cooling fan operation is necessary, the ECA grounds the electric drive fan (EDF) relay winding. This action closes the EDF relay contacts and voltage is supplied through the resistor and these contacts to the cooling fan motor. When increased cooling fan speed is required, the ECA grounds the high-speed electric drive fan (HEDF) relay winding so that full voltage is supplied through these relay contacts to the cooling fan.

When the ignition switch is turned on, a signal is sent from the ignition switch to the relay drive unit in the integrated relay control module. As a result of this signal, the relay drive unit grounds the power relay winding, which closes the relay contacts and supplies voltage to ECA terminals 37 and 57, the fuel pump relay winding, and the EDF relay windings. The ECA grounds the fuel pump relay winding, which closes the relay contacts and supplies voltage through the inertia switch to the electric fuel pump. If the engine is not cranked within 1 second, the ECA opens the fuel pump relay winding circuit to stop the pump.

Pressure Feedback Electronic EGR System

Some applications, such as the 3.0L V6 engine, have a pressure feedback electronic EGR system. The main components in this system are the pressure feedback EGR valve, electronic vacuum regulator, and the pressure feedback electronic EGR sensor, as shown in Figure 9-156.

Exhaust pressure is supplied through a metering orifice to a controlled pressure chamber under the EGR valve. A metal tube connects this chamber to the pressure feedback electronic sensor. When the EGR valve is open, the exhaust pressure in the controlled pressure chamber decreases as EGR flow increases. The pressure feedback electronic sensor sends a signal to the ECA in relation to the controlled chamber pressure. A duty cycle output is sent from the ECA to the electronic vacuum regulator, and this regulator supplies the correct

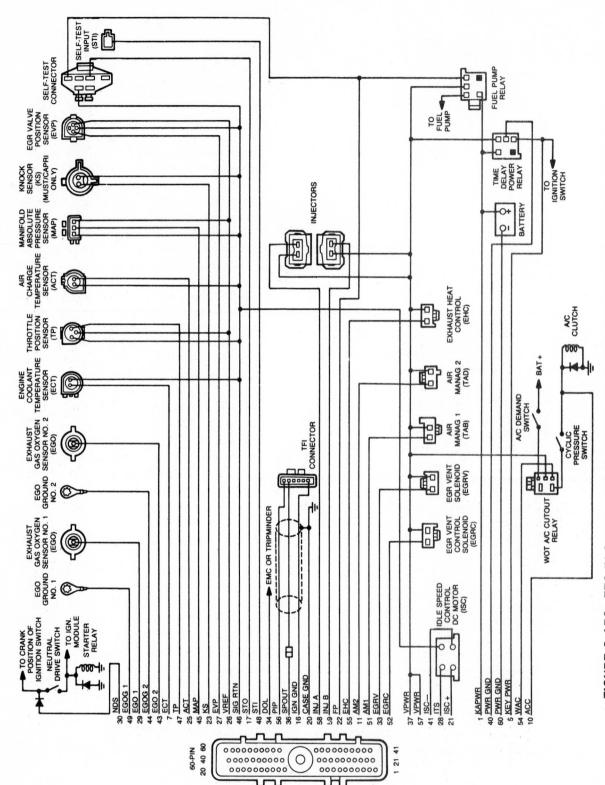

FIGURE 9-152. EEC IV System with Central Fuel Injection (CFI). (Courtesy of Ford Motor Co.)

294

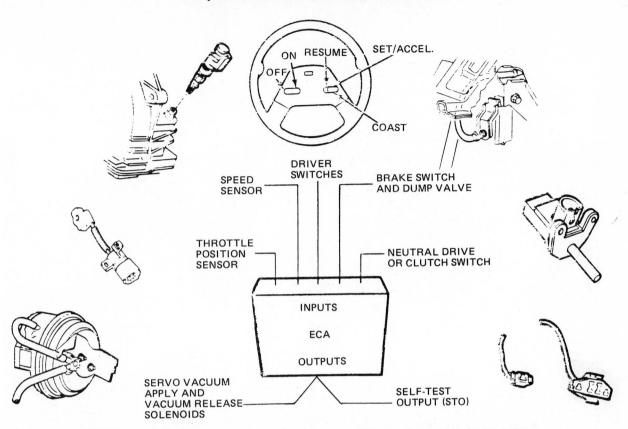

FIGURE 9-153. Integrated Vehicle Speed Control System. (Courtesy of Ford Motor Co.)

FIGURE 9-154. Integrated Relay Control Module. (Courtesy of Ford Motor Co.)

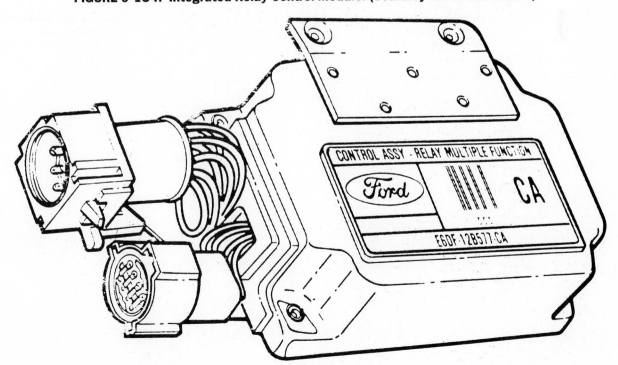

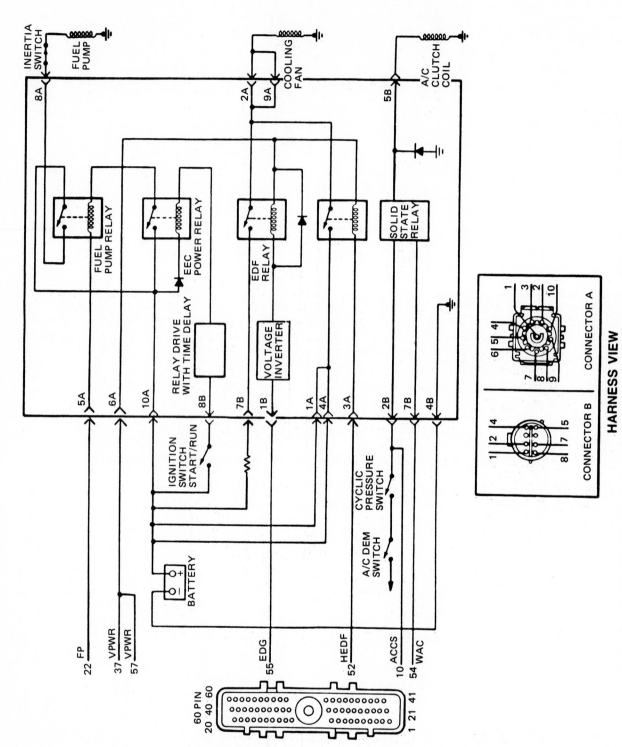

HARNESS VIEW

FIGURE 9-155. Integrated Relay Control Module Wiring Diagram. (Courtesy of Ford Motor Co.)

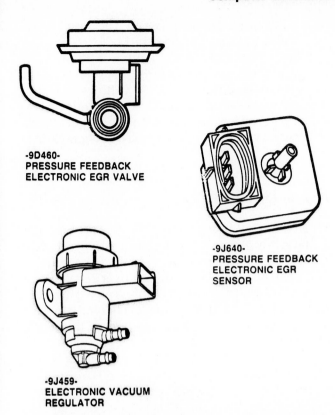

-9D460-
PRESSURE FEEDBACK
ELECTRONIC EGR VALVE

-9J640-
PRESSURE FEEDBACK
ELECTRONIC EGR
SENSOR

-9J459-
ELECTRONIC VACUUM
REGULATOR

FIGURE 9-156. Pressure Feedback Electronic EGR System Components. (Courtesy of Ford Motor Co.)

amount of vacuum to the EGR valve. The ECA matches the EGR flow rate in relation to the input sensor signals, and the ECA compares the controlled pressure chamber input signal to the amount of pressure that should exist in this chamber for a specific engine operating mode. If the pressure in the chamber does not match the required pressure, the ECA adjusts the position of the EGR valve.

A diagram of the pressure feedback electronic EGR system is illustrated in Figure 9-157.

DIAGNOSIS OF FORD ELECTRONIC ENGINE CONTROL IV (EEC IV) SYSTEMS

Fuel Pump Pressure Test

The fuel pump pressure in the EEC IV system with EFI may be diagnosed in the same way as the fuel pump pressure in the General Motors port injection system that was described earlier in this chapter.

FIGURE 9-157. Pressure Feedback Electronic EGR System. (Courtesy of Ford Motor Co.)

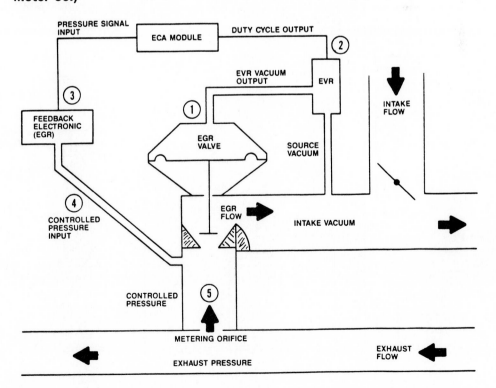

Self-Test Initiation

The same test procedure may be used to diagnose the EEC IV systems with fuel injection, or the EEC IV systems with feedback carburetors. When defects occur in the EEC IV system, service codes are stored in the ECA memory. The service codes may be obtained from the ECA memory by connecting a voltmeter or STAR tester to the self-test connector that is located in each EEC IV wiring harness. The voltmeter or STAR tester must be connected to the self-test connector as shown in Figure 9-158 or Figure 9-159, depending on the type of self-test connector.

When the STAR tester power switch is turned on, 88 will be displayed in the tester window, which will be followed by a 00 display. These displays indicate that the tester is ready to begin the self-test, which is initiated by depressing a pushbutton on the tester. The button will latch down and a colon must appear beside the 00 display before any service codes will be displayed. If the "LO BAT" indicator is displayed steadily at any time, the internal bat-

tery in the tester must be replaced. The initial STAR tester displays are indicated in Figure 9-160.

When a voltmeter and a jumper wire are connected to the self-test connector, the service codes will appear as pulsations on the voltmeter needle. Two pulsations of the voltmeter needle followed by a 2-second pause and three more needle pulsations indicate service code 23, as pictured in Figure 9-161.

Key-On Engine-Off Test

When the STAR tester or the voltmeter is connected to the self-test connector and the ignition switch is turned on, service codes caused by hard failures will be displayed first, followed by a code 10 and then the service codes that are caused by soft failures. Hard failures are defects that are present at the time of setting. Soft failures are the result of intermittent faults.

The key-on/engine-off service code format is outlined in Figure 9-162.

FIGURE 9-158. Voltmeter and STAR Tester Connections to the Self-Test Connector. (Courtesy of Ford Motor Co.)

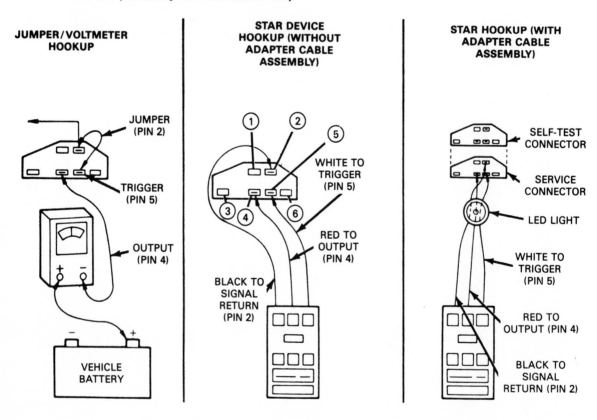

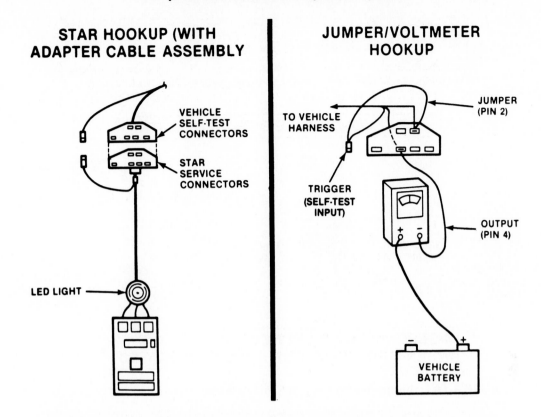

FIGURE 9-159. Voltmeter and STAR Tester Connections to Self-Test Connector with Trigger Self-Test Input. (Courtesy of Ford Motor Co.)

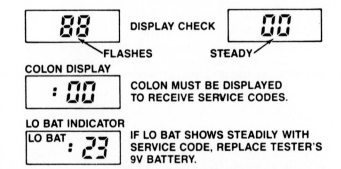

FIGURE 9-160. Initial STAR Tester Displays. (Courtesy of Ford Motor Co.)

Output Cycling Test

This test is performed in the key-on/engine-off test after the service codes have been displayed. Without disabling the self-test connections, momentarily depress the throttle to the wide-open position and then release the throttle to initiate the output cycling test. This test procedure allows the technician to force the ECA to activate the ECA output controls one after the other. The technician may listen to or

observe each output control, such as relays or motors, to determine if they are operating normally. A second throttle depression will stop the output cycling tests, or the test will shut off automatically after 10 minutes.

Engine-Running Test

Before the engine-running test is attempted, complete these procedures:

1. Run the engine at 1,500 RPM for 15 minutes to warm up the oxygen sensor.
2. Turn the engine off and depress the self-test button on the STAR tester. If a voltmeter is used for test purposes, connect the jumper wire to the self-test connector.
3. Wait 10 seconds and start the engine.

With the STAR tester or voltmeter connected to the self-test connector and the engine running, the ECA will monitor all the sensors for proper operation, energize all the output actuators, and check the corresponding results. An engine identification

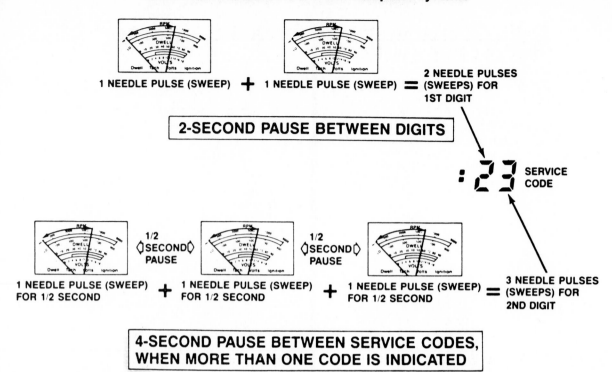

FIGURE 9-161. Voltmeter Reading of Service Codes. (Courtesy of Ford Motor Co.)

code will be displayed first in the engine-running test code format. The engine identification code pulses will be equal to half the number of engine cylinders. For example, three pulses on the voltmeter would be the engine identification code for a 6-cylinder engine. Three engine identification pulses will be displayed as 30 on the STAR tester. When dynamic response code 10 is received, the throttle must be pushed momentarily to the wide-open position. The engine-running service code format is shown in Figure 9-163.

The service code sequence for the key-on/engine-off test and the engine-running test with a STAR tester is outlined in Figure 9-164, and the service code sequence when these tests are performed with a voltmeter is shown in Figure 9-165.

FIGURE 9-162. Key-On/Engine-Off Service Code Format. (Courtesy of Ford Motor Co.)

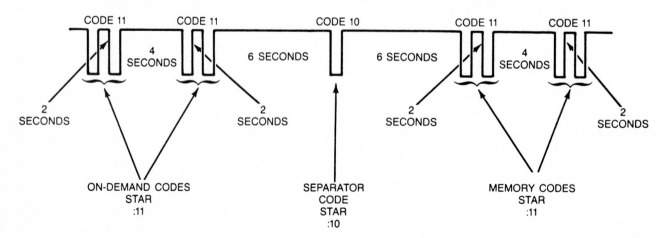

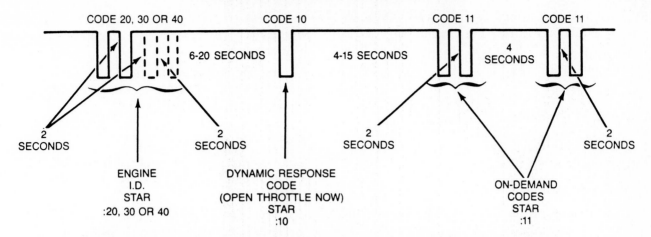

FIGURE 9-163. Engine-Running Service Code Format. (Courtesy of Ford Motor Co.)

SELF-TEST SERVICE CODE SEQUENCE	
KEY-ON ENGINE-OFF SEGMENT	**ENGINE-RUNNING SEGMENT**
1. FAST CODES—LED LIGHT FLICKERS 2. ON-DEMAND CODES—HARD FAULTS 3. SEPARATOR CODE—NUMERAL 10 4. MEMORY CODES—INTERMITTENT FAULTS	1. ENGINE I.D. CODE—NUMERALS 20, 30, OR 40 2. DYNAMIC RESPONSE CODE—NUMERAL 10 (GOOSE THE THROTTLE) 3. FAST CODES—LED LIGHT FLICKERS 4. ON-DEMAND CODES—HARD FAULTS

FIGURE 9-164. Service Code Sequence with STAR Tester. (Courtesy of Ford Motor Co.)

SERVICE CODE SEQUENCE	
KEY-ON ENGINE OFF-SEGMENT	**ENGINE-RUNNING SEGMENT**
1. FAST CODES—NEEDLE FLUCTUATES 2. ON-DEMAND CODES—HARD FAULTS 3. SEPARATOR CODE—NEEDLE SWEEPS ONCE 4. MEMORY CODES—INTERMITTENT FAULTS	1. ENGINE I.D. CODE—TWO, THREE OR FOUR NEEDLE SWEEPS 2. DYNAMIC RESPONSE CODE—NEEDLE SWEEPS ONCE (GOOSE THE THROTTLE) 3. FAST CODES—NEEDLE FLUCTUATES 4. ON-DEMAND CODES—HARD FAULTS

FIGURE 9-165. Service Code Sequence with Voltmeter. (Courtesy of Ford Motor Co.)

Wiggle Test

If a voltmeter is connected to the self-test connector, the jumper wire must be disconnected from the self-test connector during this test. When the ignition is on and the wiring harness or connectors are wiggled, the voltmeter needle will deflect each time a defect occurs and a service code will be stored. A STAR tester with a light-emitting diode (LED) light, as illustrated in Figure 9-158, may be used for this test. With the tester pushbutton not depressed and the ignition on, the LED light will go out each time a defect occurs as the wiring harness and connectors are wiggled. The LED light will stay off as long as the defect exists.

The wiggle test may be repeated with the engine running and the self-test mode activated with the jumper wire connected to the self-test connector, or with the self-test button depressed on the STAR tester. The key-on and the engine-running tests may be repeated to read any service codes that were stored in the ECA during the wiggle test.

Later model STAR testers provide an audible "beep" in place of the LED light.

Erase Code Procedure

The service code can be erased from the ECA using the following procedure:

1. Turn the ignition switch off.
2. Depress the self-test button to deactivate the STAR tester.
3. Turn the ignition switch on.
4. Depress the self-test button to reactivate the STAR tester.
5. When the STAR tester begins to display the first code, depress the self-test button to deactivate the tester. This should erase all codes from the ECA.

Service Procedure and Codes

Intermittent codes should not be erased from the ECA until the necessary repairs have been completed. Hard faults should be repaired before intermittent faults. Fast codes, indicated by a slight flutter of the voltmeter needle or flash of the LED light on the STAR tester, should be ignored, because they are used in factory test procedures. (Refer to Figure 9-164 for fast codes.)

If the basic timing is being checked, the in-line timing connector near the distributor must be disconnected. The computed timing may be checked by depressing the self-test button on the STAR tester, or by connecting the jumper wire to the self-test connector as in the voltmeter diagnosis, with the engine running. The computer timing should be advanced 20° more than basic timing. Service codes should be disregarded during this test. The service codes are listed in Figure 9-166.

Additional Service Code Abbreviations and Procedures

The following abbreviations and their explanations have not been listed previously. The abbreviations apply to 1986 and later models.

OCC—Output cycling check. This is the cycling test where the ECA cycles all the solenoids or motors in the system.

AXOD—This is the name of Ford's automatic overdrive transaxle, which has a lockup-type torque converter operated by the ECA.

LUS—Lockup solenoid. This solenoid is in the converter lockup circuit in the AXOD transaxle.

NPS—Neutral pressure switch. This switch in the AXOD transaxle is in the converter clutch lockup circuit.

BOO—Brake on/off switch. This switch is an input to the ECA for cruise control operation and other functions.

FMEM—Failure mode effects management. This is a default program in the ECA in which the ECA compensates for data loss caused by a hard sensor failure.

SCVNT—Servo control vent.

SCVAC—Servo control vacuum.

Some of the 90 series codes were introduced because dual oxygen sensors are used on later model engines. Various Ford engines have different sensors or components in the EEC IV system. Therefore, some service codes will only be available on a specific engine. Service codes vary, depending on the model year.

When repairs are made to correct the defects that are indicated by service codes, always begin repairing the defect that was indicated by the lowest service code number, and always begin the repair procedure with defect indicated by hard failures. In many cases, a defective sensor can cause more than one service code to be stored in the ECA.

If the air conditioner is turned on during the key-on or engine-running tests, code 67 will appear on the tester. Therefore, the air conditioner should be turned off during the diagnostic procedures.

During the engine-running tests on 1987 and later models, the brake pedal must be depressed and the steering wheel turned after the engine ID code is provided. Otherwise, codes will be given representing the brake on/off (BOO) switch and the power steering pressure switch (PSPS). Some 1986 and later models do not have a dynamic response code in the engine-running test sequence.

A cylinder balance test may be performed on 1986 and later models with SFI. This test is initiated by momentarily pushing the throttle wide open at

CODES

11 — "PASS"
12 — RPM/EXTENDED IDLE
13 — RPM/NORMAL IDLE (1/ 2.3L, 2.5L, 3.8L)
14 — PIP ERRATIC
15 — ROM OR RAM TEST FAILED
16 — RPM TOO LOW FOR EGO TEST
17 — IDLE STOP — TEST
18 — NO TACH (IDM)
19 — IVPWR TEST FAILURE
21 — ECT CIRCUIT OUT OF RANGE
22 — MAP/BP CIRCUIT OUT OF RANGE
23 — TPS CIRCUIT OUT OF RANGE
24 — ACT/VAT CIRCUIT OUT OF RANGE
25 — KNOCK NOT SENSED
26 — MAF/VAF CIRCUIT OUT OF RANGE 2/
29 — AXOD VSS FAILED
31 — EGR TEST 3/
32 — EGR TEST 4/
33 — EGR TEST 5/
34 — EGR TEST 6/
35 — EGR TEST 7/
39 — AXOD LUS FAILED
41 — EGO STATE ALWAYS LEAN
42 — EGO STATE ALWAYS RICH
43 — EGO COOLDOWN OCCURED
44 — THERMACTOR AIR SYS INOP.
45 — THERMACTOR AIR ALWAYS UPSTREAM
46 — AIR ALWAYS BYPASSED
47 — SCCS BUTTON(S) NOT FUNCTIONAL
51 — ECT INPUT HIGH (OPEN)
52 — PSPS CIRCUIT OPEN
53 — TPS INPUT HIGH
54 — ACT/VAT INPUT HIGH (OPEN)

CODES

55 — KEYPWR CKT OPEN
56 — MAF SHORT TO PWR (VAF HIGH)
57 — NPS FAILED OPEN — NEUTRAL
58 — ITS — CLOSED (KOER)
59 — 4/3 SW FAILED OPEN
61 — ECT INPUT LOW (GROUNDED)
62 — AXOD THS 4-3/3-2 SW CLOSED
63 — TPS INPUT LOW
64 — ACT/VAT INPUT LOW (GROUNDED)
66 — MAF/VAF — SHORTED/OPEN
67 — NDS = DRIVE/ACC = ON
68 — ITS — OPEN (KOEO)
69 — AXOD THS 3-2 SW FAILED OPEN
72 — NO MAP CHANGE IN GOOSE TEST
73 — NO TP CHANGE IN GOOSE TEST
74 — BOO SWITCH CIRCUIT OPEN/NOT ACTUATED
75 — BOO SWITCH CIRCUIT CLOSED
76 — NO VAF CHANGE IN GOOSE TEST
77 — OPERATOR DID NOT DO GOOSE TEST
78 — POWER RELAY INTERRUPT
81 — OCC CKT #1 FAILED
82 — OCC CKT #2 FAILED
83 — OCC CKT #3 FAILED
84 — OCC CKT #4 FAILED
85 — OCC CKT #5 FAILED
87 — OCC CKT #7 FAILED
88 — OCC CKT #8 FAILED
89 — OCC CKT #9 FAILED
91 — RIGHT EGO ALWAYS LEAN
92 — RIGHT EGO ALWAYS RICH
94 — RIGHT SECONDARY AIR INOP.
98 — FMEM TEST ABORT
99 — IDLE NOT LEARNED

1/ CODE 13 — CONTINUOUS-D.C. MOTOR ISC FAILURE
2/ CODE 26 — NOT APPLICABLE TO 2.3L T.C.
3/ CODE 31 — EVP EVR/PFE EVR (SENSOR INPUT LOW) — EGRC/V (EVP OUT OF LIMITS)
4/ CODE 32 — EVP EVR (EVP SENSOR EXCEEDS LOW LIMIT) — PFE EVR (EGR FLOW AT IDLE) — EGRC/V (EGR NOT CONTROLLING)
5/ CODE 33 — EVP EVR/PFE EVR (NO INFERRED EGR FLOW) — EGRC/V (EGR NOT CLOSING PROPERLY)
6/ CODE 34 — EVP EVR (VALVE EXCEEDS HIGH LIMIT) — PFE EVR (KOEO = SENSOR OUT OF LIMITS) (KOER/CONT = HIGH EXH. B.P.) EGRC/V (EGR VALVE NOT OPENING)
7/ CODE 35 — EVP EVR/PFE EVR (SENSOR INPUT HIGH) — EGRC/V (RPM TOO LOW FOR TEST)

CYLINDER BALANCE TEST CODE DESCRIPTION (5.0L SFI)
CODES 10, 20, 30...80; [10 = CYLINDER #1, 20 = CYLINDER #2, etc.]
CODE 90 IS A PASS
CODE 77 IS TEST ABORT OR THROTTLE MOVED DURING CYLINDER BALANCE TEST

FIGURE 9-166. EEC IV Service Codes. (Courtesy of Ford Motor Co.)

the end of the engine-running tests. The STAR tester disables each injector in sequence, which prevents air-fuel mixture flow to the cylinders. A tachometer may be connected to the engine to record RPM drop on each cylinder.

The wiggle test has been modified on later model vehicles. This test is now initiated by latching, unlatching, and relatching, the STAR tester at the end of the key-on or engine-running test sequence. The technician must wait two minutes after the last service code in the engine-running test before entering the wiggle test.

Cruise Control Diagnosis

When the ECA operates the cruise control, the ECA will diagnose the cruise control system. Follow this procedure for the key-on engine-off cruise control tests.

1. Place the transmission in park and unplug the self-test input connector.
2. Turn on the STAR tester and depress the self-test button.

KEY ON ENGINE OFF

CODE	DESCRIPTION
11	SYSTEM PASS
23	TP OUT OF RANGE
47	SCCS BUTTON(S) DOES NOT FUNCTION
48	A BUTTON IS STUCK
49	SIG RTN OPEN TO SWITCHES OR BUTTONS
53	TP VOLTAGE TOO HIGH
63	TP VOLTAGE TOO LOW
74	BRAKE SWITCH FAILED LOW (BRAKE OFF)
75	BRAKE SWITCH FAILED HIGH (BRAKE ON)
67	NEUTRAL DRIVE SWITCH FAILED OPEN (DRIVE INDICATION)
81	SCVNT CIRCUIT FAILED (OCC)
82	SCVAC CIRCUIT FAILED (OCC)
10	IDENTIFIER CODE (ENTERING SPEED CONTROL SELF-TEST)

KEY ON ENGINE RUNNING

11	SYSTEM PASS
27	VEHICLE SPEED CONTROL SERVO LEAKS DOWN
28	VEHICLE SPEED CONTROL SERVO LEAKS UP
36	RPM FAILS TO INCREASE DURING DYNAMIC TEST
37	RPM FAILS TO DECREASE DURING DYNAMIC TEST
10	IDENTIFIER CODE (ENTERING SPEED CONTROL SELF-TEST)

FIGURE 9-167. Cruise Control Service Codes. (Courtesy of Ford Motor Co.)

3. Turn on the ignition switch and the cruise control switch.

4. Observe code 10 on the STAR tester.

5. Press the speed control off, coast, accel, and resume buttons.

6. Depress the brake pedal once on vehicles with automatic transaxles, or depress the clutch pedal once if the vehicle has a manual transaxle.

7. Observe and record all service codes.

When the engine-running cruise control test is performed, use this procedure:

1. Connect the STAR tester and the self-test input connector.

2. Start the engine and turn on the STAR tester.

3. Turn on the cruise control switch.

4. Press the STAR tester self-test button within 15 seconds.

5. Observe code 10 and all other service codes.

All the cruise control service codes are listed in Figure 9-167.

Questions

Questions on General Motors Computer Systems

1. The amount of fuel that is injected is determined by the injector _____ _____.

2. When the ignition switch is on and no attempt is made to start the engine, the fuel pump will run until the battery is discharged.   T   F

3. In the open-loop mode, the electronic control module ignores the oxygen sensor signal.
         T   F

4. Cold fast idle speed is controlled by the idle control (IAC) motor.   T   F

5. A sticking IAC motor could result in:
a) erratic idle operation
b) detonation
c) reduced maximum speed

6. The minimum air rate should be adjusted each time an engine tune-up is performed.   T   F

7. On a port-type injection system, trouble code 33 indicates a defect in the _____ _____ _____ circuit.

8. A defective coolant sensor could result in an engine flooding complaint.   T   F

Questions on Chrysler Electronic Fuel Injection (EFI) Systems

1. The automatic shutdown (ASD) relay remains closed when the ignition switch is on and the engine is not running.   T   F

2. When the power-loss light is on, the EFI system is in the _____ mode.

3. Battery voltage is supplied to the electric fuel pump through the _____ _____ relay contacts.

4. In the single-point EFI system, the pressure regulator limits the fuel pump pressure to _____ PSI.

5. An in-line fuel filter is located in the engine compartment.   T   F

6. The power module increases the injector pulse width when engine load is increased. T   F

7. The automatic idle speed AIS motor increases the idle speed by allowing more _____ _____ to bypass the throttle.

8. In the multi-point EFI system, the power module energizes the injectors individually.   T   F

9. A fuel hose from a conventional carburetor fuel system may be used on an EFI system.   T   F

10. A 12-V test light should be used to test for power at the logic module terminals. T   F

11. When the basic timing is being checked, the system must be in the _____ mode.

12. If the power-loss light is being used to diagnose the EFI system, two light flashes followed by a pause and one light flash indicates a defect in the _____ _____ circuit.

Questions on Ford Electronic Engine Control (EEC IV) Systems

1. The vane air temperature (VAT) sensor sends a signal to the ECA in relation to:

a) intake air temperature
b) coolant temperature
c) air-fuel mixture temperature

2. While the engine is being cranked, the ECA moves the throttle air bypass valve to the _____ _____ position.

3. The high-pressure fuel pump is located in the fuel tank.   T   F

4. When a cold engine is flooded, the condition may be corrected by:
a) turning off the ignition switch and waiting 10 seconds.
b) pushing the throttle wide open while the engine is being cranked.
c) leaving the throttle in the idle position while the engine is being cranked.

5. When a signature profile ignition pickup (PIP) armature is used, the signal from the narrow blade segment is used by the ECA to:
a) supply the correct spark advance
b) control the basic timing
c) time and fire the injectors

6. In a pressure feedback electronic EGR system, an exhaust pressure signal is sent from the controlled pressure chamber to the:
a) pressure feedback electronic sensor
b) electronic vacuum regulator
c) EGR valve

7. The low-speed and high-speed electric drive fan (EDF) relays are contained in the integrated relay control module.   T   F

8. Hard faults may be defined as defects that are present when the system is being diagnosed.   T   F

9. When repairs are being made to correct defects that were indicated by service codes, the defect indicated by the service code with the highest number should be repaired first.   T   F

10. Defects that are indicated by intermittent fault codes should be repaired before defects indicated by hard fault codes.   T   F

11. When the basic timing is being checked, disconnect the:
a) coolant sensor connector
b) in-line timing connector
c) PIP wire at the distributor

OSCILLOSCOPE DIAGNOSIS OF ELECTRONIC IGNITION SYSTEMS

CHRYSLER ELECTRONIC IGNITION SCOPE PATTERNS

Scope Test Connections

The necessary test connections for a typical oscilloscope are shown in Figure 10-1. The scope connections are as follows:

Scope 1 lead is clamped over the coil secondary wire to pick up the secondary scope pattern.

Scope 2 lead is clamped over spark plug wire 1. This lead arranges the voltage waveforms in a specific order on the scope screen and also operates the timing light.

A scope lead with a red boot (1A) is attached to the negative "tach" terminal on the ignition coil.

Black lead 18 is the ground lead for the scope.

Twin-flex red lead 3 ends in voltmeter leads, which are connected to the battery terminals.

Scope leads 4 and 4A are the ammeter pickup leads.

Scope 5 connection is a vacuum hose that should be connected to the intake manifold.

Primary Pattern

A primary scope pattern in which the voltage waveforms from each cylinder are superimposed on top of each other is illustrated in Figure 10-2. The letters in Figure 10-2 represent the following events during the operation of the ignition system:

Point A indicates the instant that the module turns off the primary electron movement.

The top of the waveform at point B represents the primary induced voltage as the magnetic field collapses in the ignition coil. A primary induced voltage of 200–300 V would be normal.

The oscillations from C to D represent the primary induced voltage as it discharges back and forth through the module.

Point E indicates the instant that the module turns on the primary electron movement.

The line from point E to point A represents the dwell time or "on time" in the primary ignition circuit.

The dwell time for most electronic ignition systems is determined by the module, and therefore it is not adjustable. Dwell specifications for elec-

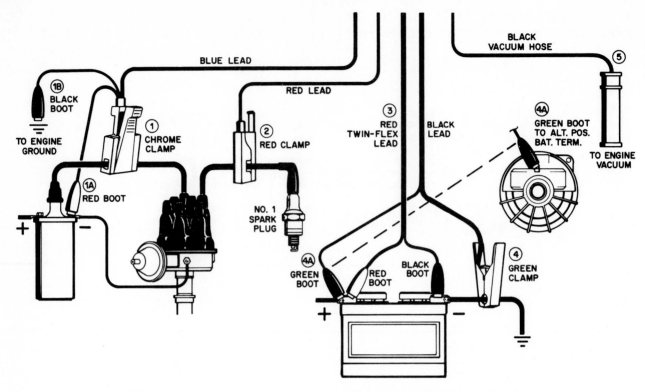

FIGURE 10-1. Scope Test Connections. (Courtesy of Sun Electric Corp.)

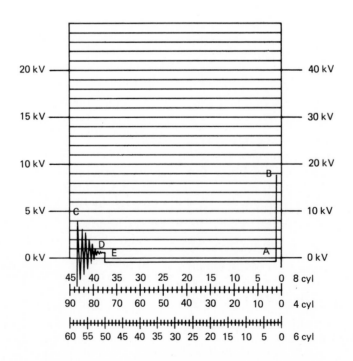

FIGURE 10-2. Chrysler Primary Scope Pattern.

tronic ignition systems are not supplied by the manufacturer. Scope patterns may vary slightly on Chrysler electronic lean burn (ELB) and electronic spark advance (ESA) systems, but the explanation of the pattern would be the same.

Secondary Pattern

An expanded secondary display scope pattern is shown in Figure 10-3. The letters in Figure 10-3 indicate the following events in the operation of the secondary ignition system:

Point A indicates the instant that the module turns off the primary electron movement.

The line from A to B is called a firing line. Point B at the top of the firing line indicates the secondary voltage that is required to start the spark plug firing.

The line from C to D is the voltage required to keep the spark plug firing. This line is referred to as a spark line.

When the spark plug stops firing, a high voltage charge of 3,000–4,000 V will remain on the center spark plug electrode and spark plug wire. This high voltage will discharge back through the coil windings to the spark plug ground electrode. The line from D to E represents this voltage discharge from the center spark plug electrode through the coil windings to the ground electrode after the plug stops firing.

Point E indicates the instant that the module turns on the primary electron movement.

When a spark plug wire is removed, the maximum secondary coil voltage should exceed 22,000 V (22 kV).

GENERAL MOTORS HIGH-ENERGY IGNITION (HEI) SCOPE PATTERNS

Primary Pattern

A primary superimposed scope pattern from an HEI ignition system is illustrated in Figure 10-4. The letters in Figure 10-4 represent the following events in the operation of the primary ignition circuit:

Point D indicates the instant that the module turns off the primary electron movement.

Point E represents the primary induced voltage, which should be 120–150 V.

From A to B the module keeps the primary electron movement shut off. The oscillations after point A indicate the primary induced voltage discharging back and forth through the module.

At point B the module turns on the primary electron movement.

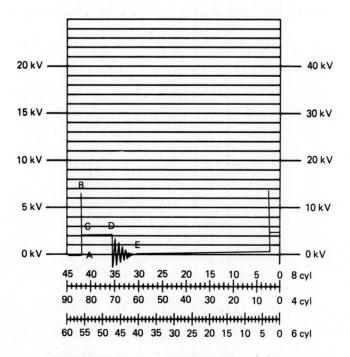

FIGURE 10-3. Chrysler Secondary Pattern.

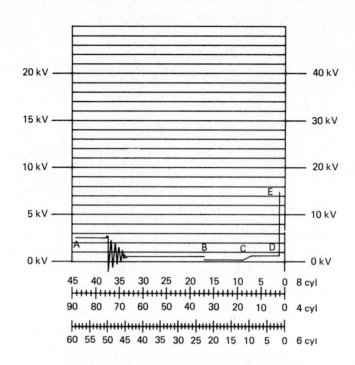

FIGURE 10-4. HEI Primary Scope Pattern.

Point C represents the instant at which the primary electron movement reaches its maximum value.

Secondary Pattern

An expanded secondary display pattern from an HEI ignition system is shown in Figure 10-5. The events represented in the operation of the secondary ignition system are as follows:

Point A indicates the instant that the module turns off the primary electron movement.

The secondary voltage that is required to fire the spark plug is illustrated by point B.

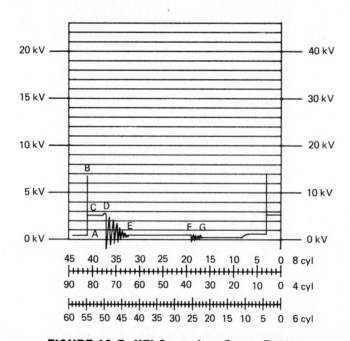

FIGURE 10-5. HEI Secondary Scope Pattern.

The voltage that is required to keep the spark plug firing is indicated by the line from C to D.

The oscillations from D to E are caused by a voltage discharge from the center spark plug electrode through the coil windings to the ground spark plug electrode after the plug stops firing.

At point F the module turns on the primary electron movement.

The oscillations from F to G indicate a low secondary induced voltage that is caused by the buildup of the primary magnetic field.

In either a primary or a secondary HEI scope pattern, the module increases the dwell time in relation to the engine speed. At idle speed the dwell should be approximately 15° and it should increase to about 30° at 2,000 engine RPM. The HEI system should provide a maximum secondary coil voltage at 34,000 V (34 kV) or more when a spark plug wire is removed.

SCOPE PATTERNS OF FORD ELECTRONIC SYSTEMS

Primary Pattern

A primary pattern of a Ford Dura-Spark ignition system is outlined in Figure 10-6. The letters in

Figure 10-6 indicate the following specific events in the operation of the primary ignition system:

Point A indicates the instant that the module shuts off the primary electron movement.

The primary induced voltage is indicated at point B. This voltage should be 200–300 V.

The oscillations from C to D represent the primary induced voltage oscillating back and forth through the module.

At point E the module turns on the primary electron movement.

The primary circuit dwell time is indicated by the line from E to A. In many Ford electronic systems such as Dura-Spark 1 and thick film integrated (TFI) systems, the module increases the dwell time in proportion to engine speed.

Secondary Pattern

A secondary scope pattern for a Ford electronic ignition system is illustrated in Figure 10-7. The letters in the pattern represent the following events:

At point A the module turns off the primary electron movement.

Point B represents the secondary induced voltage that is required to fire the spark plug.

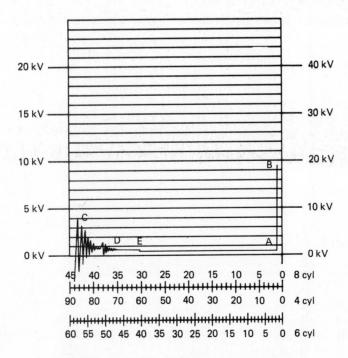

FIGURE 10-6. Ford Primary Scope Pattern.

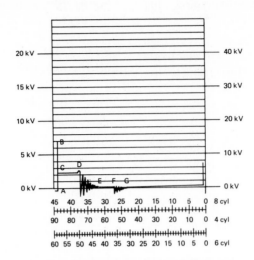

FIGURE 10-7. Ford Electronic Ignition Secondary Scope Pattern.

The line from C to D indicates the voltage that is required to keep the spark plug firing.

The oscillation from D to E is caused by a high voltage discharge from the center spark plug electrode through the coil windings to the ground plug electrode after the plug stops firing.

At point F the module turns on the primary electron movement.

The oscillation from F to G is a low induced secondary voltage thât is caused by the buildup of the primary magnetic field.

When a spark plug wire is removed, the maximum secondary coil voltage should exceed 26,000 V (26 kV).

DEFECTIVE SCOPE PATTERNS

Defective Primary Patterns

The most important primary scope pattern test for electronic ignition systems is to check the primary induced voltage. A primary scope pattern that illustrates excessively high primary induced voltage is shown in Figure 10-8. The high primary induced voltage could be caused by low resistance in the coil primary winding, or low resistance in the primary

circuit resistor. Another cause of high primary voltage would be an excessively high charging circuit voltage.

A superimposed primary pattern with a low primary voltage is shown in Figure 10-9. The low primary induced voltage could be caused by high resistance in the primary circuit resistor, coil primary winding, or the connecting wires. A low primary induced voltage could also be caused by a defective charging circuit and a partly discharged battery.

Defective Secondary Patterns

A secondary display pattern that shows reversed coil polarity is illustrated in Figure 10-10. The defective pattern in Figure 10-10 would arise when the primary coil wires are reversed; thus, all the secondary patterns would be upside down.

A secondary display pattern that illustrates a normal secondary maximum coil voltage is shown in Figure 10-11. When the coil is being tested for maximum secondary voltage, insulated pliers must be used to remove spark plug wires. The maximum coil voltage should equal or exceed the specifications provided for each electronic system in this chapter.

In a secondary display pattern, the waveform from cylinder 1 is usually positioned on the left side

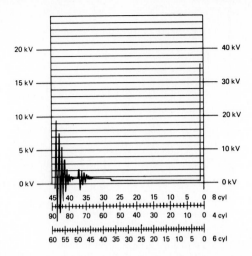

FIGURE 10-8. Excessively High Primary Induced Voltage.

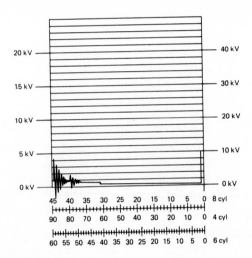

FIGURE 10-9. Low Primary Induced Voltage.

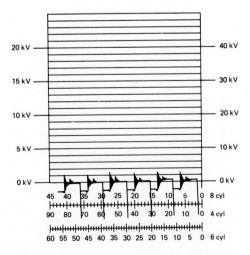

FIGURE 10-10. Reversed Coil Polarity.

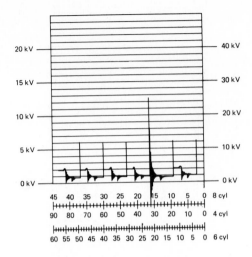

FIGURE 10-11. Normal Maximum Secondary Coil Voltage.

of the screen, and the other waveforms appear in the firing order of the engine from left to right on the screen. Some scopes place the waveform from cylinder 1 on the right side of the screen, and the other waveforms appear in the firing order of the engine reading from left to right on the screen. The voltage sweep below the zero line on the scope screen should be approximately half the length of the maximum voltage firing line, as illustrated in Figure 10-11. A low maximum secondary coil voltage could be caused by a defective coil. A defect in the primary ignition circuit that results in low primary induced voltage will also cause low maximum sec-

ondary voltage. When the maximum secondary coil voltage is lower than specified, a voltage sweep below the zero line will still be evident, as pictured in Figure 10-12.

A leakage defect may occur in the distributor cap, rotor, or spark plug wires. When testing the secondary ignition circuit for a leakage defect, remove each spark plug wire one after the other. If a leakage defect is present, the maximum secondary coil voltage will be lower than specified and there will be no downward voltage sweep below the zero line, as illustrated in Figure 10-13.

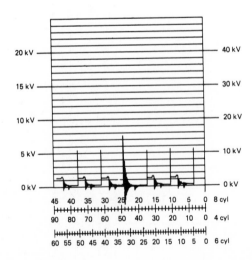

FIGURE 10-12. Low Maximum Secondary Coil Voltage.

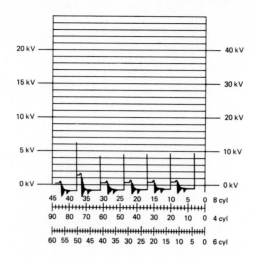

FIGURE 10-13. Secondary Leakage Defect.

If the spark plug gaps, plug wires, and the rotor gap are in satisfactory condition, the voltage required to fire each spark plug should be 6–12 kV, with no more than 3 kV variation between the lowest and the highest firing voltage. A spark plug firing voltage that is excessively high is pictured in Figure 10-14.

The high firing voltage would be caused by excessive resistance in the spark plug, spark plug wire, or rotor gap. In order to determine which component is at fault, remove the spark plug wire from the spark plug and connect it to ground with a ground probe. If the firing voltage is 8 kV or less with the spark plug

wire grounded—as illustrated in Figure 10-15—the high resistance defect exists in the spark plug.

When the firing voltage remains above 8 kV with the spark plug wire grounded, the excessive resistance is in the spark plug wire or the rotor gap. In order to find out whether the resistance defect is in the spark plug wire or the rotor gap, remove the spark plug wire from the distributor cap and connect the cap terminal to ground with a ground probe. If the firing voltage remains above 8 kV with the distributor cap terminal grounded, the rotor gap has excessive resistance. When the firing voltage with the distributor cap terminal grounded is less

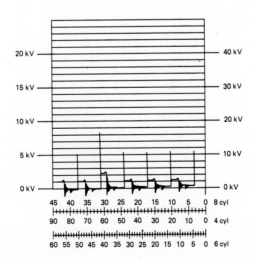

FIGURE 10-14. High Secondary Firing Voltage Caused by Excessive Resistance.

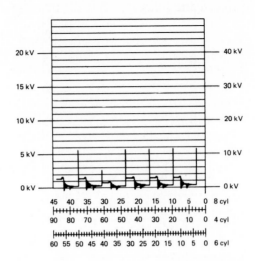

FIGURE 10-15. Secondary Firing Voltage with the Spark Plug Wire Grounded.

than 8 kV and the firing voltage with the spark plug wire grounded is more than 8 kV, the spark plug wire has excessive resistance.

General Motors HEI systems and Ford Dura-Spark ignition systems with a large diameter distributor cap have a 0.094 in (2.35 mm) rotor gap. These systems should have a firing voltage of 8 kV or less with the spark plug wire or the distributor cap terminal grounded. Other ignition systems have a 0.035 in (0.889 mm) rotor gap and they should have a firing voltage of 5 kV or less with the spark plug wire or the distributor cap terminal grounded.

A high resistance defect will be unmistakable—it will appear as a high, short spark line on a raster pattern, as shown in Figure 10-16.

In a raster scope pattern, the waveforms from each cylinder are displayed one above the other, with waveform 1 located at the bottom of the screen and the other waveforms positioned in the firing order of the engine going upward on the screen. Some scope manufacturers use a stacked pattern in place of the raster pattern. In the stacked pattern, the waveforms are positioned one above the other but waveform 1 is at the top of the screen and the other

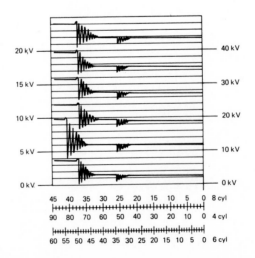

FIGURE 10-16. High Resistance Defect on a Raster Pattern.

waveforms are in the firing order of the engine going downward on the screen.

A spark line that is longer and lower than the other spark lines would be caused by a low-resistance defect such as a fouled spark plug, or a grounded spark plug wire. Low cylinder compression will also cause this type of pattern. If the oscillations after the spark line are reduced, the coil secondary wire or the coil secondary winding has a high-resistance defect. A high-resistance defect in one of these locations will cause the high voltage discharge from the center plug electrode through the coil windings to the ground electrode to be dropped across the secondary ignition circuit in one or two oscillations. Under normal conditions, there should be seven or more peaks on the oscillation that is located immediately after the spark line on the secondary scope pattern.

INFRARED EXHAUST EMISSION ANALYSIS

Carbon Monoxide (CO) and Hydrocarbon (HC) Emission Tests

The infrared analyzer passes a single beam of infrared light through the exhaust sample. A wheel that contains optical filters is rotated through the beam to obtain a signal that is proportional to the percentage of carbon monoxide (CO) and the hydrocarbon (HC) levels in the exhaust sample. The signal from the optical filters is electronically processed to separate the CO and the HC signals and apply them to meters on the analyzer face. A pump in the tester is used to move the exhaust sample through a hose that is connected from the tail pipe probe to the tester. The meters on an infrared analyzer are shown in Figure 10-17.

Carbon monoxide levels in the exhaust are proportional to the air-fuel ratio when it is between 14.7:1 and 10:1, as outlined in Figure 10-18. As the air fuel ratio becomes richer, the CO levels increase. When the air-fuel ratio is leaner than 14.7:1, CO levels remain unchanged. Therefore CO levels are a reliable indicator of rich air-fuel ratios, but a poor indicator of lean air-fuel ratios. The CO emission levels in Figure 10-18 are from a vehicle that does not have a catalytic converter.

Some of the causes of high CO emission levels are pictured in Figure 10-19. Excessive hydrocarbon (HC) levels in the exhaust of vehicles that are not equipped with catalytic converters are the result of incomplete combustion because of mechanical, electrical, or carburetor defects. Hydrocarbon levels are low at the normal operating air-fuel ratio of 14.7:1. When the air-fuel ratio is too lean or too rich,

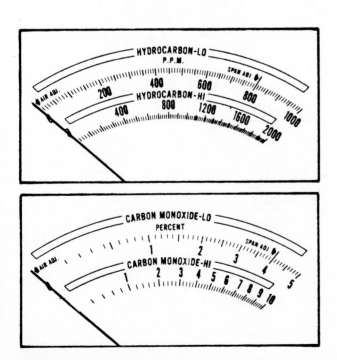

FIGURE 10-17. Infrared Analyzer. (Courtesy of Bear Automotive Inc.)

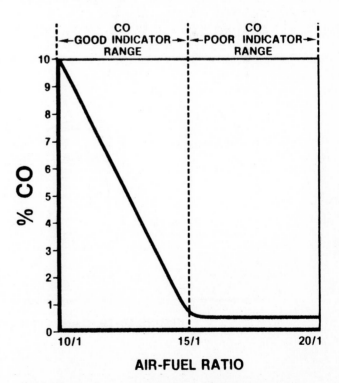

FIGURE 10-18. Relationship of Carbon Monoxide and Air-Fuel Ratio. (Courtesy of Sun Electric Corp.)

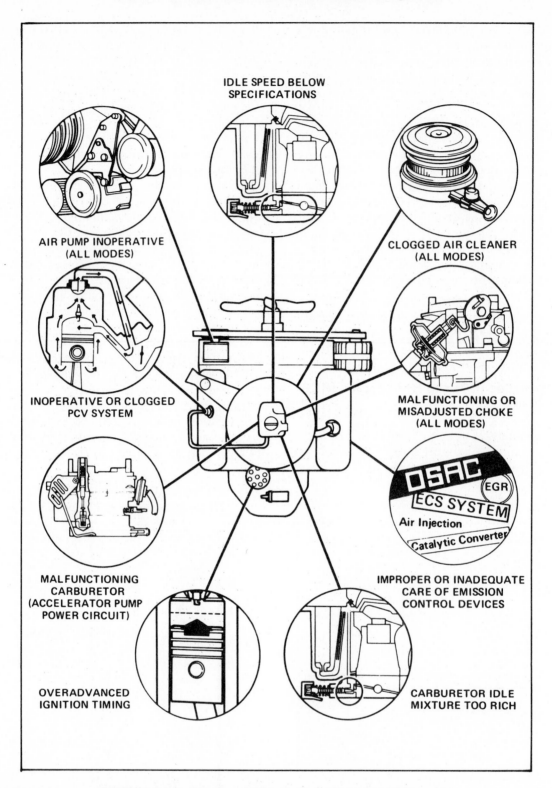

FIGURE 10-19. Causes of High Carbon Monoxide (CO) Levels.

hydrocarbon levels will increase. The HC levels for vehicles without catalytic converters increase significantly when the air-fuel ratio becomes leaner than 17:1. Therefore, HC levels are a reliable indicator of lean air-fuel ratios for precatalytic converter vehicles and they could be used to set the idle mixture screws on these vehicles. Hydrocarbon levels in relation to air-fuel ratios are illustrated in Figure 10-20.

Some of the causes of high HC levels in the exhaust are illustrated in Figure 10-21.

When an infrared tester is being used to check emission levels, the tester must be warmed up for 10–15 minutes and the engine should be at normal operating temperature. The infrared tester should be calibrated according to the manufacturer's recommended procedure. When a tester probe is inserted in the tail pipe of vehicles equipped with catalytic converters, the tester will indicate the emission levels that are coming out of the catalytic converter. In order to read the emission levels coming out of the engine, insert the tester pickup probe ahead of the catalytic converter. Some vehicles have a plug in the exhaust pipe between the converter and the engine for this purpose. A special adapter may be installed in the exhaust pipe to accommodate the pickup probe. Most manufacturers recommend that the air injection pump be disconnected when the emission level tests are being performed.

The emission label in the engine compartment of many vehicles provides the HC and CO levels recommended in each case as well as the location of the tester pickup probe. Some states have vehicle inspection and maintenance regulations pertaining to HC and CO levels that dictate the maximum emission levels for each year of vehicle. The most common test procedure is to check HC and CO levels at idle speed and then at 2,500 RPM.

Four-Gas Emission Tests

The HC and CO emission levels in Figure 10-22 are from the tail pipe of a vehicle that is equipped with a catalytic converter. The HC levels do not increase when the air-fuel ratio becomes too lean because the catalytic converter oxidizes the HC. When the air-fuel ratio is too rich, CO levels do increase, but not as much as they would on a precatalytic converter vehicle. There is no way of determining whether a mixture is lean on the basis of the CO and HC levels from a vehicle equipped with a catalytic converter.

Some emission testers have the capability to read oxygen (O_2) and carbon dioxide (CO_2) levels in the exhaust as well as HC and CO levels. The oxygen (O_2) level in the exhaust increases significantly when the air-fuel ratio becomes leaner than 14.7:1. The CO and O_2 emission levels from a vehicle with a catalytic converter are indicated in Figure 10-23.

The stoichiometric air-fuel ratio is the ratio at which the fuel burns most efficiently. For a gasoline-fueled engine, the stoichiometric air-fuel ratio is 14.7:1. For a vehicle equipped with a catalytic converter, the O_2 level is an excellent indicator of a lean air-fuel ratio. If the O_2 level is above 0.5 percent, the catalytic converter is getting enough oxygen to function properly. When the CO level is above 0.5 percent and the O_2 level 0.5 percent, the catalytic converter is usually not oxidizing the CO. Therefore the O_2 and CO levels can be used to check the catalytic converter.

Carbon dioxide (CO_2) levels in the exhaust increase as the air-fuel ratio becomes leaner, from 8:1 to 14:1. When the air-fuel ratio is leaner than 14:1, CO_2 levels decrease, as shown in Figure 10-24.

At the stoichiometric air-fuel ratio, CO_2 levels are just beginning to decrease. When the four-gas tester is being used, the emission levels should be checked at idle speed and 2,500 RPM. The emission level specifications may be obtained from the vehicle emission label or from state emission regulations. The emission specifications in Table 10-1 could be used as a guide if other specifications are not available.

If the vehicle emission levels do not meet the specifications in Table 10-1, some corrective service

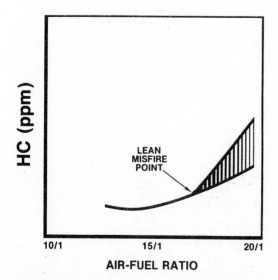

FIGURE 10-20. Hydrocarbon (HC) Levels in Relation to Air-Fuel Ratios. (Courtesy of Sun Electric Corp.)

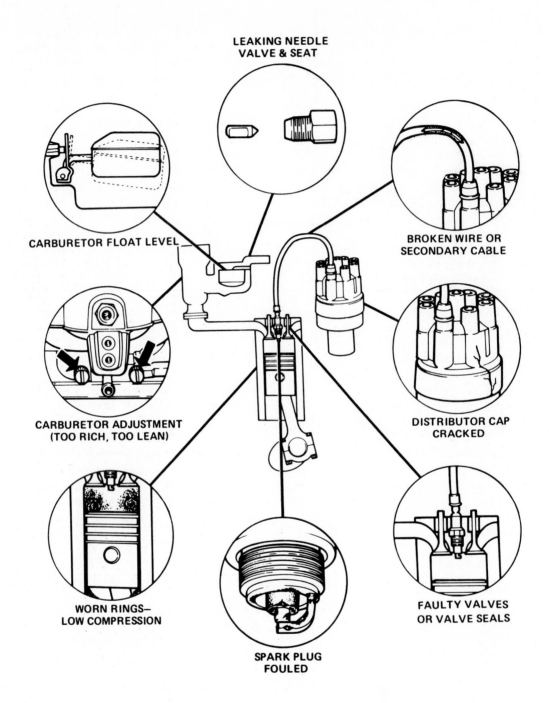

FIGURE 10-21. Causes of High Hydrocarbon (HC) Levels.

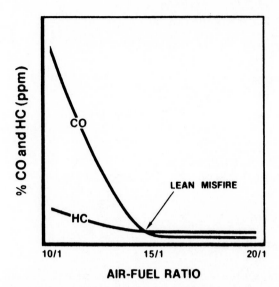

FIGURE 10-22. HC and CO Levels with Catalytic Converter. (Courtesy of Sun Electric Corp.)

is required. The O_2 levels may be used to check the air injection system. Record the O_2 level with the air injection system disabled and the engine operating at idle speed. When the air injection system is operational, the O_2 level should be 2–5 percent higher than the O_2 level with the air injection system disabled. The CO_2 and O_2 levels may be corrected by adjusting the carburetor mixture screws as follows:

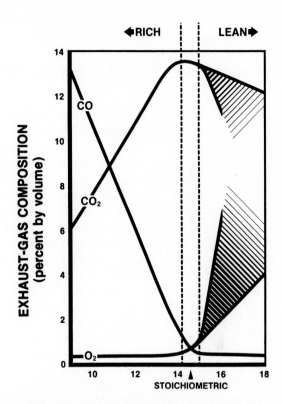

FIGURE 10-24. CO_2, O_2, and CO Emission Levels. (Courtesy of Sun Electric Corp.)

1. Run the engine at the specified idle speed and normal operating temperature.
2. Turn each mixture screw in or out until the highest CO_2 reading is obtained.
3. Rotate the mixture screws inward ¼ turn at a time. The mixture screws should be turned alternately. Rotate the mixture screws until the O_2 begins to increase and fluctuate; this pattern indicates the idle "lean point" has been reached.
4. Back the idle mixture screws out slightly to enrich the mixture. Emission levels must meet all state and federal regulations.

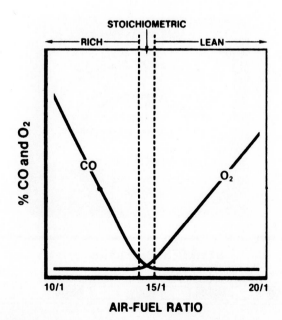

FIGURE 10-23. CO and O_2 Emission Levels with Catalytic Converter. (Courtesy of Sun Electric Corp.)

TABLE 10-1. Emission Level Specifications.

Vehicle Year	HC (ppm) Less Than	CO_2 (%) Greater Than	CO (%) Less Than	O_2 (%) in the Range of
71-75	400 ppm	8%	3%	.1-7%
76-79	200 ppm	8%	2%	.1-7%
80-81	100 ppm	8%	.5%	.1-7%

Courtesy of Sun Electric Corp.

Cylinder Output Tests

When performance or economy complaints are being diagnosed, the engine compression and the ignition system should be checked before the fuel injection system or computer-controlled carburetor system. The cylinder output tester on the oscilloscope may be used to check the cylinder output, which will give an indication of the cylinder compression.

A series of push buttons on the cylinder output tester are used to stop each spark plug from firing. The primary pattern pickup lead on the scope is connected to the negative "tach" side of the ignition coil and one scope lead is connected to ground. When the cylinder output test buttons are depressed, the scope momentarily grounds the primary ignition circuit each time the module attempts to open the primary ignition circuit, and this action prevents the spark plug from firing.

The cylinder output test buttons are arranged to stop each spark plug from firing in the firing order of the engine. For example, if the engine has a firing order of 18436572, button 2 will stop spark plug 8 from firing. When the cylinder output tester is being used, the secondary display scope pattern will indicate which cylinder is "killed" or prevented from firing, as illustrated in Figure 10-25.

When each cylinder is killed, the engine RPM should slow down a specific amount. This drop will be indicated on a tachometer on the scope. If a cylinder is not producing much power, very little drop in RPM will be indicated on the tachometer when that cylinder is "killed." The RPM drop should not vary more than 30 percent between the cylinders. The causes of low cylinder output and RPM drop are:

1. Vacuum leaks into the intake manifold.
2. The exhaust gas recirculation (EGR) valve stuck open.
3. Ignition system misfiring, which would be indicated on the scope patterns.
4. Low cylinder compression because of burned valves, worn rings, or a damaged head gasket.

The cylinder output test buttons and the tachometer on the scope are illustrated in Figure 10-26.

Spark Advance Tests

Most scopes have the capability to test the spark advance on computer-controlled spark advance systems or on distributors with conventional vacuum and centrifugal advance mechanisms. The following procedure should be used to check the spark advance:

1. Check the initial timing with the timing light on the scope. On computer-controlled advance systems, the special instructions must be followed that were provided for each system in chapters 7, 8, and 9.

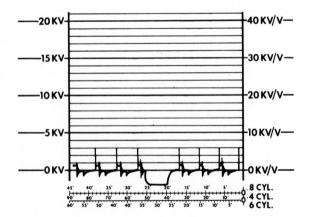

FIGURE 10-25. Cylinder Output Test Scope Pattern. (Courtesy of Sun Electric Corp.)

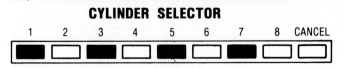

FIGURE 10-26. Cylinder Output Tester. (Courtesy of Sun Electric Corp.)

2. Operate the engine at 2,000 RPM and rotate the advance control knob on the timing light until the timing mark returns to the initial setting. At this time the advance meter on the timing light or the scope will indicate the total spark advance.

3. Compare the total spark advance to the vacuum and centrifugal advance specifications at 2,000 RPM, or to the computer advance specifications. If conventional distributor advances do not provide the specified advance, the spark advance test may be repeated with the vacuum advance hose disconnected to determine if the vacuum advance or the centrifugal advance is defective.

Many vehicles are equipped with a special device to prevent the vacuum advance from operating instantly when the throttle is opened. (See chapter 5, "Distributor Spark Retard Emission Systems.") In these vehicles, the delay device will have to be bypassed when spark advance is being tested, or time will have to be allowed for the delay device to operate. When the spark advance is being tested on computer-controlled systems, all the special instructions must be followed that were provided for each system in chapters 7, 8, and 9.

Some scopes have a magnetic timing feature rather than a timing light. A magnetic timing probe is inserted in a hole in the timing indicator. The initial timing and advance timing are indicated on a digital display or a meter on the scope. The timing light and the advance meter are shown in Figure 10-27.

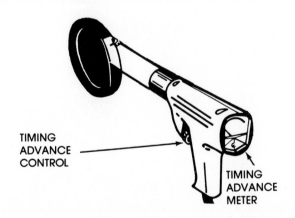

TIMING
ADVANCE
CONTROL

TIMING
ADVANCE
METER

FIGURE 10-27. Timing Light and Advance Meter. (Courtesy of Sun Electric Corp.)

Questions

1. On the primary superimposed scope pattern from a Chrysler electronic ignition system, a primary induced voltage of 100 V would be satisfactory.   T   F

2. Dwell specifications are provided by the manufacturers of electronic ignition systems. T   F

3. When a spark plug wire is removed in a high-energy ignition (HEI) system, the maximum secondary coil voltage should exceed _____ _____ kV.

4. The dwell increases in relation to engine speed in a thick film integrated (TFI) ignition system.   T   F

5. A primary induced voltage that is higher than normal could be caused by low resistance in the primary circuit ignition resistor.   T   F

6. When a spark plug wire is removed and a leakage defect exists in the distributor cap, a sweep will appear below the zero line on the scope screen.   T   F

7. Carbon monoxide (CO) emission levels are a reliable indicator of a lean air-fuel ratio.   T   F

8. For a vehicle that is equipped with a catalytic converter, hydrocarbon (HC) levels will increase when the air-fuel ratio becomes leaner than 17:1.   T   F

9. Oxygen levels in the exhaust are a reliable indicator of a lean air-fuel ratio.   T   F

10. If the oxygen level is 0.5 percent and the CO level is 1 percent, the _____ _____ is defective.

ELECTRONIC INSTRUMENT PANELS, VOICE ALERT SYSTEMS, AND TRIP COMPUTERS

TYPES OF ELECTRONIC INSTRUMENTATION DISPLAYS

Light-Emitting Diodes

A light-emitting diode (LED) is a diode that emits light as electric current flows through it. Small dotted segments are arranged in the LED display so that numbers and letters can be formed when selected segments are turned on. The LEDs usually emit red light and they can be seen easily in the dark; however, they are less visible in direct sunlight. The tachometer on the electronic instrument cluster shown in Figure 11-1 has 36 LEDs.

Liquid Crystal Displays

The liquid crystal display (LCD) uses a polarized light principle on a nematic liquid crystal to display numbers and characters. An LCD display uses extremely low electrical power, but it requires backlighting to be viewed in the dark. A complete LCD instrument display panel is illustrated in Figure 11-2.

The retainer is made of heat-resistant mineral and glass-filled polyester, and it provides a closely dimensioned support for the LCD cells. A dark background is used on the nematic LCDs, and active areas change from dark to clear when they are energized. Polarizers in the LCDs provide the proper balance of contrast, transmission, and hue. The protective matte film on the front surface of the LCDs reduces reflections and provides resistance to scratches and chemicals.

A thin polycarbonate transflector is mounted behind the LCD to provide color. The front surface of the transflector is silkscreened with translucent fluorescent inks which make use of front-incident ambient light and rear-incident backlighting to obtain proper day and night intensity levels.

Backlighting is achieved with the use of clear acrylic light pipes. Light entrance areas and special optical patterns on the rear surfaces of the light pipes provide optimum backlighting balance and intensity. White polycarbonate reflectors are placed behind the light pipes to use the light escaping from their rear surfaces.

Hundreds of individual electrical connections are made through the elastomeric connectors which make contact between the indium tin oxide conductor pads on the LCDs to corresponding contacts on the driver board.

The driver board contains eight static LCD driver integrated circuits (ICs) which are individually responsible for driving 32 LCD segments.

The microprocessor on the logic board receives input data from various sensors and switches and then determines whether each LCD segment should be on or off. A 12-pin connector is used to connect

FIGURE 11-1. Electronic Tachometer with LEDs. (Courtesy of Chrysler Canada Ltd.)

the logic board to the driver board, whereas a 24-pin and a 36-pin connector complete the electrical connections between the microprocessor and the vehicle electrical system. The inputs to the microprocessor in the LCD instrument cluster and the output control functions are outlined in block form in Figure 11-3.

The front of the LCD instrument cluster is pictured in Figure 11-4.

The switch and telltale console shown in Figure 11-5 is mounted near the LCD instrument cluster.

The LCD instrument cluster is continuously backlit by four halogen lamps. Their intensity is automatically controlled by pulse width modulation to provide proper display intensity under varied natural lighting conditions. A photocell in the upper left corner senses natural light conditions, and the microprocessor then determines the correct pulse width cycle for the halogen lamps to provide the proper LCD display intensity.

When the ignition switch is turned on, the LCD cluster is lit at full intensity for 2 seconds and all the LCD segments are activated. During this time the microprocessor measures all inputs so it can display them immediately after the initialization sequence. The driver may adjust the brilliance of the

LCD display by rotating the rheostat on the headlight switch.

LIQUID CRYSTAL DISPLAY FUNCTIONS

Vehicle Speed

The speedometer displays vehicle speed in both digital and analog bar-graph form. Metric or English values may be selected on the LCD displays with the Metric/English switch on the switch and telltale console. The yellow digital speedometer display indicates speed from 0 to 157 miles per hour (MPH), or 0 to 255 kilometers per hour (KPH). A green 41-segment bar graph graphically displays speed from 5–85 MPH (8–137 KPH). A multi-pole permanent-magnet speed sensor in the transaxle provides a speed signal to the LCD microprocessor. The odometer is updated every 0.1 mile (mi) or 0.1 kilometers (km). This odometer counts from 0 to 999.9 mi or (999.9 km), and then counts from 1,000 to 4,000 mi (1,000 to 6,436 km). The trip odometer is set on 0 when the reset button is depressed.

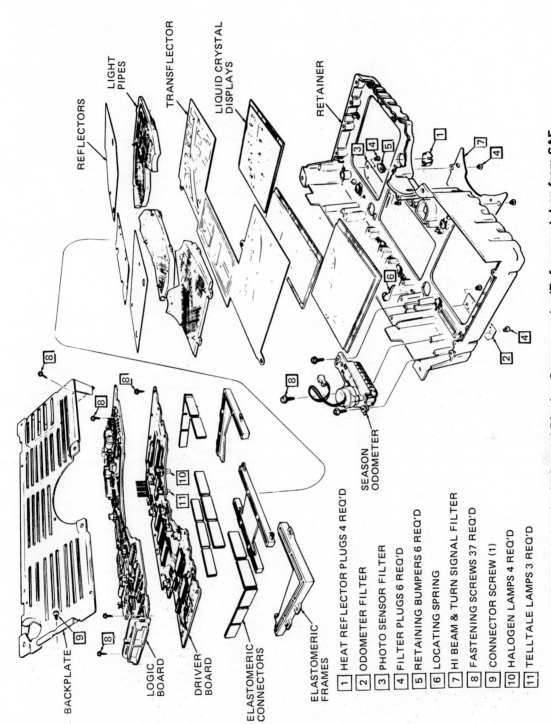

REFLECTORS

LIGHT PIPES

TRANSFLECTOR

LIQUID CRYSTAL DISPLAYS

RETAINER

SEASON ODOMETER

BACKPLATE

LOGIC BOARD

DRIVER BOARD

ELASTOMERIC CONNECTORS

ELASTOMERIC FRAMES

1 HEAT REFLECTOR PLUGS 4 REQ'D
2 ODOMETER FILTER
3 PHOTO SENSOR FILTER
4 FILTER PLUGS 6 REQ'D
5 RETAINING BUMPERS 6 REQ'D
6 LOCATING SPRING
7 HI BEAM & TURN SIGNAL FILTER
8 FASTENING SCREWS 37 REQ'D
9 CONNECTOR SCREW (1)
10 HALOGEN LAMPS 4 REQ'D
11 TELLTALE LAMPS 3 REQ'D

FIGURE 11-2. Liquid Crystal Display Components. (Reference taken from SAE paper No. 830041. Reprinted with Permission © 1983 Society of Automotive Engineers)

327

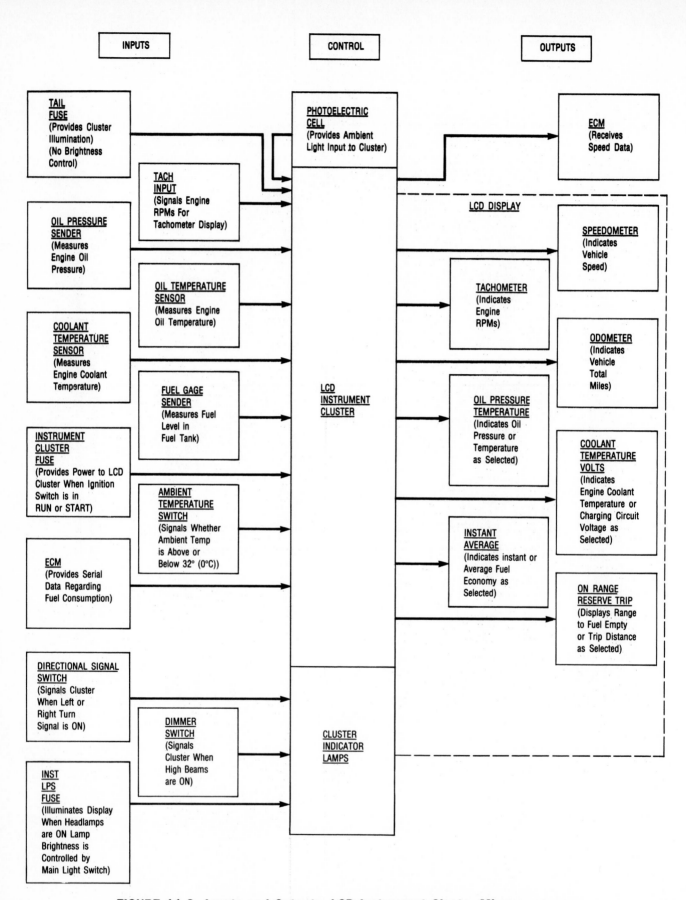

**FIGURE 11-3. Inputs and Outputs, LCD Instrument Cluster Microprocessor.
(Courtesy of Chevrolet Motor Division, General Motors Corporation)**

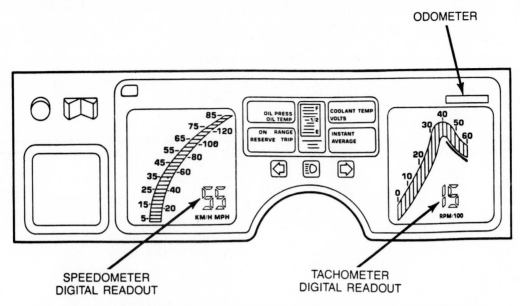

FIGURE 11-4. LCD Instrument Cluster. (Courtesy of Chevrolet Motor Division, General Motors Corporation)

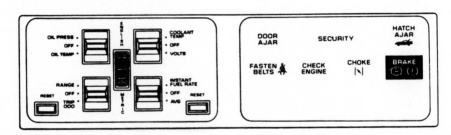

FIGURE 11-5. Switch and Telltale Console Used with LCD Instrument Cluster. (Reference taken from SAE paper No. 830041. Reprinted with Permission © 1983 Society of Automotive Engineers)

Engine Speed

The tachometer displays engine revolutions per minute (RPM) in analog and digital forms. Yellow digits display engine speed from 0 to 7,000 RPM. A 31-segment bar graph is green from 0 to 4,300 RPM, yellow to 5,100 RPM, and red to 6,000 RPM. Vehicle speed and engine speed displays are updated 16 times per second. An input signal from the high energy ignition (HEI) module is used by the microprocessor to control the tachometer displays.

Oil Pressure and Temperature

The driver can select an oil pressure or oil temperature reading on the LCD cluster with the selector switch on the switch and telltale console. Oil pressure displays range from 0 to 80 PSI (0 to 560 kPa), while oil temperature displays read from 149° to 320°F (65° to 160°C). When the oil temperature exceeds 300°F (149°C), an out-of-normal-limits warning is activated in the oil display.

Coolant Temperature and Volts

Coolant temperature or a voltage reading can be selected on the LCD display by depressing the appropriate selector button on the switch and telltale console. Coolant temperature is displayed from 104° to 302°F (40° to 150°C) and electrical system voltage is displayed in the 11.5–16.5 V range. When coolant temperature exceeds 255°F (124°C), an out-of-normal-limits warning is shown on the temperature display.

Fuel Economy

Instant or average fuel economy is displayed in the lower right LCD quadrant. This display indicates miles per gallon (MPG) or liters per 100 kilometers (L/100 km). Speed sensor pulses are counted by the LCD microprocessor to determine the distance travelled. The electronic control module (ECM) on the vehicle controls the fuel system. (Refer to Chapter 9 for an explanation of the ECM.) A serial data line from the ECM to the LCD microprocessor provides the necessary information regarding the amount of fuel consumed. This signal is updated each 0.675 second, and the instant fuel economy display is updated each 0.75 second. Average fuel economy is computed from the distance travelled and the fuel used since the reset button was depressed.

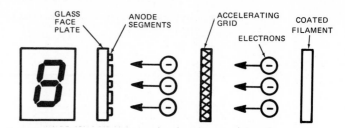

FIGURE 11-6. Vacuum Fluorescent Display Principles. (Courtesy of Chrysler Canada Ltd.)

Fuel Level

An illuminated bar-graph fuel gauge is located in the center of the LCD cluster. When the fuel tank is full, all the bars are brightly illuminated. As the fuel level is lowered in the tank, the bars gradually fade out from the top down. An amber low-fuel warning light is activated when only two bars remain illuminated. When this occurs, the lower left reading in the LCD display switches from trip distance or range to a display of distance travelled with the low-fuel warning activated. Miles or kilometers may be displayed in this reserve fuel mode. When two bars are left illuminated on the fuel gauge, the range display will read zero. At higher levels in the fuel tank, the range display indicates the distance that may be travelled on the fuel remaining in the tank. If the vehicle is operating in the reserve fuel mode, the driver can display the trip odometer reading for 5 seconds by selecting "off" or "range" and then switching to "trip odometer." After 5 seconds this display will change back to display the distance travelled since the low-fuel warning was activated.

VACUUM FLUORESCENT DISPLAYS

Operation

A vacuum fluorescent display (VFD) generates its light with the same basic principles as a television picture tube. In the VFD, a heated filament emits electrons which strike a phosphorescent material and emit a blue-green light. The filament is a resistance wire that is heated by electric current. A coating on the heated filament emits electrons

which are accelerated by the electric field of the accelerating grid. The anode is charged with a high voltage, which attracts the electrons from the grid. A VFD computer supplies high voltage to the specific anode segments which are needed to emit light for any given message. The operating principles of a VFD are shown in Figure 11-6.

The brightness of the VFD may be intensified by increasing the voltage on the accelerating grid. Another method of controlling VFD brilliance is pulse width timing. When this method is used, the VFD is turned on and off very rapidly. A shorter "on time" dims the display, whereas a longer "on time" increases the brilliance.

Shock can damage a VFD display; therefore, they must be handled with care. A typical VFD is pictured in Figure 11-7.

ELECTRONIC INSTRUMENT PANEL DIAGNOSIS

Diagnosis of Chrysler Vacuum Fluorescent Instrument Panel Display

If no display appears when the ignition switch is turned on, check the fuses that supply power to the VFD. On the 21-pin connector, pin 20 is ground, pin 21 is ignition, and pin 10 is battery feed. The 21-pin VFD connector and terminal identification is provided in Figure 11-8.

If the VFD cluster illuminates, use the following sequence for cluster self-diagnosis:

1. Press the "trip" and "trip reset" buttons with the ignition switch off. While these buttons are depressed, turn on the ignition switch. The letters EIC should appear in the odometer and the low-oil lamp will illuminate. If EIC does not appear, replace the cluster.

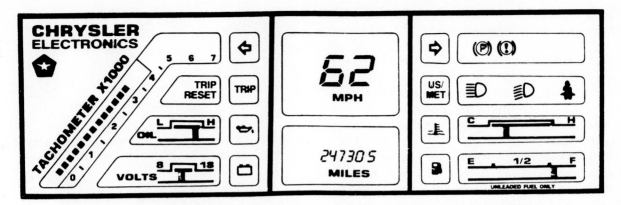

FIGURE 11-7. Vacuum Fluorescent Display. (Courtesy of Chrysler Canada Ltd.)

FIGURE 11-8. Vacuum Fluorescent Display 21-Pin Connector and Terminal Identification. (Courtesy of Chrysler Canada Ltd.)

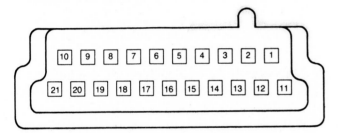

RIGHT PRINTED CIRCUIT BOARD CONNECTOR

CAV	CIRCUIT	GUAGE	COLOR	FUNCTION
1	P51	20	GY *	LAMP CHECK SWITCH
2	P2	20	BK *	PARK BRAKE SWITCH
3	E2	20	OR *	ILLUMINATION LAMPS
4	L7	20	BK/YL*	PARKING LAMPS
5	K3	20	BK/RD*	POWER LOSS LAMP
6	H4	20	BK	LAMP GROUND
7	L4	18	VT *	LOW BEAM LAMP
8	D5	20	TN	RIGHT TURN SIGNAL LAMP
9	L5	18	RD	HIGH BEAM LAMP
10	X1	20	RD *	BATTERY FEED
11				
12				
13	G7	20	WT/OR*	SPEED SENSOR
14	T11	20	GY/RD*	TACHOMETER
15	L17	20	YL *	DISPLAY DIMMING
16	G4	20	DB	FUEL SENDER
17	G60	20	GY/YL*	OIL SENDER
18				
19	G20	20	VT/YL*	TEMPERATURE SENDER
20	H40	18	BK/LG*	CHASIS GROUND
21	G5	20	DB *	IGNITION RUN/START

2. When EIC is present, press the "US/Metric" button and all the VFD displays should illuminate. If the word "fail" and a code appears and some displays do not illuminate, a fault exists in the VFD. For example, a fail code 4 indicates a defective or incorrectly installed odometer chip.

3. When 5 appears in the speedometer window, the first five tests are completed. Test six is initiated by waiting 5 seconds and depressing the "US/Metric" button, which should cause all the individual segments to appear in sequence. If any two segments in the same digit appear simultaneously, such as a horizontal line and a vertical line, replace the cluster. If more than one image appears at once in the gauge sequence, cluster replacement is also necessary.

4. After a 5-second pause, press the US/Metric button to initiate test seven. This test will illuminate the gauge scales and sequence the speedometer and odometer. If this test fails to occur, replace the cluster.

5. Wait 5 seconds and press the "US/Metric" button to begin test eight. On systems with warning lamps, the battery, temperature, and fuel lamps should light in sequence, followed by the appearance of an 8 in the speedometer. If these lamps do not light, test the individual bulbs. Replace the cluster if bulb replacement does not correct the problem. On systems without warning lamps, the 8 in the speedometer will appear immediately when the "US/Metric" button is depressed. Once the 8 appears, the VFD test sequence is completed, and the cluster is confirmed good if all the tests were satisfactory. Press the "US/Metric" button or the "trip reset" button to return the VFD to normal operation.

Vacuum Fluorescent Display Cluster or Odometer Chip Replacement

If the VFD cluster is replaced, the odometer memory chip should be removed from the old cluster and installed in the new cluster. This will retain the original miles or kilometers on the odometer. A special tool is available for odometer chip removal.

The odometer chip may be ruined by static electric charges. Therefore, extreme care must be used when this chip is handled. Do not carry the odometer chip around unless the foam retainer is installed on the chip pins. New odometer chips are supplied with the foam retainer in place, and it must not be removed until the chip is installed in the VFD. Be careful not to bend the chip pins during installation. The odometer chip must be installed as indicated in Figure 11-9.

When the fuel, temperature, and oil gauges all provide a maximum reading check, check the pull-up module connector at the back of the cluster as indicated in Figure 11-10.

Service Precautions Vacuum Fluorescent Displays

The following precautions should be observed when VFD-equipped vehicles are being serviced

1. The ignition switch should be turned on before battery booster cables are connected to the battery. When the VFD is illuminated, a protection

FIGURE 11-10. Pull-Up Module Connector. (Courtesy of Chrysler Canada Ltd.)

circuit prevents static charge damage to the odometer chip.

2. Non-resistor plugs or non-resistance type plug wires must not be used, or VFD damage may result.

3. When secondary ignition system spark is being tested, do not allow spark to jump more than 1/4 in (6.35 mm).

CHRYSLER 11-FUNCTION VOICE ALERT SYSTEM

Operation

The messages that may be provided by this system are the following:

1. Your headlights are on.
2. Don't forget your keys.
3. Your washer fluid is low.
4. Your fuel is low.
5. Your electrical system is malfunctioning. Prompt service is required.
6. Your parking brake is on.
7. A door is ajar.

FIGURE 11-9. Odometer Chip Installation. (Courtesy of Chrysler Canada Ltd.)

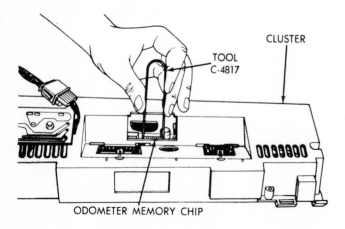

8. Please fasten your seat belts.

9. Your engine is overheating. Prompt service is required.

10. Your engine oil pressure is low. Prompt service is required.

11. All monitored systems are functioning.

12. Thank you.

A pulsating beep is provided before the audible message, and a tone follows the message. Message number 1 will be heard after the following sequence of events:

1. The driver's door is closed.

2. Headlights are on.

3. Ignition switch is turned on and off.

4. Key is removed from ignition switch.

5. Driver's door is opened.

Message number 5 occurs if the charging system is below 11.75 V and the engine is running above idle speed for several minutes. If the engine temperature is above 270°F (132°C) for 1 minute and the engine speed is above idle for 1 minute, message number 9 is heard. Message number 10 is provided if the oil pressure is low for a minimum of 2 seconds and the engine is running. Forward vehicle motion is necessary, plus the other applicable conditions, before messages 6, 8, or 11 are heard. Message number 7 is provided if a door is ajar and the vehicle is in forward or reverse motion.

The heart of the electronic voice alert system is a microprocessor or module mounted above the glove compartment. A switch or sensor located in each monitored component sends the necessary input signal to the module. If an unsatisfactory signal is received, the control module provides the appropriate message. Forward and reverse vehicle motion signals are provided by a speed sensor and the backup light switch.

A volume control is located on the underside of the module. The on/off switch on the module cancels the voice signal if the switch is moved toward the rear of the vehicle. This switch is accessible through an opening in the top right area inside the glove box. The voice alert module is accessible through the glove box opening after the glove box has been removed, as indicated in Figure 11-11.

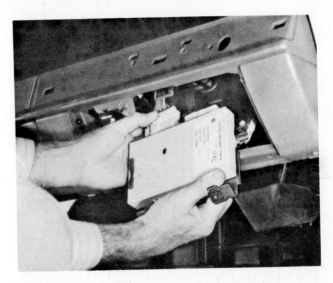

FIGURE 11-11. Voice Alert Module. (Courtesy of Chrysler Canada Ltd.)

CHRYSLER 24-FUNCTION MONITOR AND VOICE ALERT SYSTEM

Operation

The 24-function message center is referred to as an Electronic Monitor. This system provides the driver with visual and audible warnings. Visual messages are presented on a two-line, ten-character blue-green vacuum fluorescent display. Orange and yellow symbols are activated simultaneously. Audible messages are provided by a voice synthesis microprocessor.

The monitor system contains an electronic monitor module with two vacuum fluorescent displays and an electronic voice alert microprocessor or module. The monitor module is located in the center dash area, and the voice alert module is located above the glove box. This system is capable of displaying and verbalizing 24 warning conditions about the vehicle. The actual number of warning conditions varies, depending on the vehicle options. Messages displayed and verbalized include three categories as follows:

1. **Safety**—Passenger door ajar, driver door ajar, hatch ajar, fasten seat belts.

2. **Operational**—Oil pressure low, engine temperature high, fuel level low, transmission pressure low, voltage low.

3. **Convenience**—Coolant level low, brake fluid low, disc brake pads worn, washer fluid low, rear washer fluid low, engine oil level low, headlights out, brake lamp out, tail lamp out, parking brake engaged, keys in ignition, exterior lights on.

The electronic monitor module senses various defective conditions from the inputs. When a defective condition exists, this module displays a warning message and sends a tone and talk signal to the voice alert module. This causes the voice alert module to generate a short tone and provide the appropriate audible message. When more than one defect exists, the same sequence is followed for each fault and each message is displayed for 4 seconds. If the defect is driver-correctable, the monitor module signals the voice alert module to provide a thank-you tone after the defect is corrected.

The voice alert module does not generate tones through the radio speaker, as in the 11-function system. Instead, an external sound transducer in the voice alert module provides the tone signals. If the radio is on, the audible messages interrupt the radio and are delivered through the radio speaker as in the 11-function system. Two visual messages and the corresponding audible messages are shown in Figure 11-12.

Requirements to Obtain Messages

A "keys in ignition" or "exterior lamps on" message is provided if these conditions are present and the driver's door is opened with the ignition switch in the off, lock, or acc positions. The audible message is followed by a pulsating tone for "keys in ignition," or a continuous tone for "exterior lamps

on," and these tones continue until the condition is corrected.

A "fasten seat belt" message is provided for 6 seconds when the ignition switch is turned on. It will continue to be displayed if the driver's seat belt is not buckled, or until the car has been moved 16–24 in (40.6–60.9 cm), at which time the audible message is given.

The "monitored system OK" message is provided when the ignition switch has been on for 6 seconds and no defects have been found.

Some visual messages, such as "park brake engaged" and all door, trunk, and hatch ajar messages are displayed when the ignition switch is on and the corresponding fault is sensed. The audible message is provided with the visual message after the vehicle has been moved 16–24 in (40.6–60.9 cm). When the fault is corrected, a short tone is heard.

Faults in the following systems result in visual and audible messages if the ignition switch is on and the defective condition has been sensed for 15 seconds: "washer fluid low," "rear washer fluid low," "fuel level low," "coolant level low," "brake fluid low," and "disc brake pads worn." After the defect has been corrected, the ignition switch must be turned off to clear these messages from the monitor module.

Any lamp message, such as head lamp out, tail lamp out, or brake lamp out, is displayed and heard if the ignition switch is on and the fault has been sensed for 3.5 seconds. Correction of the fault clears the failure message.

A "low oil pressure" message is provided if low oil pressure is sensed and the engine RPM is between 300 and 1,500. The failure message is cleared when the fault is corrected.

The "engine temperature high" message is displayed and heard when the engine temperature is sensed high and engine RPM is above 300 for 30 seconds. A second-level audible message is heard

FIGURE 11-12. Visual Messages and Corresponding Audible Messages. (Courtesy of Chrysler Canada Ltd.)

Visual Messages		Audible Messages
	WASHER FLUID LOW	"Your washer fluid is low"
	RR WASHER FLUID LOW	"Your rear washer fluid is low"

after the same fault conditions are present for 60 seconds. The failure message is cleared when the fault is corrected or when the engine speed drops below 300 RPM.

A "voltage low" audible and visual message is given if the battery voltage is below 12.35 V and the engine speed is above 1,500 RPM for 15 seconds. The ignition switch must be turned off after the fault is corrected to clear this message.

When the transmisson pressure is sensed low for 15 seconds and the engine speed is above 300 RPM with the vehicle not in reverse, a "low transmission pressure" audible and visual message is provided. The ignition switch must be turned off after the correction procedure to clear this message.

An "engine oil level low" audible and visual message is provided if the monitor module is powered by the time delay relay (TDR) only, and the engine oil level is sensed low. The TDR supplies power to the monitor module for a specific time period after the ignition is turned off. This message is also given with the ignition on and the TDR still on. The ignition switch must be turned off after the condition is corrected to clear this message.

A system-check button is located on front of the monitor. When this button is pushed twice within a 5-second interval, the system is muted and a "mute engaged" message is displayed for 4 seconds. This message is only visual. The system can also be muted with the switch on the voice alert module, as in the 11-function system.

The complete 24-function system is illustrated in Figure 11-13. The functions of other voice alert systems would be similar to the Chrysler systems.

CHRYSLER ELECTRONIC VOICE ALERT SYSTEM DIAGNOSIS

Diagnosis of 11-Function Electronic Voice Alert System

A module test may be performed with the following procedures:

1. Remove the key from the ignition switch.

2. Open and close the driver's door.

3. Wait one minute.

4. Open the driver's door.

5. Press and hold the door-ajar switch located on the left B post.

When these steps are completed, the module should deliver all 12 audible messages. If the module misses some messages, module replacement is necessary. If there are no audible messages, be sure the on/off module switch is moved toward the front of the vehicle. If audible messages are still not heard, check battery feed to module, ground circuit to the module, left door-ajar switch, and left door courtesy-lamp switch. The left door courtesy-lamp switch is grounded and the door-ajar switch is not grounded in the test mode. If these circuits and components are satisfactory, replace the module. This test sequence determines the ability of the module to supply audible messages, but it does not test individual switches or circuits.

Individual circuits may be tested by creating the necessary conditions to provide the audible message with a known-good chimes module inserted in place of the electronic voice alert module. If the ignition switch is turned on and off and the keys are left in the switch, a "don't forget your keys" message should be heard when the driver's door is opened. If no message is heard, connect a known-good chimes module in place of the electronic voice alert module and repeat the above test procedure. If chimes are heard, replace the electronic voice alert module. If no chimes are heard, test the circuit from the module to the ignition switch. Each circuit that is monitored by the module may be tested in the same way.

The number of functions and the conditions required to provide audible messages vary, depending on the year of the vehicle.

Diagnosis of 24-Function Monitor and Voice Alert System

When the system-check button on front of the monitor is pressed once, a tone will sound and many of audible and visible messages will be presented in a demo sequence. When this sequence is completed, the system automatically returns to normal operation. The demo sequence will be aborted if the system-check button is pressed during the sequence. This sequence proves that the monitor module and the voice alert module are capable of delivering visual and audible messages.

All the system fault signals to the monitor module are provided by a ground switch except the "voltage low," "lamp out," and "oil level low" signals. If a fault signal is not given and that signal is controlled by a ground switch, proceed as follows:

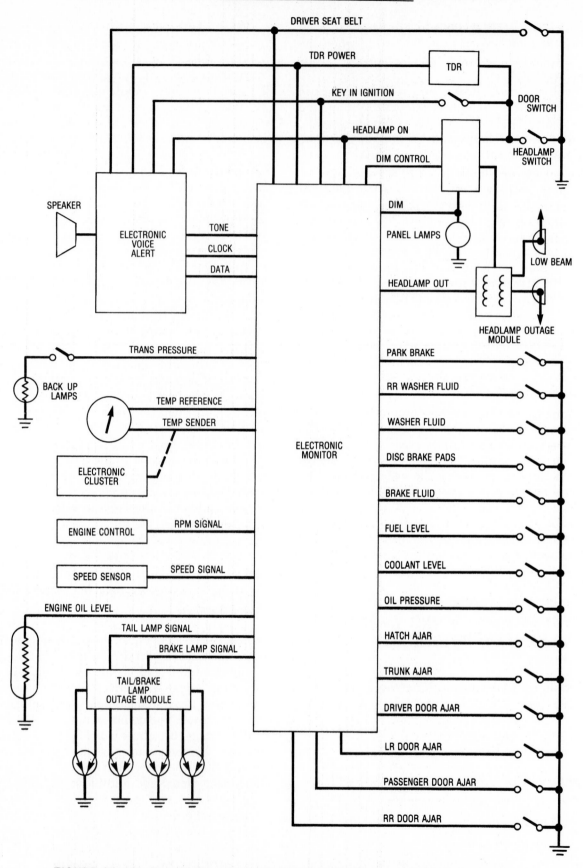

FIGURE 11-13. Twenty-Four Function Monitor and Voice Alert System. (Courtesy of Chrysler Canada Ltd.)

1. Check the operating conditions required to give the fault message.

2. Check the wiring from the monitor module to the specific switch that is not providing a message.

3. Check engine RPM if it applies to the specific message that is missing.

4. Check the distance input if it applies to the missing fault message.

5. Connect a jumper wire from the input switch terminal to ground, and meet all other requirements for the fault message being diagnosed. If a fault message is given, the switch is defective. If the fault message is not provided, the monitor module is defective.

If a fault message is still given after the fault condition has been corrected, proceed with the following diagnosis:

1. Check the conditions that are required to provide the fault message.

2. Check the wiring to the switch that provides the fault message.

3. Disconnect the switch; if the fault message stops, replace the switch. Monitor module replacement is necessary if the fault message continues.

The "lamp out" input signals are provided by front and rear lamp modules, which are shown in Figure 11-13. These modules send an alternating circuit (A/C) signal to the monitor module if a light burns out. If a "lamp out" message is not provided, proceed as follows:

1. Check the conditions that are required to provide the fault message.

2. Check the wiring.

3. Remove a tail/stop light bulb and disconnect a headlight.

If a "lamp out" message is not displayed, replace the monitor module. When only some failures occur, replace the lamp module that corresponds to the missing "lamp out" message.

The input sensor for engine oil level is a variable resistor that is built into the dipstick. If an "oil level low" message is not given, use the following diagnosis:

1. Check the conditions that are required to provide the fault message.

2. Check the wiring to the dipstick.

3. Remove the dipstick and connect a jumper wire across the wiring harness conductor. If a fault message is given, replace the dipstick. If the fault message is not given, replace the monitor module.

CHRYSLER ELECTRONIC NAVIGATOR

Functions

Most trip computers provide similar functions. The Chrysler trip computer, referred to as an electronic navigator, provides the following information:

1. Miles until empty

2. Estimated time to arrival and time of arrival

3. Distance to destination

4. Time, day of week, month, and day of month

5. Present and average miles per gallon

6. Fuel consumed

7. Average speed

8. Miles travelled

9. Elapsed driving time

The main components in the electronic navigator system are illustrated in Figure 11-14.

FIGURE 11-14. Electronic Navigator System Components. (Courtesy of Chrysler Canada Ltd.)

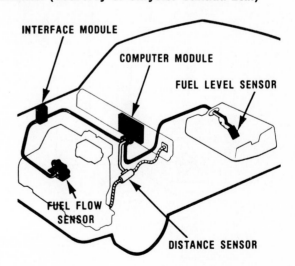

Inputs

The electronic navigator supplies the driver with trip information that is not supplied by standard instrumentation.

A vehicle speed and distance travelled signal is supplied from the speed sensor to the module in the navigator. This speed sensor signal is also used in the electronic fuel injection (EFI) system and the electronic voice alert system. (The EFI system is explained in Chapter 9.)

An input signal regarding the amount of fuel consumed is sent from the logic module in the EFI system to the navigator module. On some trip computer systems, this signal is generated by a fuel sensor in the fuel line. The fuel gauge sending unit in the fuel tank sends an input signal to the navigator module in relation to the amount of fuel in the tank. This signal is transmitted on the G4 circuit. The navigator control buttons and wiring diagrams are illustrated in Figure 11-15. The instrument panel, which contains an electronic instrument cluster, an electronic navigator, and the message center, is shown in Figure 11-16.

FIGURE 11-15. Electronic Navigator Control Buttons and Wiring Diagram. (Courtesy of Chrysler Canada Ltd.)

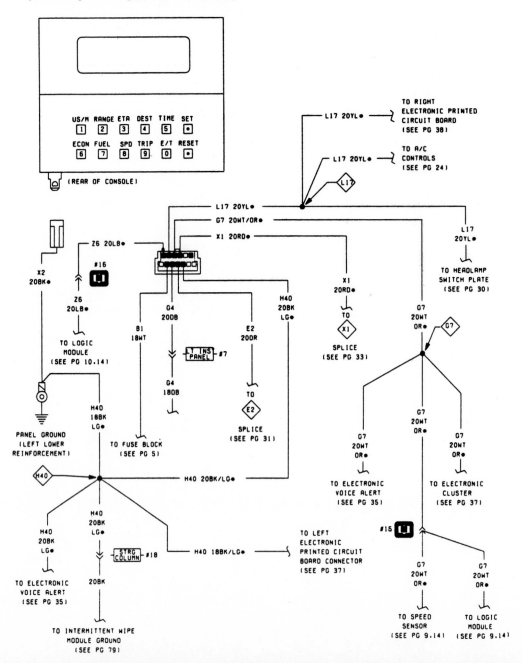

ELECTRONIC CLUSTER   NAVIGATOR

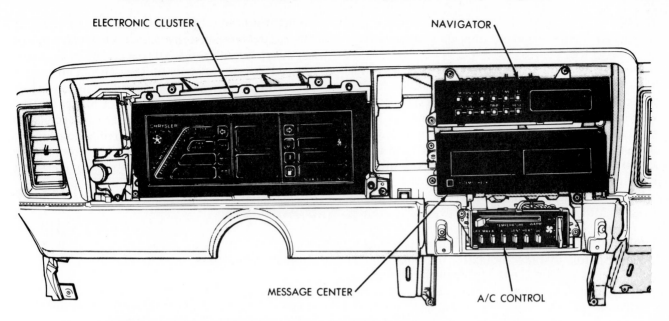

MESSAGE CENTER   A/C CONTROL

FIGURE 11-16. Instrument Panel with Electronic Cluster, Navigator, and Message Center. (Courtesy of Chrysler Canada Ltd.)

Range Function. When the RANGE button is pressed, the navigator digital display indicates the number of miles that can be driven on the fuel remaining in the tank. The navigator module multiplies the amount of fuel in the tank by the projected fuel mileage to provide this calculation. An update of the range reading is provided every few seconds by the navigator module as fuel is used out of the tank.

Distance to Destination Function. The distance to destination (DEST) must be set before this function will operate. To set the DEST, press the DEST button, and then press the SET button within 5 seconds. This action causes the navigator display to indicate 0 miles. At this time, enter the DEST with the appropriate numbered navigator buttons. When the SET button is pressed, the navigator module begins the countdown of distance travelled. The maximum distance setting is 9,999 miles (16,091 km). After the DEST is entered, the navigator displays the remaining DEST when the DEST button is pressed. When the destination is reached, the navigator displays "trip completed," followed by several audible tones.

The electronic navigator system is connected to the voice alert system. The navigator will display US or metric readings if the US/M button is pressed.

Time Function. When the engine is started or the TIME button is pressed, the navigator will con-

tinuously display the time of day, day of week, month of year, and day of month. The display can be reset by pressing the TIME button followed by the SET button within 5 seconds. When this action is taken, an arrow appears on the display to indicate that the hours are to be set. The hour display is advanced if the RESET button is pressed, or the US/M button will backup the display. After the hour display has been set, press the SET button again and the arrow will point to the minute display. When the minutes are set, repeat the procedure for day of week, month, and day of month. Any portion of the time display may be bypassed if the SET button is pressed until the arrow indicates the time function that requires setting. When the entire time display is correct, press the SET button a final time to establish the readings.

If the battery has been disconnected, the time display will require resetting.

Estimated Time of Arrival Function. When the estimated time of arrival (ETA) button is pressed, the navigator displays the estimated driving time to destination for 5 seconds. After this time, the display switches to continuous reading of time and date of arrival at a previously entered destination. If the vehicle is operating at normal or high speeds, the navigator module calculates the ETA from the current vehicle speed. At low vehicle speeds, the module uses the trip average speed to make this calculation. If the ETA exceeds 100 hours, "trip over

100 hours" will be displayed. Once the destination is reached, "trip completed" appears on the display.

Economy Function. If the economy (ECON) button is pressed, the average miles per gallon (MPG) since the last reset is displayed for 5 seconds. After this time, the display indicates the present MPG, and this reading is updated continuously. The navigator module begins a new average MPG calculation if the RESET button is pressed while the average MPG is displayed. Updating of the average MPG reading occurs every 16 seconds, whereas the present MPG display is updated every 2 seconds.

Fuel Function. When the FUEL button is pressed, the navigator display indicates the number of gallons consumed since the last reset. The highest possible display reading is 999.9 gallons, and the display is updated every few seconds. A new fuel-consumed calculation is started if the RESET button is pressed within 5 seconds after the FUEL button is pressed.

Speed Function. When the speed (SPD) button is pressed, the average miles per hour (MPH) since the last reset is displayed. This reading is updated every 8 seconds, and the highest display is 86 MPH. If the RESET button is pressed within 5 seconds after the SPD button is pressed, a new average MPH calculation is initiated. Since low-speed ETA calculations are based on the average speed, the driver may wish to reset the average speed after the DEST is entered.

Trip Function. If the TRIP button is pressed, the navigator display indicates the accumulated trip miles since the RESET button was pressed. The maximum displayed mileage is 999.9 miles, and this reading is updated every 5 seconds. When this mileage is reached, the display returns to zero. A new trip mileage calculation is initiated if the RESET button is pressed within 5 seconds after the TRIP button is pressed.

Elapsed Time Function. When the elapsed time (E/T) button is pressed, the amount of driving time since the RESET button was pressed is indicated on the navigator display. The highest reading is 99 hours and 59 minutes. After this time is reached, the reading returns to zero. During the first hour after the RESET button has been pressed, the E/T is displayed in minutes and seconds. After this time, hours, and minutes are shown on the E/T display. If the RESET button is pressed within 5 seconds after the E/T button is pressed, a new E/T calculation is initiated.

Reset Function. The reset button is used as previously described to clear various functions and begin new calculations. All trip information may be cleared simultaneously if the RESET button is pressed twice within five seconds after any of the following buttons are pressed: ECON, FUEL, SPD, TRIP, E/T. When this action is taken, the navigator display indicates "trip reset" for 5 seconds.

CHRYSLER ELECTRONIC NAVIGATOR DIAGNOSIS

General Diagnosis

When the navigator display is inoperative, fails to respond to function buttons, or goes to minimum brightness with the head lamps on, proceed with the diagnosis in Table 11-1.

Speed Diagnostic Function

If all the navigator functions illustrated in Table 11-2 fail to operate, the speed diagnostic function may be used to diagnose the system. To perform the speed diagnostic function, use the following procedure:

1. Press the SPD button and within 5 seconds simultaneously press the US/M and RESET buttons followed by the SET button.
2. Enter 122 on the navigator.
3. The display should indicate 1.11 times the vehicle speed. For example, at 10 miles per hour (MPH) the display should read 11.
4. Exit from the speed diagnostic function by pressing the SET button followed by another function button.

Fuel-Flow Diagnostic Function

If the navigator functions shown in Table 11-3 do not operate normally, use the fuel-flow diagnostic function to diagnose the system as outlined in the table. Proceed as follows:

1. Press the SPD button and within 5 seconds press the US/M and RESET buttons simultaneously, and then press the SET button.
2. Enter 123 on the navigator.
3. The display should indicate 2 to 4 with the engine idling.

TABLE 11-1. General Navigator Diagnosis

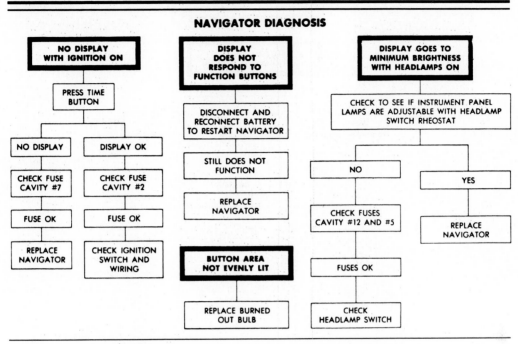

Courtesy of Chrysler Canada Ltd.

TABLE 11-2. Navigator Speed Diagnostic Function

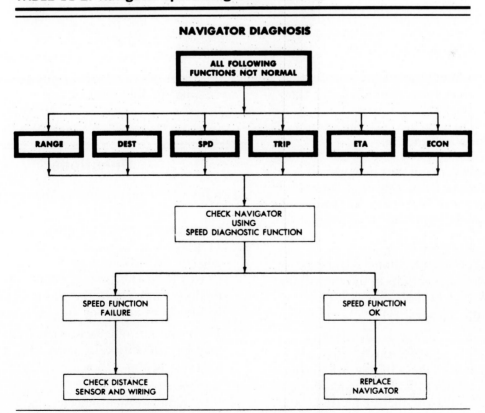

Courtesy of Chrysler Canada Ltd.

TABLE 11-3. Fuel-Flow Diagnostic Function

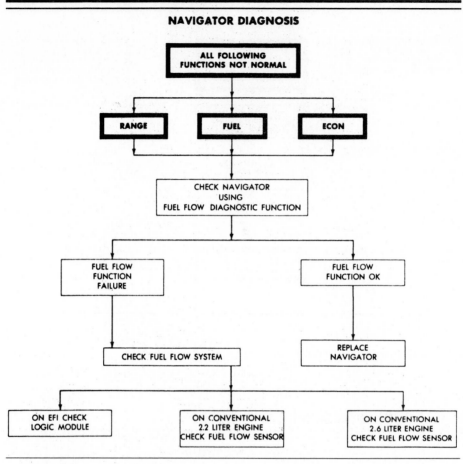

Courtesy of Chrysler Canada Ltd.

4. To exit from the fuel-flow diagnostic function, press the SET button and then any other function button.

Fuel-Level Diagnostic Function

If the range function does not operate normally, the fuel-level diagnostic function shown in Table 11-4 may be used to diagnose the system. The following procedures are used for the fuel-level diagnostic function:

1. Press the SPD button and within 5 seconds press the US/M and RESET buttons at the same time, then press the SET button.

2. Enter 124 on the navigator.

3. The display reading should be near 0 if the fuel tank is full, or above 229 if the tank is empty. The readings between these figures are proportional to the amount of fuel in the tank. For example, with 5 gallons of fuel in the tank, the display reading should be approximately 195.

4. Exit from the fuel-level diagnostic function by pressing the SET button.

Questions

1. A liquid crystal display (LCD) instrument cluster requires backlighting.   T   F

2. In the LCD instrument cluster shown in Figure 11-4, the low-fuel warning light is activated when:
   a) two bars are lit on the fuel gauge
   b) three bars are lit on the fuel gauge
   c) four bars are lit on the fuel gauge

3. The brightness of a vacuum fluorescent display may be intensified by increasing the voltage supplied to the _____ _____.

4. If the battery is disconnected, the navigator will maintain the correct time.   T   F

TABLE 11-4. Fuel-Level Diagnostic Function

NAVIGATOR DIAGNOSIS

```
        ┌─────────────────────┐
        │   RANGE FUNCTION     │
        │     NOT NORMAL       │
        └─────────────────────┘
                   │
        ┌─────────────────────┐
        │   CHECK NAVIGATOR    │
        │        USING         │
        │ FUEL LEVEL DIAGNOSTIC FUNCTION │
        └─────────────────────┘
                   │
        ┌──────────┴──────────────────────┐
        │                                  │
┌───────────────┐              ┌───────────────┐
│  FUEL LEVEL   │              │  FUEL LEVEL   │
│   FUNCTION    │              │  FUNCTION OK  │
│   FAILURE     │              └───────────────┘
└───────────────┘                      │
        │                      ┌───────────────┐
┌───────────────┐              │    REPLACE    │
│ CHECK INSTRUMENT │           │   NAVIGATOR   │
│ CLUSTER FUEL GAUGE │         └───────────────┘
│   OPERATION    │
└───────────────┘
        │
┌───────────────┐
│ TEST FUEL TANK │
│  SENDING UNIT  │
└───────────────┘
```

Courtesy of Chrysler Canada Ltd.

5. An input signal regarding the amount of fuel consumed is sent to the navigator module from the:
   a) electric fuel pump
   b) logic module or fuel flow sensor
   c) fuel injectors

6. The navigator displays the time function when the engine is started.   T   F

7. On a Chrysler vacuum fluorescent instrument panel, the diagnostic mode is entered by:
   a) depressing the US/METRIC button with the ignition switch on.
   b) holding the TRIP and TRIP RESET buttons in the depressed position and turning the ignition on.
   c) depressing the TRIP and TRIP RESET buttons with the ignition on.

8. When a Chrysler vacuum fluorescent display (VFD) is being diagnosed and the letters EIC do not appear, VFD cluster replacement is necessary.   T   F

9. An odometer chip in a VFD must not be carried unless the _____ _____ is in place on the chip.

10. Non-resistor type spark plugs may be used with VFD instrument clusters.   T   F

11. In a 24-function monitor and voice alert system, the diagnostic mode is entered by:
    a) depressing the system-check button once with the ignition on.
    b) depressing the system-check button twice in a 5-second interval with the ignition on.
    c) depressing the system-check button three times in a 5-second interval with the ignition on.

12. During the speed diagnostic function, the number that must be entered with the navigator button is:
    a) 111
    b) 122
    c) 229

12

COMPUTER-CONTROLLED DIESEL INJECTION AND EMISSION EQUIPMENT

DIESEL EXHAUST EMISSIONS

Diesel Emission Standards

Diesel-powered passenger cars must meet the same emission standards as gasoline-powered cars. However, diesel-powered cars and trucks are subjected to increasingly stringent particulate emission standards. (Refer to Chapter 1 for emission standards.)

Origin and Content of Particulate Emissions

Particulate emissions are formed during the diesel combustion process. In the diesel cycle, fuel is injected directly into the combustion chamber as a fine spray of high-pressure droplets.

The average compression ratio on a passenger car diesel engine is 23:1, which creates enough heat on the compression stroke to enable fuel droplets to reach their self-ignition temperature. Each droplet burns down rapidly to a microscopic carbon site. This process is referred to as evaporative droplet burning. These sites then form embryonic nuclei that are 0.001 in (0.025 mm) in diameter or smaller. Following the nucleation stage, aggregation takes place; this is the process by which the embryonic nuclei form simple chain structures known as intermediate particles.

When the piston moves down on the power stroke, the internal energy of the gas in the cylinder is lowered. The opening of the exhaust valve near the end of the power stroke allows instantaneous expansion of the gas in the cylinder. This sudden expansion causes further agglomeration, condensation, and absorption, so that eventually a fine particulate of chain carbon is formed, which is often referred to as "soot." The basic diesel particulate emission formations are shown in Figure 12-1.

Diesel particulate emissions appear as gray particles under an electron microscope. The large, interconnected agglomerated particulate in the center of Figure 12-2 is surrounded by smaller particles.

DIESEL ENGINE PRINCIPLES

Diesel Ignition and Compression Ratio

There are many similarities between the four-stroke cycle diesel engine and the four-stroke cycle gasoline engine. Here we consider only the major differences between the two engines.

The diesel engine does not require an ignition system to ignite the air-fuel mixture in the cylinders. Instead, the diesel engine uses a very high compression ratio, which creates enough heat on

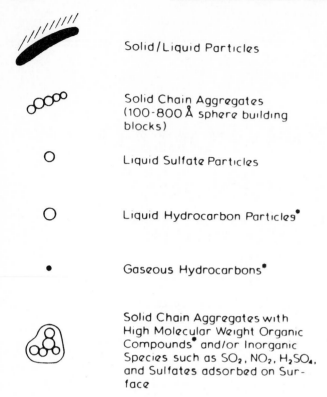

Solid/Liquid Particles

Solid Chain Aggregates (100-800 Å sphere building blocks)

Liquid Sulfate Particles

Liquid Hydrocarbon Particles*

Gaseous Hydrocarbons*

Solid Chain Aggregates with High Molecular Weight Organic Compounds* and/or Inorganic Species such as SO_2, NO_2, H_2SO_4, and Sulfates adsorbed on Surface

*Unburned Hydrocarbons, Oxygenated Hydrocarbons (Ketones, Esters, Organic Acids) and Polynuclear Aromatic Hydrocarbons

FIGURE 12-1. Particulate Emission Formations. (Reprinted with Permission © 1979 Society of Automotive Engineers)

FIGURE 12-2. Diesel Particulates under an Electron Microscope. (Reprinted with Permission © 1980 Society of Automotive Engineers)

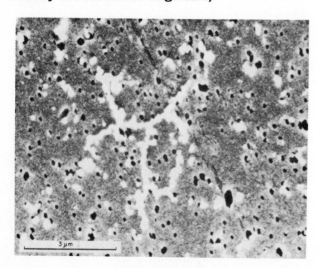

the compression stroke to ignite the air-fuel mixture. The average automotive diesel engine would have a compression ratio of 22.5:1 and a cranking compression pressure of 400 PSI (2,758 kPa). For every PSI of compression pressure, the temperature of the air in the cylinder increases approximately 2°F. Therefore, a compression pressure of 400 PSI (2,758 kPa) would create a cylinder temperature of approximately 800°F (426°C). The ignition temperature of diesel fuel is about 750°F (393°C).

The diesel engine does not use a throttle to control engine speed. The air intake system is unrestricted, and the speed of the diesel engine is controlled by regulating the amount of fuel that is injected into the cylinders by the injection pump and injectors.

Most of the components in a diesel engine are more heavily constructed than comparable gasoline engine parts because of the higher pressure encountered in the diesel engine. A 4-cylinder automotive diesel engine is illustrated in Figure 12-3.

ELECTRONICALLY CONTROLLED DIESEL INJECTION PUMP

Electronic Control System

As diesel particulate emission standards become increasingly stringent, especially in California, it has been necessary to introduce diesel injection pumps that are electronically controlled. These pumps are similar to conventional pumps, but the injection advance is controlled by an electronic control unit which controls a solenoid valve in the injection pump. Input signals to the electronic control unit are the following:

1. Coolant temperature sensor
2. Speed and crankshaft position sensor
3. Altitude switch
4. Injection nozzle with inductive transmitter

On the basis of the input signals that it receives, the electronic control unit controls the pump solenoid valve to provide the precise injection advance required by the engine. The complete injection pump electronic control system is shown in Figure 12-4.

The electronic control unit is mounted in the luggage compartment. A power relay supplies voltage to this control unit when the ignition switch is

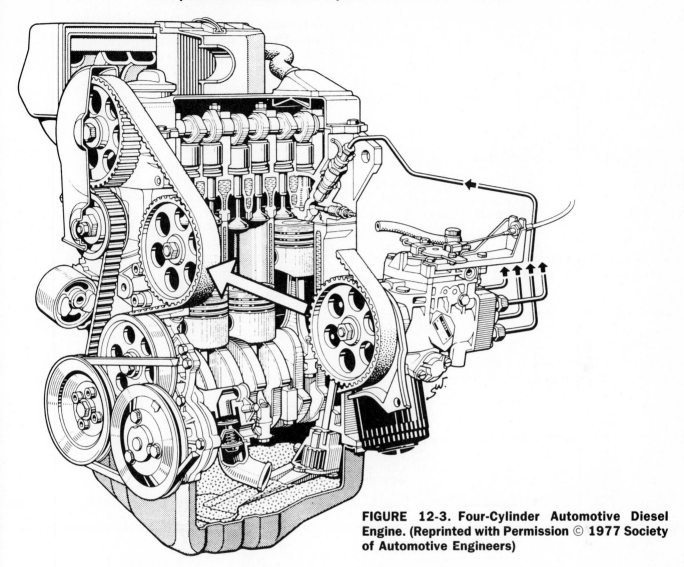

FIGURE 12-3. Four-Cylinder Automotive Diesel Engine. (Reprinted with Permission © 1977 Society of Automotive Engineers)

FIGURE 12-4. Diesel Injection Pump Electronic Control System. (Courtesy of Ford Motor Co.)

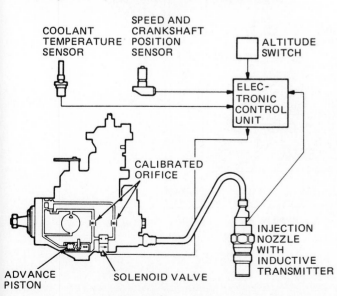

turned on. If the battery polarity is reversed, a diode in the power relay prevents the relay from closing and thus protects the electronic control unit. The power relay is illustrated in Figure 12-5.

FUEL SYSTEM

Operation

An electric lift pump delivers fuel from the fuel tank to the injection pump. A fuel conditioner is connected in the fuel line between the tank and electric pump. This conditioner contains a filter which removes contaminants such as dirt and water from the fuel. A transfer pump in the injection pump de-

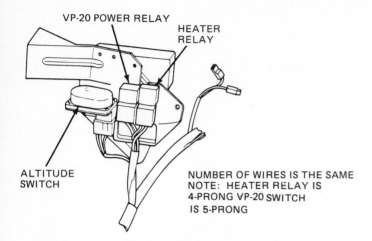

FIGURE 12-5. Electronic Control Unit Power Relay. (Courtesy of Ford Motor Co.)

livers fuel to the injection pump plunger. High-pressure fuel is forced from the pump plunger to the injection nozzles, which inject fuel into the cylinders at the right instant. Excess fuel is returned from the pump and the injectors to the fuel

tank. The internal pump components are lubricated and cooled by the diesel fuel.

A complete diesel fuel system is pictured in Figure 12-6.

INJECTION PUMP

Transfer Pump

The transfer pump contains four spring-loaded vanes mounted in rotor slots. An eccentric pump cavity surrounds the outer edges of the vanes. Fuel enters the transfer pump from the lift pump, and the transfer pump forces fuel past the pressure regulating valve to the injection pump plunger. The pressure regulating valve opens at a specific pressure and returns some of the fuel to the pump inlet, which limits transfer pump pressure. When engine speed and transfer pump speed increase, the

FIGURE 12-6. Complete Diesel Fuel System. (Courtesy of Ford Motor Co.)

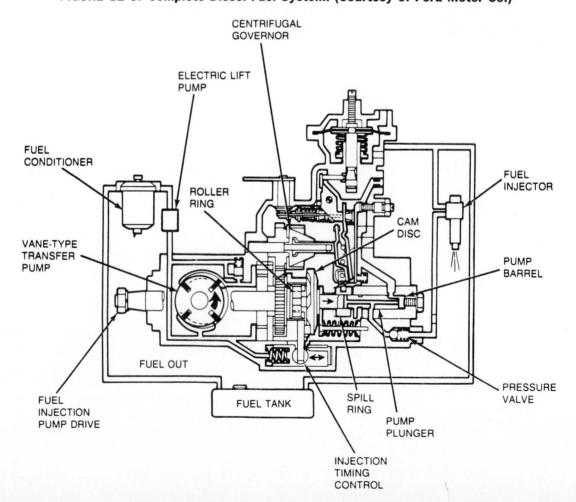

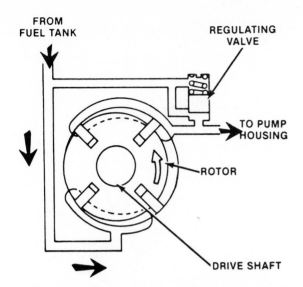

FIGURE 12-7. Transfer Pump and Pressure Regulating Valve. (Courtesy of Ford Motor Co.)

transfer pump pressure gradually increases due to the design of the pressure regulating valve.

A transfer pump with a pressure regulating valve is shown in Figure 12-7.

High-Pressure Pump Plunger

When the ignition switch is turned on, the electric fuel shut-off plunger is lifted, which allows transfer pump pressure to force fuel into the passages in the high-pressure plunger. The transfer pump and the high-pressure plunger are rotated by the pump drive. A cam disc is mounted on the high-pressure plunger, and the cams on this disc contact rollers in a roller ring. When the low points on the cam are in contact with the rollers, the fuel flows from the transfer pump through one of the plunger fill slots into the center of the plunger. The number of rotor fill slots corresponds to the number of engine cylinders.

A return spring holds the plunger and cam disc against the rollers. As the disc high points contact the rollers, the plunger is forced forward and the plunger seals the inlet port from the transfer pump. Forward plunger movement creates a very high fuel pressure in the center of the plunger, which forces fuel from the outlet port through the pressure valve to the injection nozzles. This extremely high fuel pressure opens the injection nozzle at the right instant, and the nozzle sprays fuel into the combustion chamber.

The clearance between the plunger and the barrel is in millionths of an inch to prevent fuel leaks

between the components. A series of equally spaced outlet ports is located around the plunger, and there is always one outlet port for each engine cylinder. The plunger stroke is shown in Figure 12-8, and the plunger and barrel are illustrated in Figure 12-9.

The amount of fuel injected is controlled by the spill ring position, which determines the effective plunger stroke. When the plunger is moved forward, the vertical fuel passage in the plunger is uncovered at the edge of the spill ring, which allows fuel to be spilled from the center of the plunger. When this occurs, fuel pressure decreases instantly in the plunger and the injection line, which causes the injection nozzle to close. The actual plunger stroke remains constant, but the effective plunger stroke that determines the amount of fuel delivered by the pump is controlled by the spill ring position. If the spill ring is moved toward the cam disc, the effective pump stroke, fuel delivery, and engine speed is reduced. When the spill ring is moved away from the cam disc, the effective pump stroke, fuel delivery, and engine speed are increased.

Governor

The position of the spill ring is controlled by the accelerator pedal and the governor in the pump. A linkage is connected between the accelerator pedal

FIGURE 12-8. Plunger with Cam Discs, Roller, and Roller Ring. (Courtesy of Ford Motor Co.)

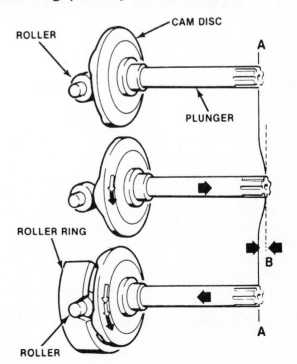

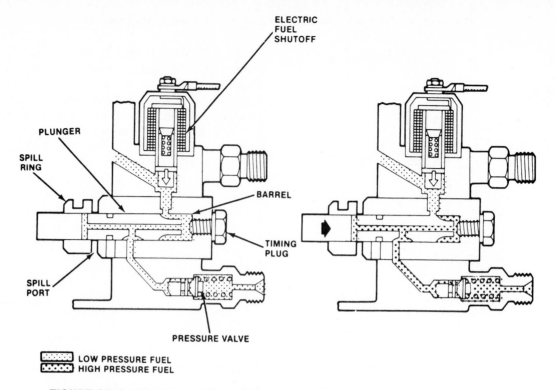

FIGURE 12-9. Plunger and Barrel Fuel Passages. (Courtesy of Ford Motor Co.)

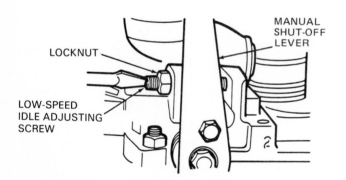

FIGURE 12-10. Idle Speed Adjustment. (Courtesy of Ford Motor Co.)

and the governor lever. The governor contains a group of pivoted flyweights which are rotated by a set of gears. A governor drive gear is mounted on the pump driveshaft, and the driven gear rotates the flyweights and retainer. The accelerator pedal is connected to the governor lever through the linkage and governor spring. When the engine is idling, the governor weights, spring, and linkage position the spill ring to inject the correct amount of fuel to obtain the correct idle speed. A low-speed idle adjusting screw on the injection pump is used to adjust the idle speed, as indicated in Figure 12-10.

As the accelerator pedal is depressed, the linkage and governor spring move the spill ring away from the cam disc. This movement increases the effective plunger stroke, so that more fuel is in-

jected and engine speed increases. When the engine speed reaches the governed revolutions per minute (RPM), the outward weight movement overcomes the governor spring tension and forces the lever away from the pump drive. This action moves the spill ring toward the cam disc, which reduces the effective plunger stroke and the amount of fuel injected to reduce engine speed. The governor operation is shown in Figure 12-11.

Pressure Valve and Injection Lines

A pressure valve is located in each pump outlet fitting. When the spill port opens at the end of the effective plunger stroke, pressure drops rapidly in the injection line. When this occurs, the pressure valve closes, which controls the negative pressure in the injection line. The pressure valve allows enough sudden pressure decrease in the injection line to cause rapid closure of the injection nozzle to prevent fuel dribbling at the injector.

Excessive negative pressures in the injection lines can actually tear the fuel apart and create vapor cavities in the fuel. When these vapor cavities are formed, very small particles of metal can be torn from the inside surface of the injection lines. This process, known as cavitation, can severely deteriorate injection lines in a short time. The pressure valve controls negative injection line pressure and prevents cavitation.

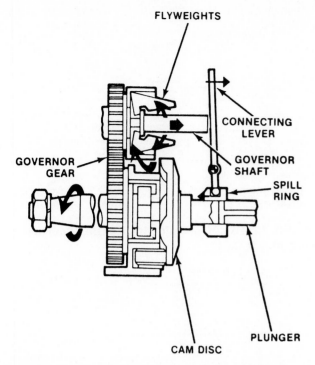

FIGURE 12-11. Governor Operation. (Courtesy of Ford Motor Co.)

Injection lines are manufactured from high-pressure steel tubing. These lines are all the same length on each engine. Injection lines of different lengths on the same engine would cause a slight variation in injection timing. Therefore, injection lines must never be altered in any way.

Electronic Injection Advance

A timing control piston is mounted in a bore below the roller ring, and a pin is connected from the roller ring to the piston. Transfer pump pressure is applied to the right side of the timing control piston, and a return spring is positioned on the opposite end of the piston. When engine speed and transfer pump pressure increase, the increase in transfer pump pressure forces the timing control piston to the left against the spring tension. This piston movement rotates the roller ring in the opposite direction to plunger and cam disc rotation, which causes the cams to contact the rollers sooner and advance the injection timing. The advance mechanism is illustrated in Figure 12-12.

FIGURE 12-12. Electronically Controlled Injection Advance Mechanism. (Courtesy of Ford Motor Co.)

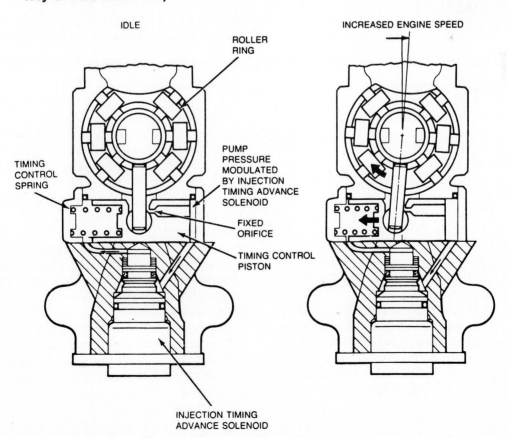

The injection timing advance solenoid is controlled by the electronic control unit. Some fuel moves from the advance side of the timing control piston through the advance solenoid to the return spring side of the timing control piston. When the input signals inform the electronic control unit that increased injection advance is required, the electronic control unit allows the advance solenoid plunger to move upward. When this occurs, fuel flow past the advance solenoid plunger is reduced, which increases the transfer pump pressure on the timing control piston. This pressure increase moves the timing control piston and roller ring to provide more injection advance.

If a reduction in injection timing is necessary, the electronic control unit moves the timing advance solenoid plunger downward. This action allows more fuel to flow from the advance side of the timing control piston, past the advance solenoid plunger and to the return spring side of the timing control piston. When this occurs, transfer pump pressure applied to the timing control piston is reduced, which allows the roller ring and timing control piston to move to the right. This movement reduces injection timing advance.

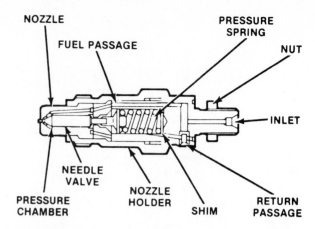

FIGURE 12-13. Diesel Injection Nozzle. (Courtesy of Ford Motor Co.)

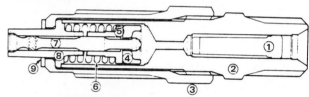

1. EDGE FILTER
2. INLET FITTING (NOZZLE HOLDER BODY)
3. BODY (CAPNUT)
4. RETAINER (COLLAR)
5. SPRING SEAT
6. SPRING
7. PINTLE VALVE (NOZZLE VALVE)
8. NOZZLE BODY
9. SEALING WASHER

FIGURE 12-14. Poppet Nozzle Design. (Courtesy of Pontiac Motor Division, General Motors Corporation)

INJECTION NOZZLES

Design and Operation

Many automotive diesel nozzles are threaded into the cylinder head. A sealing washer seals the end of the nozzle into the cylinder head. The injection line fitting is threaded to the nozzle. A needle valve with a precision tapered tip is positioned in the spray orifice at the nozzle tip. This needle valve is seated by a pressure spring. When the injection pump delivers high pressure fuel to the nozzle, the needle valve is lifted against the spring tension, and fuel is sprayed from the injector tip into the combustion chamber. The needle valve closes immediately when the injection pump pressure decreases. Diesel injection nozzle construction is shown in Figure 12-13.

The clearance between the needle valve and the nozzle body is measured in millionths of an inch. Therefore, nozzles are precision devices that require careful handling. Some nozzles have several spray orifices in the tip. The actual number and size of the spray orifices vary, depending on combustion chamber size.

Some nozzles have a valve with a tapered seat around the circumference of the valve. This type of valve opens outward into the combustion chamber and fuel sprays out around the entire seat circumference. These nozzles may be referred to as poppet nozzles. The internal design of a poppet nozzle is pictured in Figure 12-14.

An internal stop limits the maximum nozzle valve opening. Some nozzles have a return fuel line which returns excess fuel to the fuel tank. In many automotive diesel engines, the nozzles inject fuel into a small precombustion chamber. Combustion in this chamber begins with a swirling motion because of the chamber's design. The burning air-fuel mixture expands with a strong swirling motion into the main combustion chamber. This action creates increased turbulence in the main combustion chamber to provide more complete burning of the air-fuel mixture. The glow plugs are usually located in the precombustion chamber with the nozzles, as indicated in Figure 12-15.

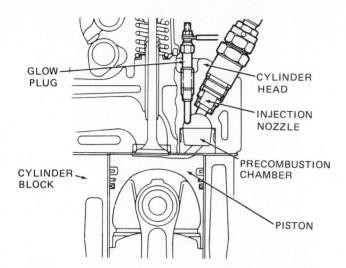

FIGURE 12-15. Precombustion Chamber with Nozzle and Glow Plug. (Courtesy of Ford Motor Co.)

GLOW PLUG CIRCUITS

Design and Operation

Current flow through the glow plugs heats the precombustion chamber, which provides easier starting. An electronic control unit operates two relays which supply voltage to the glow plugs. The electronic control unit receives a coolant temperature signal. When the ignition switch is turned on with the coolant temperature below 86°F (30°C), the electronic control unit closes both relays. Under this condition full voltage is supplied to the glow plugs, which heat the precombustion chambers very quickly so the engine can be started immediately.

A signal from the alternator informs the electronic control unit when the engine is started. This signal causes the electronic control unit to open relay 1 and keep relay 2 closed. A dropping resistor between relay 2 and the glow plugs reduces the voltage and current flow through the glow plugs. Relay 2 may remain closed for 30 seconds after the engine has started. The operation of relay 2 during this time, referred to as an after-glow period, maintains precombustion chamber temperature to prevent engine stalling. The complete glow plug circuit is shown in Figure 12-16.

The electronic control unit will only keep relay 1 closed for 3 to 6 seconds after the ignition switch is turned on, whereas the length of the after-glow period varies, depending on engine temperature. If the ignition switch is turned on and the engine is not cranked, the electronic control unit opens both relays in 3 to 6 seconds to prevent battery discharge and glow plug damage. If the coolant temperature is above 86°F (30°C), the electronic control unit will not close relay 1.

The glow plug circuit shown in Figure 12-16 does not have an instrument panel indicator light. Some glow plug circuits have a wait period before the engine is started. These circuits have an instrument panel indicator lamp that informs the operator

FIGURE 12-16. Glow Plug Circuit. (Courtesy of Ford Motor Co.)

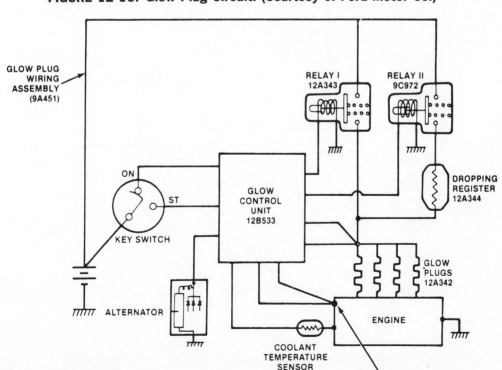

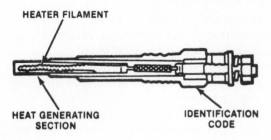

HEATER FILAMENT

HEAT GENERATING
SECTION

IDENTIFICATION
CODE

FIGURE 12-17. Glow Plug Design. (Courtesy of Ford Motor Co.)

when to start the engine. A pulsating voltage is supplied to the glow plugs during the after-glow period in some glow plug circuits. The internal design of a glow plug is illustrated in Figure 12-17.

COMPUTER-CONTROLLED EXHAUST GAS RECIRCULATION SYSTEM

Electric Circuit and Vacuum Hose Routing

The computer-controlled exhaust gas recirculation (EGR) system is shown in Figure 12-18.

A diesel engine produces high emissions of nitrogen oxide (NOx) during idle and off-idle opera-

tion. Therefore, the EGR valve is open during these operating conditions. The purpose of the components in the EGR system may be summarized as follows:

1. **Temperature switch**—senses coolant temperature. This switch is closed above 172°F (78°C).

2. **Vacuum regulator**—controls the EGR flow in relation to fuel delivery. It is mounted on the injection pump and is operated by the pump linkage.

3. **Vacuum reservoir**—prevents vacuum pulsations at the vacuum regulator.

4. **Vacuum delay valve**—delays vacuum supplied to the EGR valve to prevent sudden valve opening and improve driveability.

5. **Idle speed switch**—informs the injection timing module when the engine is idling. The module uses this signal to control the EGR valve at idle. This switch is mounted on the injection pump.

6. **EGR control solenoid**—the injection timing module opens and closes this solenoid to supply vacuum to the EGR valve.

7. **Vacuum pump**—provides vacuum to operate the EGR valve. It is mounted beneath the rocker arm cover on the engine.

FIGURE 12-18. Computer-Controlled Exhaust Gas Recirculation System. (Courtesy of Ford Motor Co.)

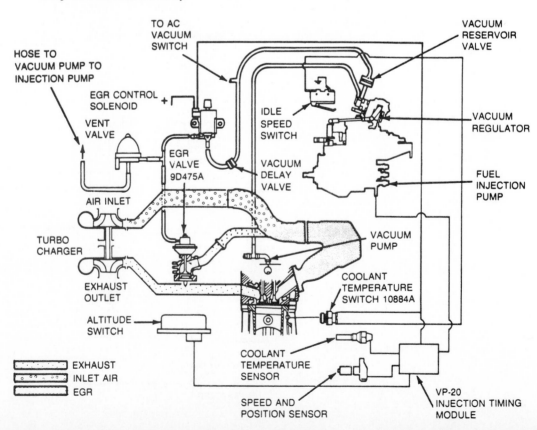

8. **Injection timing module**—controls EGR valve operation. This module also controls injection advance, as illustrated previously in Figure 12-4.

9. **EGR valve**—when this valve is open, it recirculates exhaust into the intake manifold.

10. **Speed and position sensor**—supplies an engine RPM signal to the injection timing module. This sensor is positioned on the lower left corner of the engine block.

11. **Altitude switch**—informs the injection timing module when the vehicle is operating above 9,800 ft (3,000 m) elevation.

12. **EGR vent valve**—if the EGR vacuum supply exceeds 8.6 in Hg (29 kPa), this valve vents the vacuum hose from the vacuum pump to the atmosphere.

The computer-controlled EGR system shown in Figure 12-18 and the glow plug system illustrated in Figure 12-16 are used on the Ford 2.4L turbocharged diesel engine.

DIESEL INJECTION PUMP WITH COMPUTER CONTROL OF FUEL DELIVERY AND INJECTION ADVANCE

Design

The electronic programmable injection control (EPIC) injection pump has internally mounted sensors and actuators and an external microprocessor. Compared to a similar conventional diesel injection pump, the EPIC pump provides lower combustion noise with improved economy and performance. The microprocessor could easily be expanded to control the turbocharger wastegate, intercooler, and exhaust gas recirculation (EGR).

The EPIC injection pump is lighter and more compact than a similar conventional pump. There are only 100 parts in the EPIC injection pump, compared to 220 components in a similar conventional pump. The EPIC injection pump is designed primarily for engines in passenger cars and light trucks.

Fuel Control

In many conventional injection pumps, transfer pump pressure supplies fuel past a metering valve to the chamber between the pumping pistons in the rotor. The position of the metering valve controls the amount of fuel injected by regulating fuel delivery to the rotor pumping chamber. Rotation of the metering valve is controlled by the accelerator pedal and the governor. The pumping pistons are forced together by cams on an internal cam ring as the rotor assembly rotates.

The EPIC injection pump has angled ramps on the shoes that fit against the outer ends of the pumping pistons. These ramps match similar ramps on the pump drive shaft. Axial rotor movement varies the pumping piston position, and therefore determines the pumping chamber volume to control fuel delivery. Transfer pump pressure supplies fuel to the rotor pumping chamber, and this same pressure also delivers fuel through an electrohydraulic actuator to the front of the rotor. In this way, the electronic control unit and the actuator control the rotor axial position to provide precise fuel delivery. A position sensor at the front of the rotor sends a signal to the electronic control unit in relation to rotor axial position.

The fuel metering control in a conventional injection pump is compared to the EPIC pump in Figure 12-19.

FIGURE 12-19. Conventional and Electronic Fuel Metering Control. (Reprinted with Permission © 1985 Society of Automotive Engineers. Reference taken from *Automotive Engineering*, March 1985)

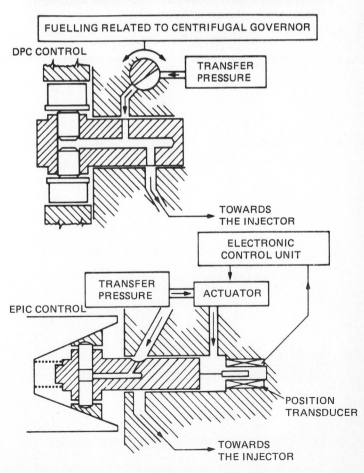

Injection Timing Control

Transfer pump pressure is also supplied through another electrohydraulic actuator to the advance piston. When the input signals indicate that more injection advance is required, the electronic control unit operates the actuator to increase the transfer pump pressure supplied to the advance piston. This pressure increase moves the advance piston against the return spring pressure, and the advance piston movement rotates the cam ring in the opposite direction to rotor rotation. Rotation of the cam ring causes the pumping piston mechanisms to strike the internal cam sooner, so the injection timing advances.

A position transducer on the return spring side of the advance mechanism sends an input signal to the electronic control unit in relation to advance piston movement. In some injection systems, a sensor in the injector informs the electronic control unit when the injector plunger begins to lift. This type of sensor may be used in place of the advance piston position transducer. Both types of sensors are illustrated in the injection advance systems shown in Figure 12-20.

FIGURE 12-20. Electronic Injection Advance Control. (Reprinted with Permission © 1985 Society of Automotive Engineers. Reference taken from *Automotive Engineering*, March 1985)

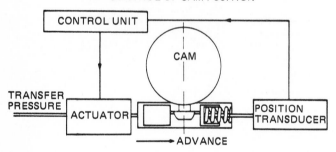

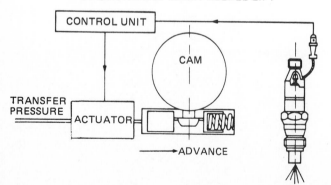

DIESEL FUEL

Types and Ratings

The engine manufacturer's recommendations should be followed when a diesel fuel is selected. Grade number 2D diesel fuel is usually recommended for most diesels that operate under medium to heavy loads and uniform speeds in warm climates. If number 2D diesel fuel is used in cold weather below 0°F (−18°C), the fuel may thicken and plug the fuel system. This process is referred to as fuel waxing. Grade number 1D diesel fuel is recommended in cold climates, and it may also be required in high-speed diesel engines that are often subjected to variations in speed and load.

The pour point of diesel fuel is the temperature below which the fuel will no longer flow. The cloud point of the fuel is the temperature at which wax crystals start to form in the fuel. The cloud point is higher than the pour point. Number 1D diesel fuel has the lowest cloud and pour points available.

The cetane number of the diesel fuel indicates the ability of the fuel to self-ignite in the combustion chamber. The cetane number of the fuel should meet the engine manufacturer's minimum requirements. There is no benefit in using a higher cetane number fuel than called for by the engine manufacturer.

When handling or storing diesel fuel, it is extremely important that water and other contaminants be kept out of the fuel. Extreme cold weather temperatures of −25°F (−31°C) or lower require special fuels such as arctic or kerosene-treated products, which often are detrimental to the injection pump during extended use.

DIESEL SERVICE

Filter Service

Because nozzles and injection pumps have precision clearances that are measured in millionths of an inch, dirt in the fuel can damage these components. Therefore, it is very important to keep dirt out of the fuel system and change the fuel filter at the manufacturer's recommended intervals. The fuel filter replacement procedure is pictured in Figure 12-21.

FUEL CONDITIONER ASSEMBLED

FUEL FILTER REMOVED FROM HOUSING

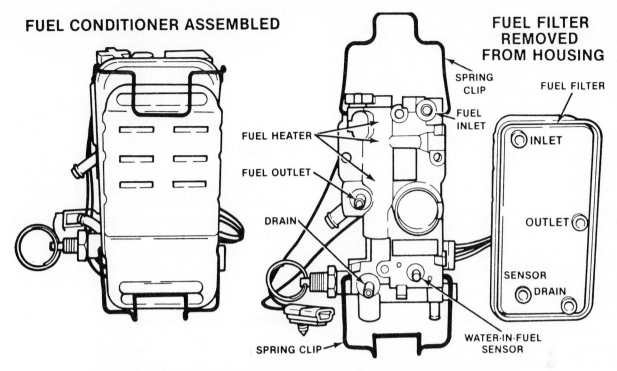

FIGURE 12-21. Fuel Filter Replacement. (Courtesy of Ford Motor Co.)

Some fuel filter assemblies contain a water separator to separate water from the diesel fuel. A water-in-fuel sensor turns on an indicator light on the instrument panel when 50 cubic centimeters (cc) of water is collected in the filter. To purge water from the filter assembly, proceed as follows:

1. Turn on the ignition switch to activate the electric lift pump.
2. Place a container below the filter drain.
3. Pull the pull ring to open the drain valve.
4. Release the pull ring to close the drain valve when diesel fuel appears a light amber color.

Air is purged from the filter assembly when the vent screw is loosened with the ignition switch on and the lift pump running. Close the vent screw when air bubbles no longer appear at the screw. Air purging will be necessary after fuel system repairs, or when the system has run out of fuel. A fuel heater in the filter assembly heats the fuel to prevent fuel waxing when fuel temperature is below 46°F (8°C). Current flow through the fuel heater is controlled by a set of contacts that are closed by a bimetal strip. A fuel filter assembly is shown in Figure 12-22.

Nozzle Testing

The injection nozzles can be tested on a nozzle tester. This tester contains a hand pump and a pressure gauge. A special test fluid is used in the

FIGURE 12-22. Fuel Filter Assembly. (Courtesy of Ford Motor Co.)

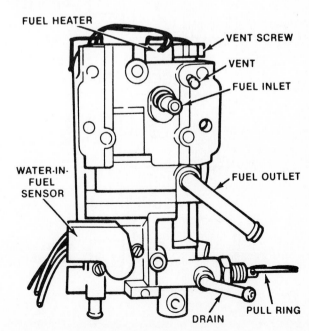

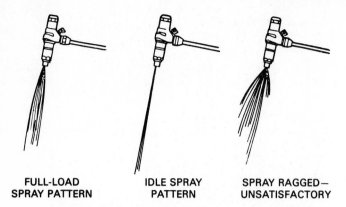

FULL-LOAD
SPRAY PATTERN

IDLE SPRAY
PATTERN

SPRAY RAGGED—
UNSATISFACTORY

FIGURE 12-23. Acceptable and Unacceptable Spray Patterns. (Courtesy of Ford Motor Co.)

tester. The following nozzle tests should be performed:

1. The spray pattern should be tested by operating the hand pump until the pressure opens the nozzle. The nozzle spray must be directed into a container. Never direct the nozzle spray against human flesh. The spray pattern will vary, depending on the type of nozzle. Acceptable and unacceptable spray patterns are illustrated in Figure 12-23.

2. When the hand pump on the tester is operated, the nozzle opening pressure on the tester gauge should equal manufacturer's specifications. On many automotive nozzles, the opening pressure is not adjustable.

3. When the tester hand pump is operated rapidly, the nozzle should chatter as it opens and closes. If the nozzle fails to chatter, the nozzle is probably sticking.

4. Nozzle seat leakage should be checked by applying a tester pressure of 300 PSI (2,068 kPa) below the nozzle opening pressure to the nozzle for 10 seconds. Acceptable and unacceptable seat leakage conditions at the end of the 10-second test are pictured in Figure 12-24.

Injection Pump Timing

On the Ford 2.4L diesel engine, the injection pump is driven by a cogged belt. This engine also has two silent shafts which are driven by a separate belt. The correct timing procedure for the silent shaft belt and the injection pump belt is illustrated in Figure 12-25.

FIGURE 12-24. Acceptable and Unacceptable Nozzle Seat Leaking Conditions. (Courtesy of Ford Motor Co.)

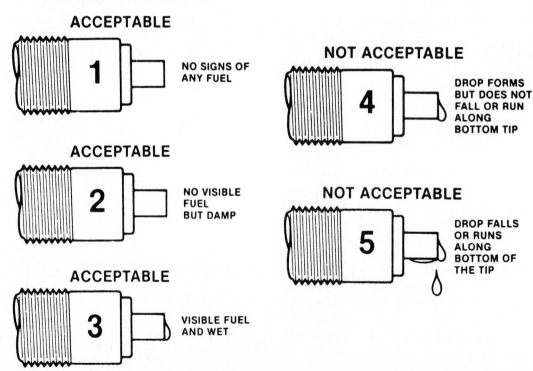

ACCEPTABLE

1   NO SIGNS OF ANY FUEL

ACCEPTABLE

2   NO VISIBLE FUEL BUT DAMP

ACCEPTABLE

3   VISIBLE FUEL AND WET

NOT ACCEPTABLE

4   DROP FORMS BUT DOES NOT FALL OR RUN ALONG BOTTOM TIP

NOT ACCEPTABLE

5   DROP FALLS OR RUNS ALONG BOTTOM OF THE TIP

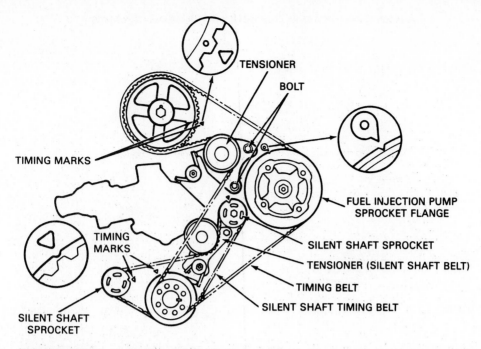

FIGURE 12-25. Injection Pump Belt and Silent Shaft Belt Timing Procedure. (Courtesy of Ford Motor Co.)

The bolt pattern on the injection pump sprocket flange is offset so the sprocket can only be installed in one position, as indicated in Figure 12-26.

The static injection pump timing may be checked with a special adaptor attached to a dial indicator. To check the static injection pump timing, proceed as follows:

1. Position number 1 piston at top dead center (TDC) on the compression stroke.

2. Remove the plug and sealing washer in the injection pump head.

3. Mount the dial indicator and adaptor in the plug opening in the pump head. There must be a minimum of 0.100 in (0.25 mm) preload on the dial indicator.

4. Rotate the crankshaft clockwise until the lowest reading is obtained on the dial indicator, and zero the dial indicator in this position.

5. Continue to rotate the crankshaft clockwise until number 1 piston is at TDC on the compression stroke. The movement on the dial must equal manufacturer's specifications.

6. To adjust the dial indicator reading, loosen the pump mounting bolts and rotate the injection pump clockwise to increase the reading, or counterclockwise to decrease the reading. Never start the engine with the injection pump mounting bolts loose, or severe pump damage and personal injury may result.

7. When the dial indicator reading is correct, remove the dial indicator and install the plug in the pump head. The static timing procedure is shown in Figure 12-27.

Never clean the outside of an injection pump with steam or hot water, because plunger seizure may occur. A dynamic timing meter can be used to check injection pump timing with the engine running. This meter is connected to a diagnostic connector in the wiring harness, as indicated in Figure 12-28.

FIGURE 12-26. Injection Pump Sprocket Mounting Bolts. (Courtesy of Ford Motor Co.)

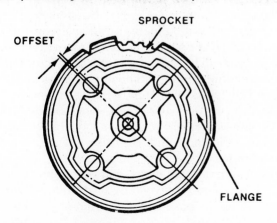

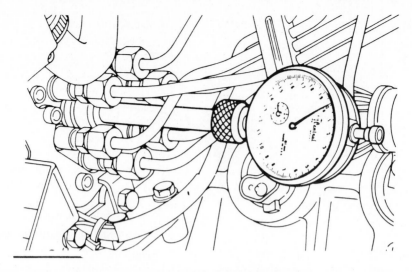

FIGURE 12-27. Static Injection Pump Timing. (Courtesy of Ford Motor Co.)

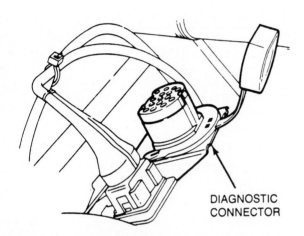

FIGURE 12-28. Dynamic Timing Meter Connector. (Courtesy of Ford Motor Co.)

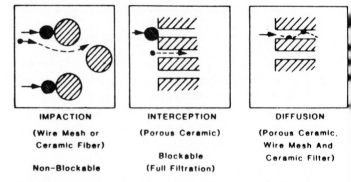

IMPACTION
(Wire Mesh or
Ceramic Fiber)

Non-Blockable

INTERCEPTION
(Porous Ceramic)

Blockable
(Full Filtration)

DIFFUSION
(Porous Ceramic,
Wire Mesh And
Ceramic Filter)

FIGURE 12-29. Particulate Trap Principles. (Reprinted with Permission © 1981 Society of Automotive Engineers)

DIESEL PARTICULATE TRAPS

Design

In future years, diesel particulate traps may be required to lower particulate emissions. The requirements for diesel particulate traps are:

1. Adequate collection efficiency.

2. Low pressure drop to ensure minimum loss in fuel economy and performance.

3. Minimum volume for packaging.

4. High-temperature (2500°F) capability to withstand regeneration.

5. Extended durability.

Most particulate traps consist of wire mesh or ceramic fibers and operate on the principle of impaction and diffusion (see Figure 12-29). The larger particulates strike the filaments of the mesh and adhere to the surface of the filaments or to the particulate material collected previously on the filaments. This type of trap is sometimes called a nonblockable trap because normally the accumulation of particulate matter will be unable to block the exhaust flow path. Although these traps tend to have relatively low pressure drops, some disadvantages have been observed, including a moderately low collection efficiency and the blow-off of collected particulates.

Some particulate traps operate on the principle of interception. That is, any particulates that are larger than the pore size of the material in the trap are intercepted and prevented from passing

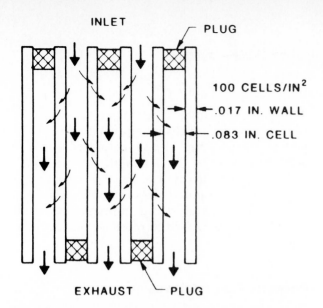

INLET

PLUG

100 CELLS/IN^2

.017 IN. WALL

.083 IN. CELL

EXHAUST   PLUG

FIGURE 12-30. Porous Ceramic Particulate Trap. (Reprinted with Permission © 1981 Society of Automotive Engineers)

through the material, as illustrated in Figure 12-29. As additional particulate material accumulates on the surface of the trapping material, the effective pore size may be reduced, and thus more particulates may be collected. This type of trap is often referred to as a blockable trap. Although the blow-off of collected particles is not a problem for this type of trap, the pressure drop across the trap and the exhaust back pressure tend to increase more rapidly than in the case of a nonblockable particulate trap. Particulate traps that use the interception principle require frequent regeneration to clean out accumulated particulates and to avoid excessive pressure drop across the trap. A porous ceramic honeycomb particulate trap that uses the interception principle is shown in Figure 12-30.

The open channels of the honeycomb are alternately blocked with high-temperature ceramic cement at the top and bottom so that all of the inlet flow must pass through the porous ceramic walls before exiting from the trap. This honeycomb trap provides a very high filtration surface area per unit volume. The initial layer of particulates on the surface of the ceramic walls will be collected by the interception method, and successive layers of collected particulates will provide an additional filtration medium. Figure 12-31 is a photograph of a honeycomb trap that is partly loaded with particulates.

A porous ceramic trap would be mounted in the exhaust manifold, as pictured in Figure 12-32.

FIGURE 12-31. Porous Ceramic Trap Partly Loaded with Particulates. (Reprinted with Permission © 1981 Society of Automotive Engineers)

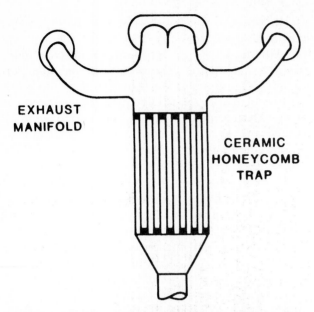

FIGURE 12-32. Exhaust Manifold with Porous Ceramic Particulate Trap. (Reprinted with Permission © 1981 Society of Automotive Engineers)

PARTICULATE TRAP REGENERATION METHODS

High-Speed, High-Load Regeneration

During the regeneration process, the collected particulates are burned off in the porous ceramic trap. Trap inlet temperature must be 900°F (482°C) to initiate regeneration. The oxygen concentration in the exhaust must be at least 2 percent before regeneration will take place. Under normal cruising speed conditions in a diesel-powered passenger car, exhaust temperatures are not high enough to initiate regeneration of a particulate trap. In a passenger car diesel engine trap, inlet temperatures will exceed 900°F (482°C) only at moderate to high speeds and under high-load conditions. Therefore regeneration of the particulate trap will take place only under these operating conditions.

If a passenger car diesel is driven continuously at low and medium speeds, particulate trap regeneration will not take place, and excessive particulate loading of the trap could cause pressure across the trap to drop excessively. In turn, exhaust back pressure would increase and engine power would be lost. Therefore, the high-speed/high-load method of trap regeneration is not practical for passenger car diesels.

Catalyzed Trap

Platinum and lead have been tested as catalysts in porous honeycomb traps impregnated with these metals. The problem with the catalyzed trap is that only the first layer of particulate material will be in contact with the catalyst. Since successive layers of collected particulate material will not be in contact with the catalyst, an impregnated catalyst would not be expected to offer significant advantages.

Catalytic Fuel Additives

Inorganic elements such as lead and copper are capable of reducing the ignition temperature of carbon by more than 300°F (149°C). When 1 g/gal of tetraethyl lead is added to diesel fuel, the regeneration temperature of the particulate trap is reduced by more than 300°F (149°C). This allows trap regeneration to occur at moderate cruising speeds in a passenger car diesel. Although catalytic fuel additives have shown promising results in particulate trap regeneration, the effects of the fuel additive on the engine, injection system, and the particulate trap have not been fully evaluated.

FIGURE 12-33. Exhaust-Fed Burner Trap Regeneration. (Reprinted with Permission © 1981 Society of Automotive Engineers)

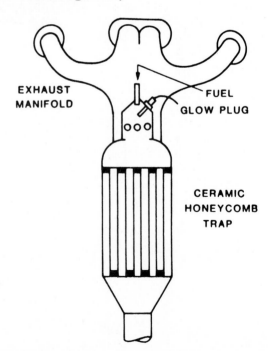

Exhaust-Fed Burner Regeneration

In this type of particulate trap regeneration, additional fuel is injected near the trap inlet, and a glow plug is used to ignite the fuel and initiate regeneration of the trap, as illustrated in Figure 12-33.

There are sufficient levels of oxygen in the diesel exhaust to initiate trap regeneration. This type of trap regeneration could be controlled by the driver, or automatic controls could be developed for this purpose. The disadvantages of this type of trap are that it consumes additional fuel and also increases hydrocarbon (HC) and carbon monoxide (CO) emission levels.

Questions

1. The ignition temperature of diesel fuel is approximately:
   a) 600°F
   b) 675°F
   c) 750°F

2. The effective pump stroke is controlled by the
   a) spill ring position
   b) roller ring position
   c) speed of plunger rotation

3. Excessive negative pressure in diesel injection lines can result in _____ of the lines.

4. When the ignition switch is turned on and the engine coolant is cold, the electronic control unit in the glow plug circuit closes:
   a) both relays
   b) relay 1 only
   c) relay 2 only

5. If the ignition switch is left on and a cold engine is not started, the electronic control unit in the glow plug circuit will:
   a) keep the glow plugs on until the battery is discharged
   b) keep relay 2 closed
   c) shut off both relays in 3 to 6 seconds

6. A number 2D diesel fuel should be used when the temperature is below 0°F (−18°C).
   T   F

7. The injection pump mounting bolts should be loosened with the engine running when the injection timing is being adjusted.   T   F

8. The outside of the injection pump should be cleaned with a steam cleaner.   T   F

9. Excessive fuel waxing in cold weather may be caused by a defective fuel heater.   T   F

10. When the static injection pump timing is checked, number 1 piston should be positioned at:
    a) TDC on the compression stroke
    b) BDC on the intake stroke
    c) BDC on the exhaust stroke

11. Nozzle seat leakage and dribbling can result in engine _____.

12. The two emission levels that tend to be high for diesel engines are:
    a) _____
    b) _____

13. A particulate trap that uses the impaction principle may be referred to as a _____ _____ trap.

14. Particulate traps that use the impaction or diffusion principle have a moderately low collection efficiency.   T   F

15. Particulate traps that use the interception principle require frequent regeneration.   T   F

16. High-speed/high-load regeneration of the particulate trap is practical in a passenger car diesel.   T   F

17. Catalytic fuel additives such as tetraethyl lead reduce the regeneration temperature of the particulate trap by _____ degrees.

18. The two disadvantages of the exhaust-fed burner regeneration methods are:
    a) _____
    b) _____

13

IMPORT EMISSION CONTROL AND COMPUTER SYSTEMS

THROTTLE POSITIONER

Design and Operation

The throttle positioner (TP) reduces hydrocarbon (HC) and carbon monoxide (CO) emissions on deceleration by holding the throttle open. A linkage is connected from the TP diaphragm to the throttle. Manifold vacuum is applied to the TP diaphragm by a vacuum switching valve (VSV). When the speed sensor that is mounted in the speedometer head signals the computer that the vehicle speed is above 16 MPH (26 KPH), the computer energizes the VSV, which then applies vacuum to the TP diaphragm, as illustrated in Figure 13-1.

When vacuum is applied to the TP diaphragm, the TP linkage holds the throttle open and prevents high emission levels on deceleration. If the vehicle speed drops below 7 MPH (11 KPH), the computer will deenergize the VSV, which will then shut off the vacuum to the TP diaphragm. With this vacuum closed, the throttle is able to return to the idle position. When the VSV is deenergized, the TP diaphragm is vented to the atmosphere through the VSV, as shown in Figure 13-2.

Venting the TP diaphragm to the atmosphere prevents the vacuum from being trapped in the diaphragm. The throttle positioner may be adjusted by applying 15 in Hg (49.5 kPa) to the diaphragm with the engine operating at idle speed and normal

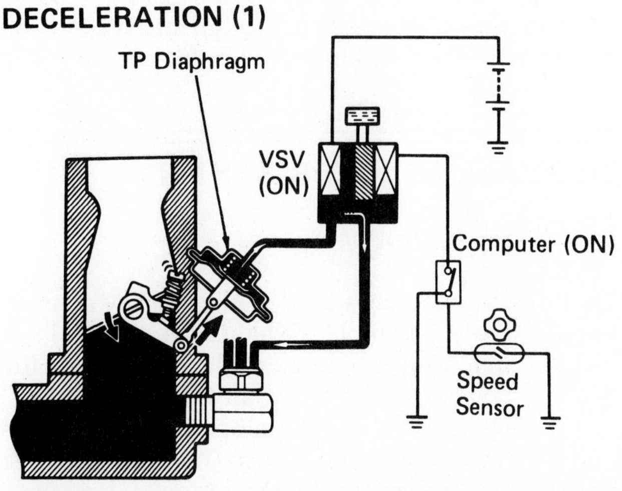

DECELERATION (1)

DECELERATION (2)

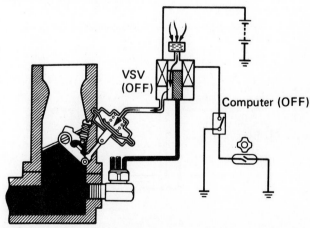

FIGURE 13-1. Operation of Throttle Positioner on Deceleration. (Courtesy of Toyota Motor Corporation, Japan)

FIGURE 13-2. Throttle Positioner Deenergized. (Courtesy of Toyota Motor Corporation, Japan)

temperature. The screw on the TP linkage should then be adjusted to the specified RPM.

MIXTURE CONTROL (MC) VALVE

Design and Operation

A mixture control (MC) valve may be used to prevent high CO and HC emissions when the engine is decelerated. The MC valve is mounted in the air cleaner and it contains a vacuum-operated diaphragm. When the engine is operating at idle speed or cruising speed, the manifold vacuum is not high enough to move the MC valve diaphragm and the MC valve remains closed, as indicated in Figure 13-3.

When the engine is decelerated, the sudden increase in manifold vacuum will move the diaphragm against the spring tension and open the MC valve. Clean air from inside the air cleaner will then be drawn through the MC valve into the intake manifold, as indicated in Figure 13-4.

The flow of air through the MC valve reduces HC and CO emissions on deceleration. In a few seconds the orifice in the MC valve diaphragm will allow the manifold vacuum to equalize on both sides of the diaphragm, and the spring will close the valve.

CONSTANT RPM

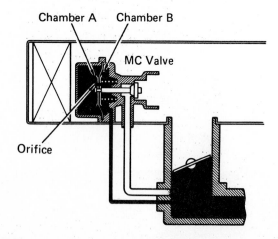

FIGURE 13-3. Mixture Control (MC) Solenoid, Closed Position. (Courtesy of Toyota Motor Corporation, Japan)

SUDDEN DECELERATION, STEP (1)

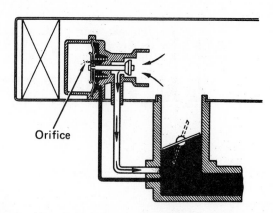

FIGURE 13-4. Mixture Control (MC) Valve, Open Position. (Courtesy of Toyota Motor Corporation, Japan)

SPARK CONTROL (SC) SYSTEM

Design and Operation

The spark control (SC) system retards the spark advance when the engine is cold and this action speeds the warm-up of the catalytic converter. A bimetal vacuum switching valve (BVSV) is connected in the distributor advance. The BVSV will be open at coolant temperatures above 104°F (40°C) and vacuum will be applied through the BVSV to the vacuum advance, as illustrated in Figure 13-5.

THERMAL REACTOR SYSTEM

Design and Operation

The thermal reactor is used in place of the conventional exhaust manifold. Exhaust gases remain at higher temperatures because of the insulation in the thermal reactor, as illustrated in Figure 13-6.

The thermal reactor system operates in conjunction with the air injection system. When the exhaust gas temperature in the thermal reactor increases, the mixture efficiency of exhaust gas and injected air also increases. As a result, HC and CO emission levels are reduced.

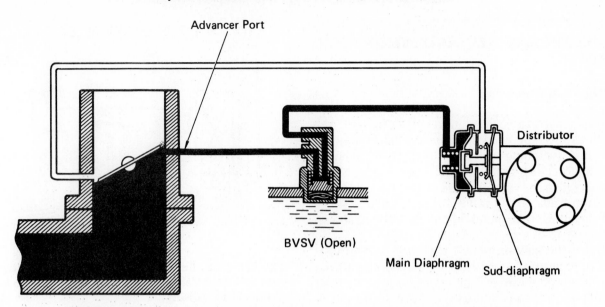

FIGURE 13-5. Spark Control (SC) System. (Courtesy of Toyota Motor Corporation, Japan)

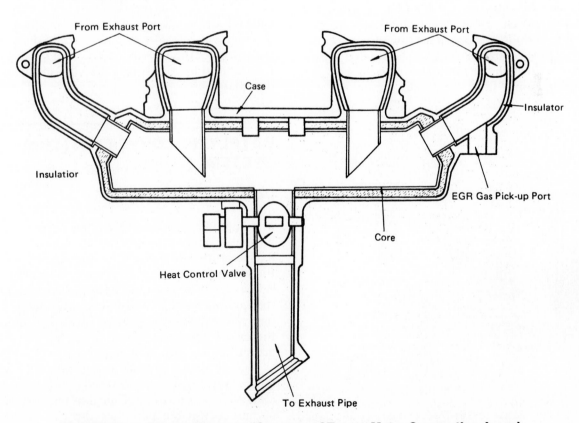

FIGURE 13-6. Thermal Reactor. (Courtesy of Toyota Motor Corporation, Japan)

EXHAUST GAS RECIRCULATION (EGR) SYSTEM

Design and Operation

The exhaust gas recirculation (EGR) valve is the same as the type used in domestic cars. A bimetal vacuum switching valve (BVSV), EGR vacuum modulator, and a vacuum switching valve (VSV) are connected in the vacuum hose to the EGR valve. The vacuum-transmitting valve is connected in the hose between the intake manifold and the VSV, as shown in Figure 13-7.

When the ignition switch is turned on, the VSV will be energized and the vacuum port from the vacuum modulator will be connected to the EGR valve port. If the coolant temperature is below 86°F (30°C), the BVSV will be closed and vacuum will not be applied to the modulator or to the EGR valve. When the coolant temperature is above 86°F (30°C), the BVSV will be open and vacuum will be applied to the modulator if the throttle is opened to uncover the EGR port to manifold vacuum. If the vehicle is operating at low speed, the exhaust pressure applied to the modulator is low and the modulator diaphragm remains in the downward position, where it bleeds off the vacuum in the EGR system. An increase in vehicle speed and exhaust pressure will force the modulator diaphragm upward and close the bleed port. Thus vacuum will be applied to the EGR valve through the modulator and the VSV, as pictured in Figure 13-8.

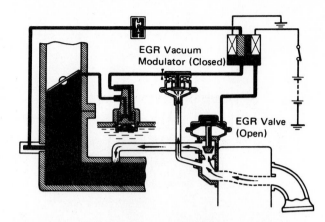

FIGURE 13-8. Exhaust Gas Recirculation (EGR) System with EGR Valve Open. (Courtesy of Toyota Motor Corporation, Japan)

When vacuum is applied to the EGR valve, the valve will open and allow exhaust gas to recirculate into the intake manifold. If the ignition switch is turned off, the VSV will be deenergized and the manifold vacuum port will be connected to the EGR valve through the VTV and VSV. As a result, the EGR valve will open momentarily when the engine is shut off and positive shutdown will occur. The VTV is a one-way valve that allows manifold vacuum to pass through the VTV to the VSV, but it is impossible for vacuum to go through the VTV in the opposite direction.

CATALYTIC CONVERTER (CCo) SYSTEM

Design and Operation

The catalytic converter (CCo) system uses a thermosensor and computer to activate an exhaust-temperature light on the instrument panel whenever the catalytic converter overheats. When the engine is being cranked, a signal from the start terminal of the ignition switch causes the computer to illuminate the exhaust-temperature light by completing the circuit from the light through the computer to ground, as illustrated in Figure 13-9.

This circuit is used to prove that the exhaust-temperature light is not burned out. If the catalytic converter is operating at normal temperature, the computer will not complete the circuit from the exhaust-temperature light to ground. When the

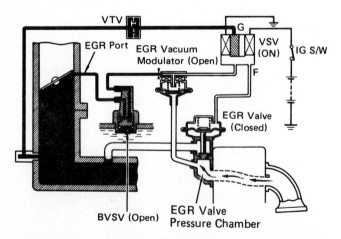

FIGURE 13-7. Exhaust Gas Recirculation (EGR) System. (Courtesy of Toyota Motor Corporation, Japan)

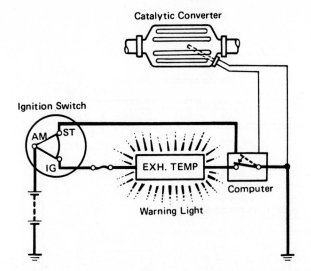

FIGURE 13-9. Proving Circuit of Catalytic Converter (CCo) System. (Courtesy of Toyota Motor Corporation, Japan)

temperature of the catalytic converter reaches 1,688°F (920°C), the thermosensor signal will cause the computer to activate the exhaust-temperature light, as indicated in Figure 13-10.

The illuminated exhaust-temperature light warns the driver that the catalytic converter has overheated. This overheating could be caused by an excessively rich air-fuel mixture or a defective air injection system.

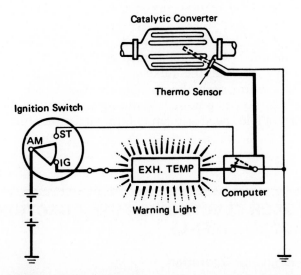

FIGURE 13-10. Catalytic Converter (CCo) System, Converter Overheated. (Courtesy of Toyota Motor Corporation, Japan)

BOSCH ELECTRONIC FUEL INJECTION MODEL D (EFI-D)

Design and Operation

The EFI-D system is illustrated in Figure 13-11. The numbered components in Figure 13-11 are:

1. Fuel pump
2. Fuel filter
5. Fuel tank
6. Pressure regulator
10. Intake manifold
12. Pressure sensor
16. Auxiliary air device
17. Cylinder head
18. Injector fuel lines
19. Throttle valve switch
20. Ignition distributor with trigger contacts
21. Thermotime switch
22. Electronic control unit (ECU)
23. Temperature sensor 1 in air intake
24. Temperature sensor 11 in cylinder head
25. Cold-start valve
26. Idle adjustment screw
27. Connection from starting motor

The ECU receives input signals from the pressure sensor, throttle valve switch, temperature sensors, and the distributor. The "on time" of the injectors is controlled by the ECU in response to the input signals. This system is referred to as a pressure sensitive system because the pressure sensor signal is the main one used by the ECU to control the air-fuel ratio at normal operating temperatures. The pressure sensor would be comparable to the manifold absolute pressure (MAP) sensor used in other systems.

A special set of contact points in the distributor relay an engine speed signal to the ECU. On receiving this signal, the ECU triggers the injectors. If the EFI-D system is used in a four-cylinder engine, the ECU can operate two injectors simultaneously because two pairs of injectors are electrically connected in parallel. When the system is used in a six-cylinder engine, two groups of three injectors will be connected in parallel. An eight-cylinder engine will

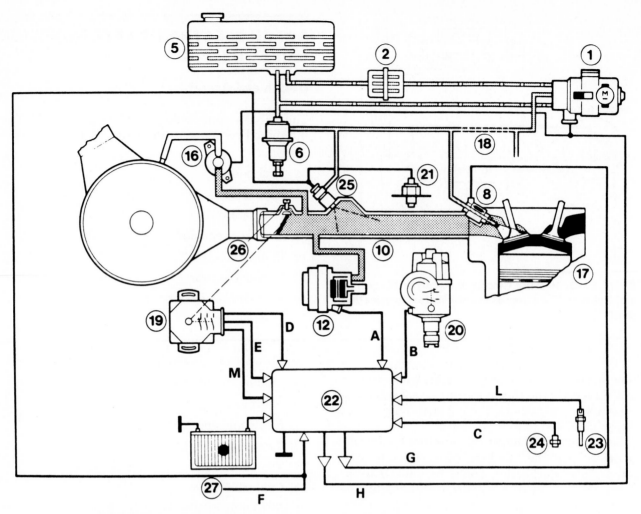

FIGURE 13-11. Electronic Fuel Injection Model D (EFI-D). (Courtesy of Robert Bosch Sales Corporation)

have four pairs of injectors connected in parallel. The special contacts are located under the breaker plate in the distributor.

When the engine is cold, the ECU will respond to the signals from the temperature switches by enriching the air-fuel ratio. Temperature sensor 11 will be located in the cylinder head on an air-cooled engine or in the cooling system of a water-cooled engine. The throttle valve switch signal to the ECU provides momentary acceleration enrichment and full load enrichment. When the engine is being cranked, the signal from the starting motor will operate the cold-start valve if the engine is cold and the thermotime switch is closed. The cold-start valve will inject additional fuel into the intake manifold to provide faster cold starting.

The auxiliary air device allows additional air into the intake manifold during the engine warm-up

period so that the correct fast idle speed can be maintained. A bimetallic strip operates a variable airflow port in the auxiliary air device. As the temperature rises, the opening through which the air can flow is gradually reduced. In this way, the correct idle speed is maintained at all temperatures.

BOSCH ELECTRONIC FUEL INJECTION MODEL L (EFI-L)

Design and Operation

The EFI-L system is shown in Figure 13-12. The numbered components in Figure 13-12 are:

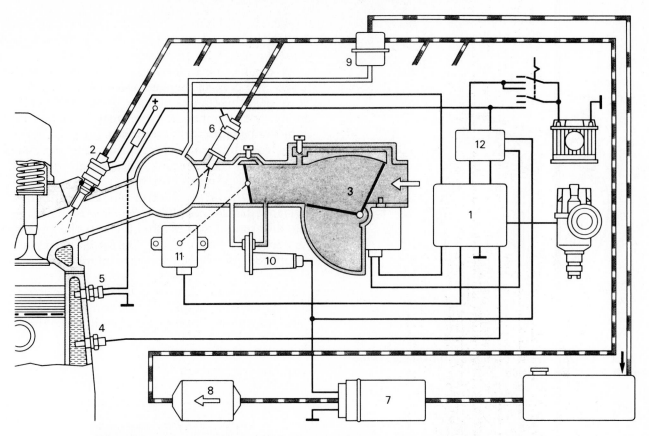

FIGURE 13-12. Electronic Fuel Injection Model L (EFI-L). (Courtesy of Robert Bosch Sales Corporation)

1. Electronic control unit (ECU)
2. Injection valve
3. Airflow sensor
4. Temperature sensor
5. Thermotime switch
6. Start valve
7. Electric fuel pump
8. Fuel filter
9. Fuel pressure regulator
10. Auxiliary air device
11. Throttle valve switch
12. Relay set

The main difference between the EFI-D and EFI-L systems is that the EFI-L system uses an airflow sensor in place of the pressure sensor. The voltage signal generated by the airflow sensor is directly proportional to the amount of air entering the engine. This is the main signal used by the ECU to control the "on time" of the injectors and the air-fuel ratio. (See Chapter 9, "Ford . . . EEC IV Systems in Turbocharged 2.3L Engines," for a complete description of the airflow sensor.) The airflow sensor is pictured in Figure 13-13.

In the EFI-L system, the ECU energizes all the injectors at the same instant. The ECU controls the "on time" of the injectors to provide the correct air-fuel ratio of 14.7:1.

BOSCH KE-JETRONIC FUEL INJECTION SYSTEM

Design and Operation

The KE-Jetronic system is a mechanically operated continuous injection system. An airflow sensor operates a control plunger in the fuel distributor, as shown in Figure 13-14.

Movement of the airflow sensor and the control plunger is proportional to the airflow through the air intake. The fuel is forced from the electric fuel pump through the fuel accumulator and filter to the control plunger, as illustrated in Figure 13-15.

As engine speed increases, the airflow sensor moves the control plunger upward and more fuel flows past the control plunger to the upper differential valve chambers, and then to the injectors. Fuel is injected continuously from the injectors. Fuel also flows from the control plunger through the pressure regulator to the fuel tank. The pressure regulator maintains a primary pressure in the system. Fuel also flows from the control plunger through the electrohydraulic actuator to the lower chamber of the differential valves and then to the pressure regulator and the fuel tank. The same fuel pressure is applied to the top of the control plunger and to the electrohydraulic actuator.

The electronic control unit (ECU) operates the electrohydraulic actuator in response to input signals from the temperature switch and the throttle

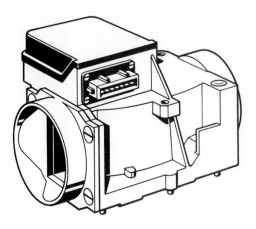

FIGURE 13-13. Airflow Sensor. (Courtesy of Robert Bosch Sales Corporation)

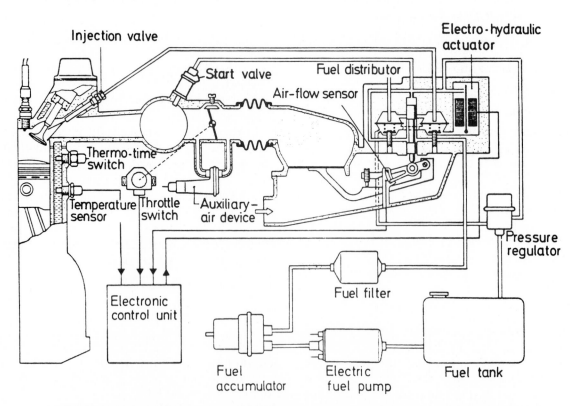

FIGURE 13-14. KE-Jetronic Fuel Injection System. (Courtesy of Robert Bosch Sales Corporation)

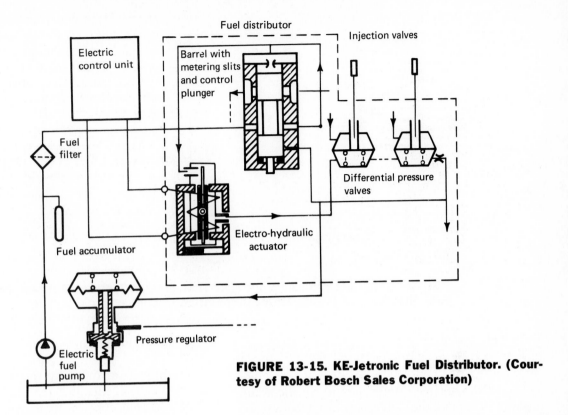

FIGURE 13-15. KE-Jetronic Fuel Distributor. (Courtesy of Robert Bosch Sales Corporation)

switch. The ECU controls the current flow through the electrohydraulic actuator winding, and the actuator varies the pressure in the lower chambers of the differential valves. Under normal conditions, the pressure in the upper differential valve chambers is higher than the pressure in the lower chambers, and the differential valve diaphragms are held in the downward position. When the engine is decelerated, the ECU reverses the current flow through the electrohydraulic actuator winding and this reversal causes the actuator to supply a higher pressure to the lower chambers of the differential valves. Under this condition, the differential valve diaphragms are forced upward and they completely shut off fuel flow to the injectors to increase fuel economy.

The ECU can assist in the control of the air-fuel ratio by varying the current flow through the electrohydraulic actuator, which controls the pressure difference between the inlet and outlet ports of the control plunger. Some KE-Jetronic injection systems also transmit an oxygen sensor signal to the ECU. This type of system is referred to as a lambda-controlled system. If some of the electronic components fail, the system will still operate mechanically, but the air-fuel ratio will not be controlled accurately. This is referred to as a "limp-home" mode.

TOYOTA ELECTRONICALLY CONTROLLED TRANSMISSION

Purpose

The electronically controlled transmission (ECT) is designed to improve fuel economy and performance. Since shift points and converter clutch lockup are controlled by the electronic control unit (ECU), these functions are optimized for economy and performance. The ECT system is available on four-speed automatic overdrive transmission in rear-wheel-drive cars, and also on four-speed automatic transaxles in front-wheel-drive cars.

Electronic Control System

A separate ECU is used for the ECT system. The inputs to this ECU are the following:

1. Driving pattern selector
2. Neutral start switch
3. Throttle position sensor
4. Brake light switch

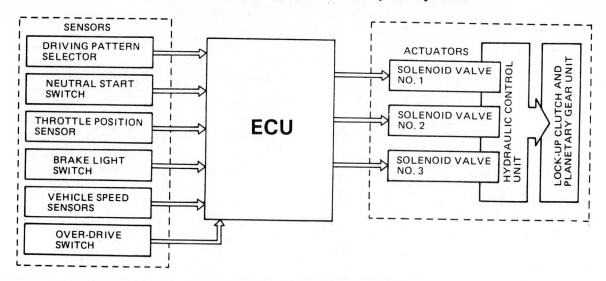

FIGURE 13-16. Electronically Controlled Transmission System Inputs and Outputs. (Courtesy of Toyota Motor Corporation, Japan)

5. Vehicle speed sensors

6. Overdrive switch

The ECU operates three solenoids in the transmission to control the converter clutch lockup and transmission shift points. A block diagram of the ECU inputs and outputs is shown in Figure 13-16, and the location of the ECT components is pictured in Figure 13-17.

Throttle Position Sensor

A direct-type or an indirect-type throttle position sensor may be used in the ECT system. The direct-type throttle position sensor contains an idle (IDL) contact which indicates a fully closed throttle, while three other contacts, L1, L2, and L3, provide signals at intermediate throttle positions. An E2 contact is a ground terminal.

The direct-type sensor is illustrated in Figure 13-18, and Figure 13-19 indicates the percentage of throttle opening at which the various contacts provide signals to the ECU.

The indirect-type throttle position sensor contains a variable resistor. A 5-V reference signal is sent from the Toyota Computer-Controlled System (TCCS) ECU to the VC terminal in the sensor. The variable signal from the sensor TA terminal is sent to the TCCS ECU. A signal converter in the TCCS ECU relays this signal to the ECT ECU. The IDL terminal in the sensor sends a fully closed throttle signal to the ECT ECU, and the E terminal provides

a ground connection between the sensor and the TCCS ECU.

Outputs controlled by the TCCS ECU include electronic fuel injection (EFI) and spark advance. (These output control functions are similar to the ones explained in Chapters 8 and 9.) An indirect-type throttle position sensor is shown in Figure 13-20.

Speed Sensors

The main, number 1 speed sensor is located in the transmission extension housing. A magnet attached to the output shaft rotates past the sensor. This rotating magnet operates a reed switch in the sensor which sends a vehicle speed signal to the ECT ECU. The signals from the throttle position sensor and the main speed sensor are used by the ECU to determine the transmission shift points and the converter clutch lockup schedule. The main, number 1 speed sensor is illustrated in Figure 13-21.

The backup, number 2 speed sensor is mounted in the speedometer. This sensor also contains a rotating magnet and a reed switch which generates four pulses for each speedometer cable revolution. When both speed sensors are operating normally, the ECU uses the signal from the main speed sensor. If the main speed sensor is defective, the ECU is programmed to use the signal from the backup speed sensor. The speed sensor signal replaces governor pressure in a conventional, hydraulically operated automatic transmission. A backup, number 2 speed sensor is pictured in Figure 13-22.

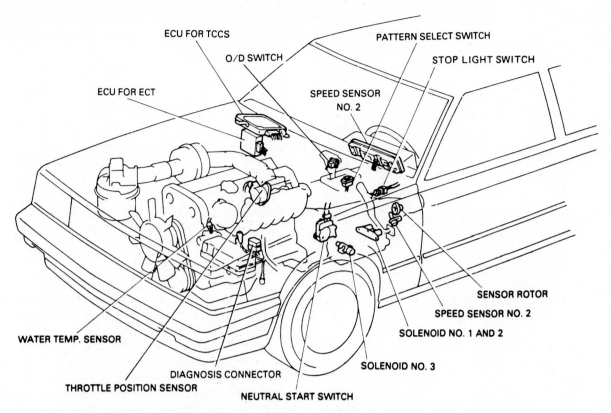

FIGURE 13-17. Electronically Controlled Transmission System Component Location. (Courtesy of Toyota Motor Corporation, Japan)

FIGURE 13-18. Direct-Type Throttle Position Sensor. (Courtesy of Toyota Motor Corporation, Japan)

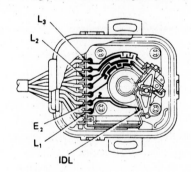

FIGURE 13-19. Direct-Type Throttle Position Sensor Signals in Relation to Throttle Opening. (Courtesy of Toyota Motor Corporation, Japan)

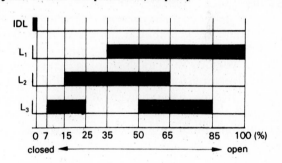

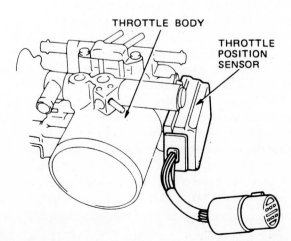

FIGURE 13-20. Indirect-Type Throttle Position Sensor. (Courtesy of Toyota Motor Corporation, Japan)

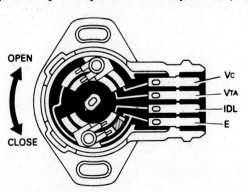

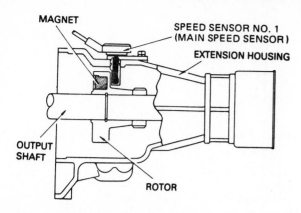

FIGURE 13-21. Main Speed Sensor. (Courtesy of Toyota Motor Corporation, Japan)

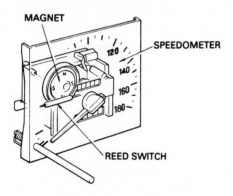

FIGURE 13-22. Backup Speed Sensor. (Courtesy of Toyota Motor Corporation, Japan)

Pattern-Select Switch

The pattern-select switch is mounted in the transmission shift console. This switch allows the driver to select various ECU modes which determine different transmission shift points. Some pattern-select switches have power or normal modes, whereas other switches provide economy, power, and normal modes. Pattern-select switches are shown in Figure 13-23.

FIGURE 13-23. Pattern-Select Switches. (Courtesy of Toyota Motor Corporation, Japan)

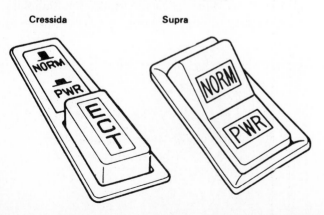

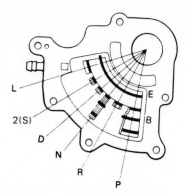

FIGURE 13-24. Neutral/Start Switch. (Courtesy of Toyota Motor Corporation, Japan)

Neutral/Start Switch

The neutral/start switch performs the conventional starter solenoid control function. However, this switch also informs the ECT ECU regarding the transmission selector lever position. When the transmission selector is moved from neutral (N) to drive (D), the ECU uses the signal from the neutral/start switch N contact to initiate a shift to third gear and then to first gear, which prevents sudden transmission shocks. This action is only initiated if the vehicle is stopped, the brake pedal is depressed, and the throttle is in the idle position. The neutral/start switch is illustrated in Figure 13-24.

Overdrive Switch

The overdrive (O/D) switch is located in the instrument panel. When this switch is in the on position, the switch contacts are open. Under this condition, a 12-V signal is supplied through the O/D off lamp to the ECU. This signal informs the ECU to allow the transmission to shift into O/D when the correct input signals are available.

When the O/D switch is moved to the off position, the switch contacts close and current flows through the O/D off lamp and the switch contacts to ground. Under this condition, the O/D off lamp is illuminated and a 0 V signal is applied to the ECU, which informs the ECU to cancel the transmission O/D operation. The operation of the O/D switch and off lamp is outlined in Figure 13-25.

The Toyota Computer-Controlled System ECU and the cruise control computer signal the ECT ECU to prevent overdrive operation if the coolant temperature is below 122°F (50°C). This action also occurs if the cruise control is operating and the actual vehicle speed exceeds the preset vehicle speed by 4–6 MPH (6–10 KPH).

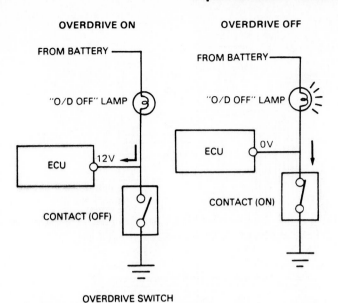

FIGURE 13-25. Overdrive Switch and Off Lamp Operation. (Courtesy of Toyota Motor Corporation, Japan)

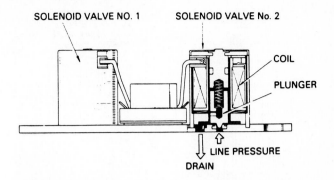

FIGURE 13-26. Number 1 and Number 2 Solenoid Valves. (Courtesy of Toyota Motor Corporation, Japan)

TABLE 13-1. Number 1 and Number 2 Solenoid Valve Application

Solenoid valve \ Gear	1st	2nd	3rd	OD
No. 1	ON	ON	OFF	OFF
No. 2	OFF	ON	ON	OFF

Courtesy of Toyota Motor Corporation, Japan

Brake Light Switch

The brake light switch is mounted on the brake pedal bracket. When the brake pedal is depressed, this switch signals the ECT ECU that the brakes have been applied, and the ECU disengages the converter lockup clutch. This action prevents engine stalling if the rear wheels lock up during a brake application.

Solenoid Valves

The number 1 and number 2 solenoid valves are mounted in the transmission valve body. These solenoids contain an electromagnet and a moveable plunger. Each plunger tip closes a bleed port in the line pressure hydraulic system. When the solenoids are not energized, the plunger tips keep the bleed ports closed. If either solenoid winding is energized by the ECT ECU, the plunger is lifted and the bleed port is opened. This action reduces the line pressure applied to one of the shift valves, which results in shift valve movement and a transmission shift. The number 1 and number 2 solenoid valves are pictured in Figure 13-26, and the application of these valves in each transmission gear is outlined in Table 13-1.

The solenoid valves are shown with the transmission shift valves in Figure 13-27.

The number 3 solenoid valve is located in the transmission valve body on some models, or mounted in the transmission case on other models. This solenoid also contains an electromagnet and a moveable plunger, and the plunger tip opens and closes a bleed port in the line pressure hydraulic system. However, the action of the number 3 solenoid valve controls a lockup clutch. (The operation of this lockup torque converter is similar to the lockup converter explained in Chapter 8.) Solenoid number 3 is shown in Figure 13-28.

When the number 3 solenoid valve is not energized, the lockup control valve is positioned so it directs fluid through the hollow input shaft into the converter. This action moves the lockup clutch away from the front of the converter, and the clutch remains disengaged. When the ECT ECU energizes the number 3 solenoid valve, the plunger opens the bleed port, which reduces line pressure on one end of the lockup control valve. When this occurs, the lockup control valve moves upward and fluid is then directed through the converter hub into the converter. This fluid forces the lockup clutch against the front of the converter, which provides a locking action between the front of the converter and the transmission input shaft. The lockup clutch is applied in second, third, and O/D gears with the

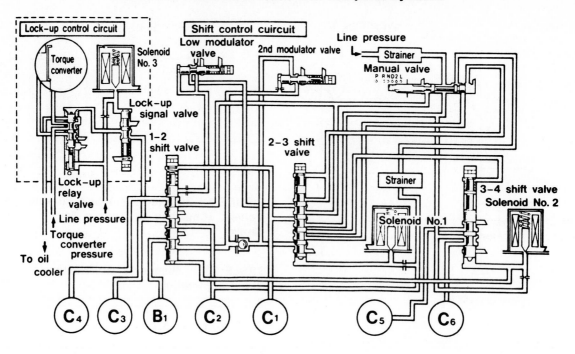

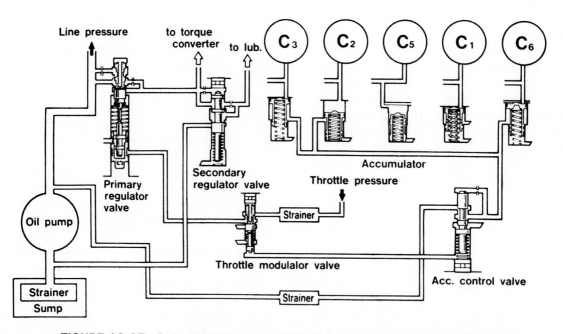

FIGURE 13-27. Solenoid Valves and Shift Valves. (Reference taken from SAE paper no. 840049. Reprinted with Permission © 1984 Society of Automotive Engineers)

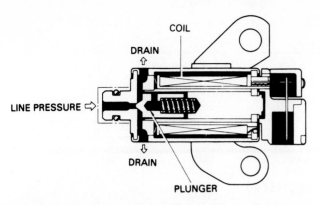

FIGURE 13-28. Number 3 Solenoid Valve. (Courtesy of Toyota Motor Corporation, Japan)

transmission selector in drive (D). The ECU controls lockup timing to prevent shock during transmission shifts.

Electronically Controlled Transmission Electronic Control Unit

The electronic control unit (ECU) is a microprocessor with a memory which contains all the shift schedules and the lockup converter schedule. The normal, power, or economy shift schedules that may be selected by the driver are also contained permanently in the ECU memory. The normal shift schedule is a well-balanced shift and lockup schedule for general driving require-

ments, whereas the economy schedule provides increased fuel efficiency, especially for city driving. When mountainous driving is encountered or when faster acceleration is desired, the power schedule changes the transmission shift points and converter lockup timing to take full advantage of torque multiplication in the converter, which improves engine performance and power.

The normal and power shift schedules are provided in Figure 13-29 and Figure 13-30. The shift schedules vary, depending on the application. The lockup converter economy shift schedule is shown in Figure 13-31.

The transmission is controlled electronically while the vehicle is moving ahead. If the transmission selector is in neutral, park, or reverse, the transmission is controlled mechanically.

Fail-Safe Functions

If solenoid number 1 or number 2 fails to operate, the ECT ECU will continue to operate the solenoid that is working normally. Therefore, some transmission shifts will occur, and the vehicle may be driven in the forward gears. When a major electronic malfunction occurs and none of the solenoid valves will operate, the transmission may be shifted with the selector lever. When the transmission selector is moved to the low (L), second (S), and drive (D) positions, the transmission operates in first gear and overdrive (O/D).

FIGURE 13-29. Normal Shift Schedule. (Reference taken from SAE paper no. 820740. Reprinted with Permission © 1982 Society of Automotive Engineers)

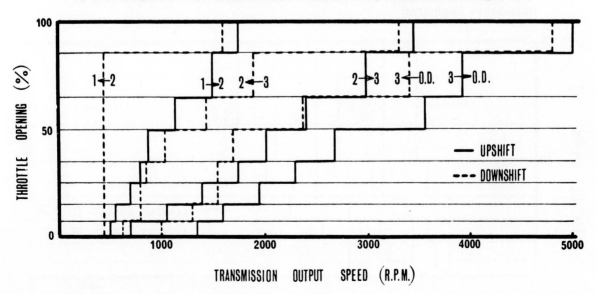

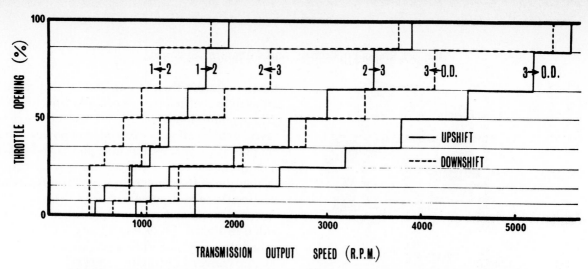

TRANSMISSION OUTPUT SPEED (R.P.M.)

FIGURE 13-30. Power Shift Schedule. (Reference taken from SAE paper no. 820740. Reprinted with Permission © 1982 Society of Automotive Engineers)

Electrical Circuit

Electrical circuits vary considerably, depending on the model and year of vehicle. A complete ECT wiring diagram for a 1985 Cressida is shown in Figures 13-32 and 13-33.

DIAGNOSIS OF ELECTRONICALLY CONTROLLED TRANSMISSION

Self-Diagnosis

The ECT ECU will cause the O/D lamp to flash on and off if a defect occurs in the speed sensors or solenoid valves. When the ignition switch is turned off, the defect is stored in the ECU memory. Diagnostic codes representing system defects may be obtained when the ignition switch is turned on and the ECT terminal is connected to the E1 terminal in the diagnostic connector. On Supra models, a single diagnostic terminal must be grounded to obtain the diagnostic codes. Both diagnostic connections are indicated in Figure 13-34.

When the diagnostic connection is completed with the ignition switch on, the O/D off lamp indicates various codes. If the O/D off lamp flashes every 0.25 seconds, the ECT system is in normal condition. When the O/D off lamp flashes six times followed by a 1.5 second pause and two more flashes, a code 62 is indicated. A normal code is compared to a code 62 in Figure 13-35, and the five defective codes which apply to the ECT systems are illustrated in Figure 13-36.

FIGURE 13-31. Lock-up Converter Economy Shift Schedule. (Reference taken from SAE paper no. 820740. Reprinted with Permission © 1982 Society of Automotive Engineers)

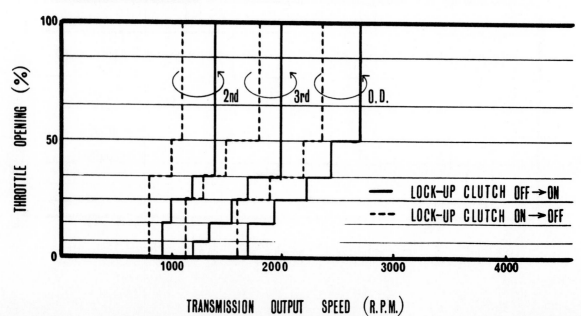

TRANSMISSION OUTPUT SPEED (R.P.M.)

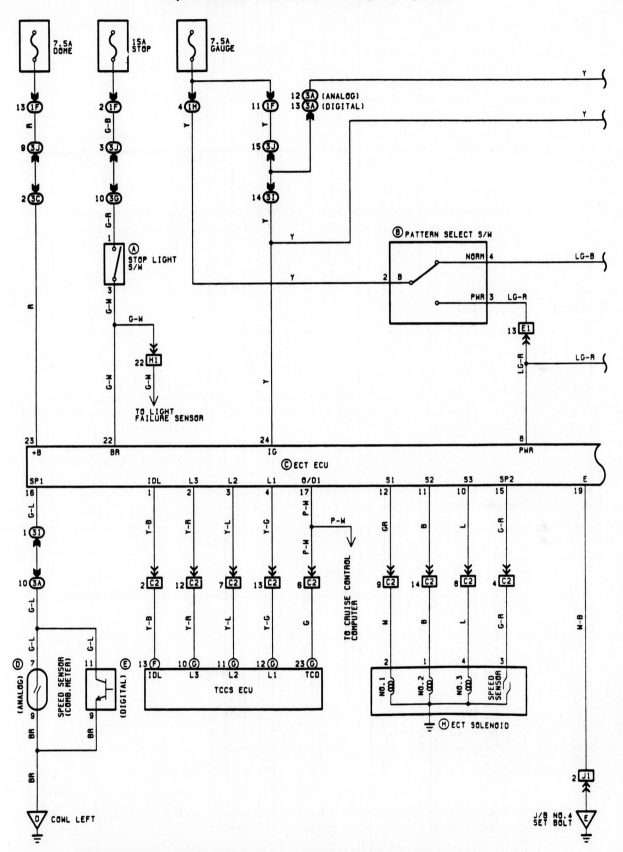

FIGURE 13-32. Electronically Controlled Transmission Wiring Diagram, Part One. (Courtesy of Toyota Motor Corporation, Japan)

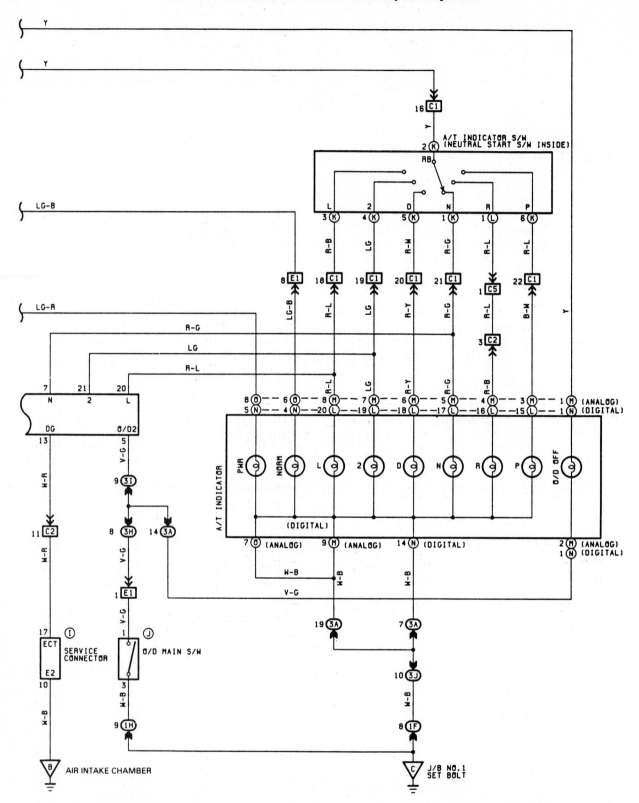

FIGURE 13-33. Electronically Controlled Transmission Wiring Diagram, Part Two. (Courtesy of Toyota Motor Corporation, Japan)

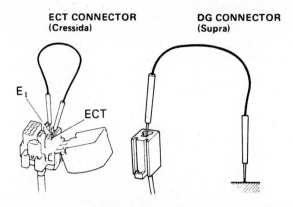

FIGURE 13-34. Diagnostic Test Connections. (Courtesy of Toyota Motor Corporation, Japan)

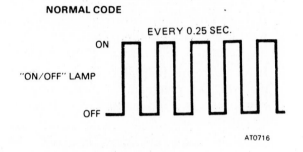

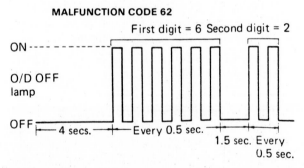

FIGURE 13-35. Normal Code and Code 62. (Courtesy of Toyota Motor Corporation, Japan)

TABLE 13-2. Voltmeter Diagnosis of Electronically Controlled Transmission

Voltage reading	Implication
0 V	Normal
4 V	Speed sensor defective
8 V	Solenoid valves defective

Courtesy of Toyota Motor Corporation, Japan

If a defect occurs in the number 3 converter clutch lockup solenoid, the ECU will not flash the O/D off lamp, but a code 64 is stored in the ECU memory.

The diagnostic codes are obtained from the flashing O/D lamp on Cressida and Supra models. On other models, a voltmeter must be connected from the ECT or DG terminal to ground with the ignition switch on. The voltmeter reading indicates normal or defective conditions, as shown in Table 13-2.

OPERATION CHECK FUNCTION

Throttle Position Sensor Check

When the ignition switch is turned on and a voltmeter is connected from the ECT or DG terminal to ground, the throttle position sensor, brake switch signal, and shift timing can be checked. If the throttle is opened from the idle position, the voltmeter readings shown in Figure 13-37 should be obtained if the throttle position sensor is working normally.

FIGURE 13-36. Electronically Controlled Transmission Defective Codes. (Courtesy of Toyota Motor Corporation, Japan)

Code No.	Light Pattern	Diagnosis System
42		Defective No. 1 speed sensor (in combintion meter) Severed wire harness or short circuit
61		Defective No. 2 speed sensor (in ATM) Severed wire harness or short circuit
62		Severed No. 1 solenoid or short circuit Severed wire harness or short circuit
63		Severed No. 2 solenoid or short circuit Severed wire harness or short circuit
64		Severed No. 3 solenoid or short circuit Severed wire harness or short circuit

THROTTLE OPENING

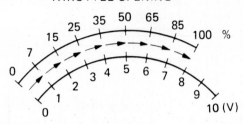

FIGURE 13-37. Voltage Readings from Throttle Position Sensor. (Courtesy of Toyota Motor Corporation, Japan)

Do not touch the brake pedal during the throttle position sensor tests.

Brake Switch Tests

With the ignition switch on and the throttle wide open, the voltmeter connected from the ECT or DG terminal to ground should read 8 V. When the brake pedal is depressed, the voltmeter reading should be 0 V if the brake switch is working normally.

Shift Timing Tests

When the vehicle is driven on the road with the voltmeter connected from the ECT or DG terminal to ground, the voltages provided in Table 13-3 should be obtained in each transmission gear.

The converter clutch lockup may be cycled on and off in second and third gear by depressing and releasing the accelerator pedal. If the correct transmission gear is not available at the specified voltage, one of the solenoid valves may be sticking or a defect may exist in the transmission valve body. The ECU cannot diagnose a sticking solenoid valve.

TABLE 13-3. Voltage Readings in Each Transmission Gear

Gear	1st	2nd		3rd		OD	
lock-up	—	OFF	ON	OFF	ON	OFF	ON
DG term. voltage	0	2	3	4	5	6	7

Courtesy of Toyota Motor Corporation, Japan

Questions

1. The throttle positioner is activated when vehicle speed reaches 30 MPH (48 KPH).   T   F

2. The mixture control valve is opened by manifold vacuum when the engine is operating at idle speed.   T   F

3. The spark control system provides faster warm-up of the _____.

4. The modulator in the exhaust gas recirculation (EGR) system is operated by _____.

5. If the exhaust-temperature light is illuminated with the engine running, the _____ is overheated.

6. In an electronic fuel injection model D system (EFI-D), the electronic control unit (ECU) energizes a pair of injectors when it receives a signal from the _____ _____.

7. In an electronic fuel injection model L (EFI-L) system, the main signal that the ECU uses to control the air-fuel ratio is the _____ signal.

8. In the KE-Jetronic fuel injection system, the airflow sensor operates the _____.

9. The two main input signals used by the electronic control unit (ECU) in the electronically controlled transmission (ECT) system to determine the transmission shift points are the:
   a) overdrive switch and neutral/start switch signals
   b) throttle position sensor and vehicle speed sensor signals
   c) neutral/start switch and driving pattern selector signals

10. If the main speed sensor in the transmission is defective, the ECU is programmed to use a vehicle speed signal from the:
   a) overdrive switch
   b) throttle position sensor
   c) backup speed sensor in the speedometer

11. If a major electronic defect occurs such as a defective ECU in the electronically controlled transmission (ECT) system, the car cannot be driven in the forward gears.   T   F

12. When the power mode is selected, the ECU causes the transmission upshifts to occur at a lower engine speed.   T   F

13. On a Toyota Cressida with the ignition switch on and the ECT terminal connected to the E1 terminal in the diagnostic connector, if the overdrive (O/D) off lamp flashes steadily every 0.25 seconds, the:
    a) number 3 solenoid valve is defective

b) number 1 solenoid valve is defective
c) ECT system is in normal condition

14. A flashing overdrive (O/D) off lamp when the vehicle is driven at 50 MPH (80 KPH) indicates:
    a) that the transmission is cycling from overdrive to third gear
    b) a defect in the ECT system
    c) a pulsating overdrive switch

INDEX